BIOPHYSICS

BIOPHYSICS
Concepts and Advances

Frank Glaser
Anita Roy
Steven Park

Kruger Brentt
Publishers
2023

Kruger Brentt Publishers UK. LTD.
Company Number 9728962

Regd. Office: 68 St Margarets Road, Edgware, Middlesex HA8 9UU

ISBN: 9781787150478

For information on all our publications visit our website at http://krugerbrentt.com/

PREFACE

Biophysics is the study of physical phenomena and physical processes in living things, on scales spanning molecules, cells, tissues and organisms. Biophysicists use the principles and methods of physics to understand biological systems. It is an interdisciplinary science, closely related to quantitative and systems biology. Biophysics has been critical to understanding the mechanics of how the molecules of life are made, how different parts of a cell move and function, and how complex systems in our bodies—the brain, circulation, immune system, and others— work. Biophysics is a vibrant scientific field where scientists from many fields including math, chemistry, physics, engineering, pharmacology, and materials sciences, use their skills to explore and develop new tools for understanding how biology—all life—works.

Physical scientists use mathematics to explain what happens in nature. Life scientists want to understand how biological systems work. These systems include molecules, cells, organisms, and ecosystems that are very complex. Biological research in the 21st century involves experiments that produce huge amounts of data. How can biologists even begin to understand this data or predict how these systems might work? This is where biophysicists come in. Biophysicists are uniquely trained in the quantitative sciences of physics, math, and chemistry and they are able tackle a wide array of topics, ranging from how nerve cells communicate, to how plant cells capture light and transform it into energy, to how changes in the DNA of healthy cells can trigger their transformation into cancer cells, to so many other biological problems. Biophysicists work to develop methods to overcome disease, eradicate global hunger, produce renewable energy sources, design cutting-edge technologies, and solve countless scientific mysteries. In short, biophysicists are at the forefront of solving age-old human problems as well as problems of the future.

The present book contains sixteen chapters covering all related disciplines. These chapters include Biophysics: An Introduction, Biophysics in Drug Discovery, Biophysics and Chromatography, Biophysics and Electrophoresis, Biophysics and Spectroscopy, Mass Spectrometry, Biophysics, Centrifugation and Microscopy, Concepts of Energy and

Metabolism, Bioelectrical and Engineering of Cell Biology, Molecular Biology Techniques, Biofluid Mechanics of Special Organs, Spatial Organization And Biochemistry In Building Blocks of Cellular Signalling Pathways, Structure And Function of Cells, Modelling Radiation Fluxes in Simple and Complex Environments, Concepts of Radiation and Temperature and Heat.

Related terminology is given at the end for ready reference. This book deals with some of the emerging topics on advanced concepts in biophysics, which would an invaluable asset for the scientists, students and teachers working in the field of physics, biophysics and medical physics. This book is also a valuable reference source for anyone interested in knowing more about advanced concepts in biophysics.

We are grateful to all those persons as well as various books, manuals, periodicals, magazines, journals etc. that helped in the preparation of this book. In spite of the best efforts, it is possible that some errors may have occurred into the compilation and editing of the book. Further queries, constructive suggestions and criticisms for the improvement of the book are always welcome and shall be thankfully acknowledged.

Frank Glaser

Anita Roy

Steven Park

CONTENTS

1

BIOPHYSICS: AN INTRODUCTION

1.1 INTRODUCTION

Biophysics is the field that applies the theories and methods of physics to understand how biological systems work. Biophysics has been critical to understanding the mechanics of how the molecules of life are made, how different parts of a cell move and function, and how complex systems in our bodies—the brain, circulation, immune system, and others— work. Biophysics is a vibrant scientific field where scientists from many fields including math, chemistry, physics, engineering, pharmacology, and materials sciences, use their skills to explore and develop new tools for understanding how biology—all life—works.

Biology is complicated, really, really complicated. This should not surprise you if you think that ultimately the laws of physics explain how the world works, because biological activity is far beyond the usual realm of simple physical phenomena. It is easy to simply turn away from a true physical explanation of biological phenomena as simply hopeless. Perhaps it is hopelessly complex at some level of detail. The ghosts of biological "stamp collecting" are still alive and well and for a good reason.

However, it is possible that in spite of the seemingly hopeless complexity of biology, there are certain emergent properties that arise in ways that we can understand quantitatively. An emergent property is an unexpected collective phenomenon that arises from a system consisting of interacting parts. You could call the phenomenon of life itself an emergent property. Certainly no one would expect to see living systems arise directly from the fundamental laws of quantum mechanics and the Standard Model that have been discussed in the first seven units of this course.

The danger is that this concept of "emergent properties" is just some philosophical musing with no real deeper physics content, and it may be true that the emergent properties of life viewed "bottom up" are simply too complex in origin to understand at a quantitative level. It may not be possible to derive how emergent properties arise from microscopic physics. In his book *A Different Universe: Reinventing Physics from the Bottom Down*, the physicist Robert Laughlin compares the local movement of air molecules around an airplane wing to the large-scale turbulent hydrodynamic flow of air around the airfoil that gives rise to lift. Molecular motion is clearly the province of microscopic physics and statistical mechanics, while turbulent flow is an emergent effect. As Laughlin puts it, if he were to discover that Boeing Aircraft began worrying about how the movement of air molecules collectively generates hydrodynamics, it would be time to divest himself of Boeing stock. Perhaps the same should have been said when banks started hiring theoretical physicists to run stock trading code. In biology, we have a much greater problem than with the airplane, because the air molecules can be described pretty well with the elegant ideas of statistical mechanics. So, while it is a long stretch to derive the emergence of turbulence from atomic motion, no one would say it is impossible, just very hard.

Fig. 1: Colored smoke marks the hydrodynamic flow around an aircraft, an emergent phenomenon.

Source: *NASA Langley Research Center (NASA-LaRC).*

In biology, even the fundamentals at the bottom may be impossibly hard for physics to model adequately in the sense of having predictive power to show the pathways of emergent behavior. A classic example is the signaling that coordinates the collective aggregation of the slime-mold *Dictyostelium* cells in response to the signaling molecule cyclic AMP (cAMP). In the movie shown in Figure 1, the individual *Dictyostelium* cells signal to each other, and the cells stream to form a fruiting body in an emergent process called "chemotaxis." This fairly simple-looking yet spectacular process is a favorite of physicists and still not well understood after 100 years of work.

So, perhaps in a foolhardy manner, we will move forward to see how physics, in the discipline known as biological physics, can attack some of the greatest puzzles of them all. We will have to deal with the emergence of collective phenomena from an underlying complex set of interacting entities, like our *Dictyostelium* cells. But that seems still within the province of physics: really hard, but physics. But there are deeper questions that seem to almost be beyond physics.

1.2 EMERGENT PHYSICS RULES IN LIFE

The amazing array of knowledge in previous units contains little inkling of the complex, varied phenomena of life. Life is an astonishingly emergent property of matter, full-blown in its complexity today, some billions of years after it started out in presumably some very simple form. Although we have many physical ways to describe a living organism, quantifying its state of aliveness using the laws of physics seems a hopeless task. So, all our tools and ideas would seem to fail at the most basic level of describing what life is.

Biology has other incredible emergent behaviors that you can hardly anticipate from what you have learned so far. British physicist Paul Dirac famously said that, "The fundamental laws necessary for the mathematical treatment of a large part of physics and the whole of chemistry are thus completely known, and the difficulty lies only in the fact that application of these laws leads to equations that are too complex to be solved." Our question is: Is the biological physics of the emergent properties of life simply a matter of impossible complexity, or are there organizing principles that only appear at a higher level than the baseline quantum mechanics?

So far, we have talked about the emergent nature of life itself. The next astonishing emergent behaviour we'll consider is the evolution of living organisms to ever-higher complexity over billions of years. It is strange enough that life developed at all out of inanimate matter, in apparent conflict with the Second Law of Thermodynamics. The original ur-cell, improbable as it is, proceeded to evolve to ever-greater levels of complexity, ultimately arriving at *Homo sapiens* several million years ago. Thanks to Darwin and Wallace and their concept of selection of the fittest, we have a rather vague hand-waving idea of how this has happened. But the quantitative modeling of evolution as an emergent property remains in its infancy.

The building blocks of evolution

Modeling evolution is a difficult task; but nevertheless, we can try. So let's start at the top and work down. Physicists believe (and it is somewhat more of a belief than a proven fact) that all life began with some ur-cell and that life evolved from that ur-cell into the remarkable complexity of living organisms we have today, including *Homo sapiens.* We will never know what path this evolution took. But the remarkable unity of life (common genetic code, common basic proteins, and common basic biological pathways) would indicate that, at its core, the phenomenon of life has been locked in to a basic set of physical modes and has not deviated from this basic set. At that core, lies a very long linear polymer, deoxyribonucleic acid (or DNA), which encodes the basic self-assembly information and control information. A related molecule, ribonucleic acid (RNA), has a different chemical group at one particular position, and that profoundly changes the three-dimensional structure that RNA takes in space and its chemical behavior.

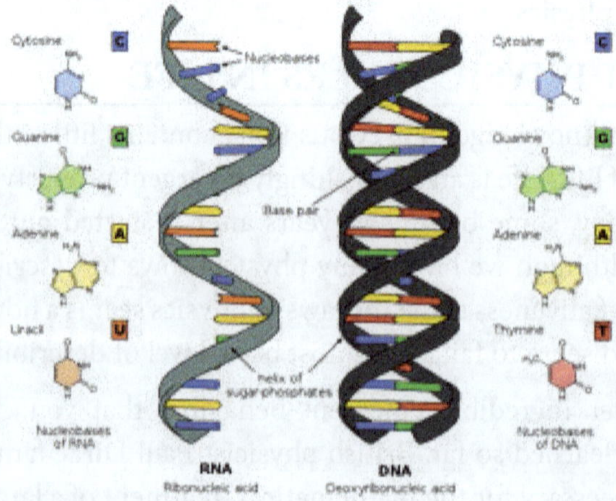

Fig. 2: Molecules of life: RNA (left) and DNA (right).

Source: *Wikimedia Commons, Creative Commons Attribution- Share Alike 3.0 license. Author: Sponk, 23 March 2010*

Although evolution has played with the information content of DNA, its basic core content, in terms of how its constituent molecules form a string of pairs, has not obviously changed. And while there is a general relationship between the complexity of an organism and the length of its DNA that encodes the complexity, some decidedly simpler organisms than *Homo sapiens* have considerably longer genomes.

So, from an information perspective, we really don't have any iron-clad way to go from genome to organismal complexity, nor do we understand how the complexity evolved. Ultimately, of course, life is matter. But, it is the evolution of information that really lies at the unknown heart of biological physics, and we can't avoid it.

1.3 THE EMERGENCE OF THE MIND IN LIVING SYSTEMS

Biology possesses even deeper emergent phenomena than the evolution of complexity. The writer and readers of this document are sentient beings with senses of identity and self and consciousness.

Presumably, the laws of physics can explain the emergent behavior of consciousness, which certainly extends down from *Homo sapiens* into the "lower" forms of life (although those lower forms of life might object to that appellation). Perhaps the hardest and most impossible question in all of biological physics is: What is the physical basis behind consciousness? Unfortunately, that quest quickly veers into the realm of the philosophical and pure speculation; some would say it isn't even a legitimate physics question at all.

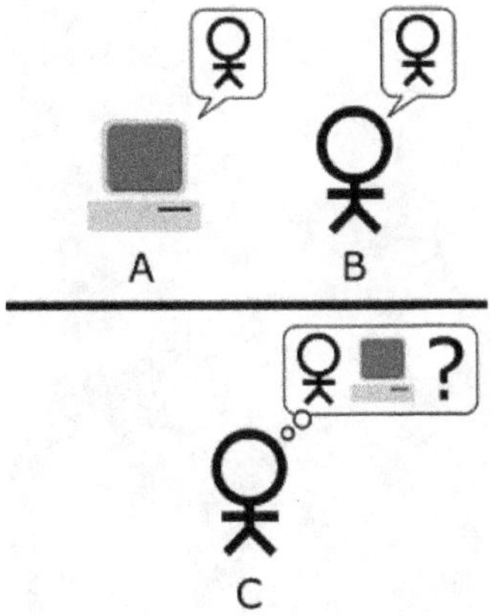

Fig. 4: Player C is trying to determine which player—A or B—is a computer and which is human.

Source: *Wikimedia Commons, Public Domain. Author: Bilby, 25 March 2008.*

There is even an argument as to whether "machines," now considered to be computers running a program, will ever be able to show the same kind of intelligence that living systems such as human beings possess. Traditional reductionist physicists, I would imagine, simply view the human mind as some sort of a vastly complicated computational machine. But it is far from clear if this view is correct. The mathematician Alan Turing, not only invented the Turing machine, the grandfather of all computers, but he also asked a curious question: Can machines think? To a physicist, that is a strange question for it implies that maybe the minds of living organisms somehow have emergent properties that are different from what a manmade computing machine could have. The answer to Turing's question rages on, and that tells us that biology has very deep questions still to be answered.

1.4 PHYSICS AND LIFE

Here's a question from a biologist, Don Coffey at Johns Hopkins University: Is a chicken egg in your refrigerator alive? We face a problem right away: What does being alive actually mean from a physics perspective? Nothing. The concept of aliveness has played no role in anything you have been taught yet in this course. It is a perfectly valid biological question; yet physics would seem to have little to say about it. It is an emergent property arising from the laws of physics, which presumably are capable of explainingthe physics of the egg.

The chicken egg is a thing of elegant geometric beauty. But its form is not critical to its state of aliveness (unless, of course, you smash it). However, you can ask pertinent physical questions about the state of the egg to determine whether it is alive: Has the egg been cooked? It's pretty easy to tell from a physics perspective: Spin the egg around the short axis of the ellipse rapidly, stop it suddenly, and then let it go.

Fig. 4: A chicken egg, is it alive or dead?

If it starts to spin again, it hasn't been cooked because the yolk proteins have not been denatured by heat and so remain as a viscous fluid. If your experiment indicates the egg hasn't been cooked it might be alive, but this biological physics experiment wouldn't take you much closer to an answer.

Assuming you haven't already broken the egg, you can now drop it. If you were right that it has not been cooked, the egg will shatter into hundreds of pieces. Is it dead now? If this were your laptop computer, you could pick up all the pieces and—if you are good enough—probably get it working again. However, all of the king's horses and all the king's men can't put Humpty Dumpty back together again and make him alive once more; we don't know how to do it. The egg's internal mechanical structure is very complex and rather important to the egg's future. It, too, is part of being alive, but surely rather ancillary to the main question of aliveness.

Aliveness is probably not a yes-no state of a system with a crisp binary answer, but rather a matter of degree. One qualitative parameter is the extent to which the egg is in thermodynamic equilibrium with its surroundings. If it is even slightly warmer, then I would guess that the egg is fertilized and alive, because it is out of thermodynamic equilibrium and radiating more energy than it absorbs. That would imply that chemical reactions are running inside the egg, maintaining the salt levels, pH, metabolites, signalling molecules, and other factors necessary to ensure that the egg has a future some day as a chicken.

Wait, the egg has a future? No proton has a future unless, as some theories suggest, it eventually decays. But if the egg is not dropped or cooked and is kept at exactly the right temperature for the right time, the miracle of embryonic development will occur: The fertilized nucleus within the egg will self assemble in an intricate dance of physical forces and eventually put all the right cells into all the right places for a chick to emerge. Can the laws of physics ever hope to predict such complex emergent phenomena?

1.4.1 Emergent and Adaptive Behavior in Bacteria

Here's an explicit example of what we are trying to say about emergent behavior in biology. Let's move from the complex egg where the chick embryo may be developing inside to the simple example of bacteria swimming around looking for food. It's possible that each bacterium follows a principle of every bug for itself: They do not interact with each other and simply try to eat as much food as possible in order to reproduce in an example of Darwinian competition at its most elemental level. But food comes and food goes at the bacterial level; and if there is no food, an individual bacterium will starve and not be able to survive. Thus, we should not be surprised that many bacteria do not exist at the level as rugged individuals but instead show quite startling collective behavior, just like people build churches.

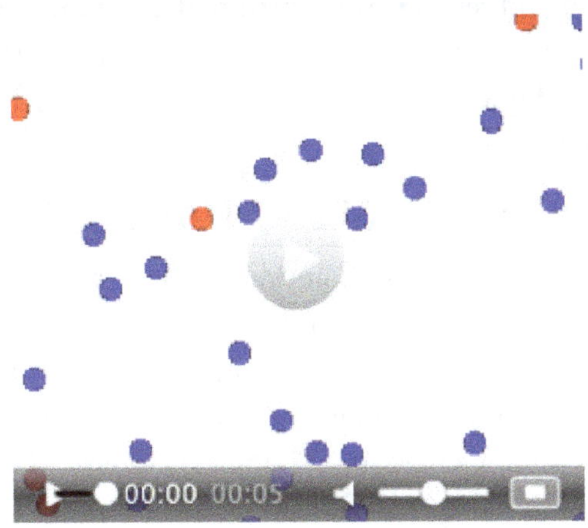

Fig. 5: Random motion of gas molecules—bottom up.

Source: *Wikimedia Commons, GNU Free Documentation License. Author: Greg L., 25 August 2006.*

If bacteria acted as rugged individuals, then we would expect their movement through space looking for food to resemble what is called a random walk, which is different from the Brownian motion that occurs due to thermal fluctuations. In a random walk there is a characteristic step size L, which is how far the bacterium swims in one direction before it tumbles and goes off randomly in a new direction. Howard Berg at Harvard University has beautiful videos of this random movement of bacteria. The effect of this random motion is that we can view individual bacteria rather like the molecules of a gas, as shown in Figure 5. If that were all there is to bacterial motion, we would be basically done, and we could use the mathematics of the random walk to explain bacterial motion.

However, bacteria can be much more complicated than a gas when viewed collectively. In the Introduction, we discussed the chemotaxis of a population of individual *Dictyostelium* cells in response to a signal created and received by the collective population of the *Dictyostelium* cells. Bacteria do the same thing. Under stress, they also begin signaling to each other in various ways, some quite scary.

For example, if one bacterium mutates and comes up with a solution to the present problem causing the stress, in a process called "horizontal gene transfer" they secrete the gene and transfer it to their buddies. Another response is to circle the wagons: The bacteria signal to each other and move together to form a complex community called a "biofilm."

1.4.2 Keller-Segel Equations

The mystery is how the supposedly simple bacteria communicate with each other to form such a complex and adapted structure. There is a set of equations, called the "Keller-Segel equations," which are usually the first steps in trying to puzzle out emergent behavior in a collection of swimming agents such as bacteria. These equations are not too hard to understand, at least in principle. Basically, they take the random walk we discussed above and add in the generation and response of a chemoattractant molecule. A sobering aspect of these equations is that they are very difficult to solve exactly: They are nonlinear in the density of the bacteria, and one of the great secrets of physics is that we have a very hard time solving nonlinear equations.

1.5 PRINCIPLES OF A COMPLEX ADAPTIVE SYSTEM

We are just skimming the surface of a monumental problem in biological physics: How agents that communicate with each other and adapt to the structures that they create can be understood. A biological system that communicates and adapts like the film-forming bacteria is an example of a complex adaptive system. In principle, a complex adaptive system could appear almost anywhere, but biological systems are the most extreme cases of this general phenomenon.

The computer scientist John Holland and the physicist Murray Gell-Mann, who played a major role in the physics developments you read about in Units 1 through 4, have tried to define what makes a complex adaptive system. We can select a few of the key properties as presented by Peter Freyer that are most germane to biological systems:

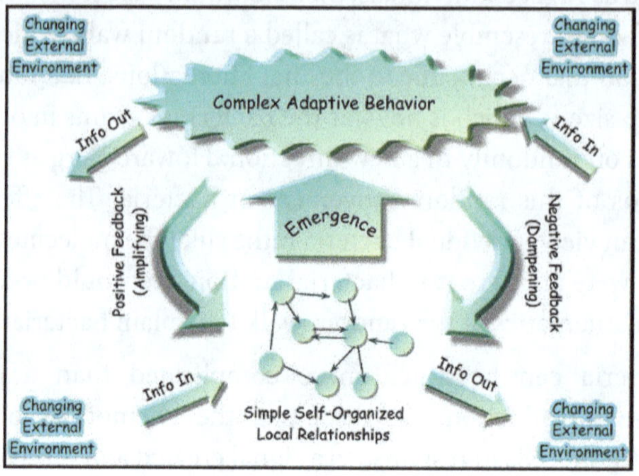

Fig. 6: A schematic view of what constitutes a complex adaptive system.

Source: *Wikimedia Commons, Creative Commons Attribution ShareAlike 3.0.*

- ***Emergence:*** We have already discussed this concept, both in this unit and in Unit 8.
- Co-evolution: We will talk about evolution later. Coevolution refers to how the evolution of one agent (say a species, or a virus, or a protein) affects the evolution of another related agent, and vice versa.
- ***Connectivity:*** This is concerned with biological networks, which we will discuss later.
- ***Iteration:*** As a system grows and evolves, the succeeding generations learn from the previous ones.
- ***Nested Systems:*** There are multiple levels of control and feedback.

These properties will appear time and again throughout this unit as we tour various complex adaptive systems in biology, and ask how well we can understand them using the investigative tools of physics.

1.6 THE EMERGENT GENOME

The challenge of biological physics is to find a set of organizing principles or physical laws that governs biological systems. It is natural to start by thinking about DNA, the master molecule of life. This supermolecule that apparently has the code for the enormous complexity seen in living systems is a rather simple molecule, at least in principle. It consists of two strands that wrap around each other in the famous double helix first clearly described by physicist Francis Crick and his biologist colleague James Watson. While the structure of DNA may be simple, understanding how its structure leads to a living organism is not.

We will use the word "emergent" here to discuss the genome in the following sense: If DNA simply had the codes for genes that are expressed in the organism, it would be a rather boring large table of data. But there is much more to the story than this: Simply knowing the list of genes does not explain the implicit emergence of the organism from this list. Not all the genes are expressed at one time. There is an intricate program that expresses genes as a function of time and space as the organism develops. How this is controlled and manipulated still remains a great mystery.

As Figure 9 shows, the DNA molecule has a helicity, or twist, which arises from the fundamental handedness, or chirality, of biologically derived molecules. This handedness is preserved by the fact that the proteins that catalyze the chemical reactions are themselves handed and highly specific in preserving the symmetry of the molecules upon which they act. The ultimate origin of this handedness is a controversial issue. But we assume that a right-handed or left-handed world would work equally well, and that chiral symmetry breaking such as what we encountered in Unit 2 on the scale of fundamental particles is not present in these macroscopic biological molecules.

It is, however, a mistake to think that biological molecules have only one possible structure, or that somehow the right-handed form of the DNA double helix is the only kind of helix that DNA can form. It turns out that under certain salt conditions, DNA can form a left-handed double helix, as shown in Figure 10. In general, proteins are built out of molecules called "amino acids." DNA, itself a protein, contains the instructions for constructing many different proteins that are built from approximately 20 different amino acids. We will learn more about this later, when we discuss proteins. For now, we will stick to DNA, which is made of only four building blocks: the nitrogenous bases adenine (A), guanine (G), cytosine (C), and thymine (T). Adenine and guanine have a two-ring structure, and are classified as purines, while cytosine and thymine have a one-ring structure and are classified as pyrimidines. It was the genius of Watson and Crick to understand that the basic rules of stereochemistry enabled a structure in which the adenine (purine) interacts electrostatically with thymine (pyrimidine), and guanine (purine) interacts with cytosine (pyrimidine) under the salt and pH conditions that exist in most biological systems.

Not only does the single-stranded DNA (ssDNA) molecule like to form a double-stranded (dsDNA) complex, but the forces that bring the two strands together result in remarkably specific pairings of the base pairs: A with T, and G with C. The pyrimidine thymine base can form strong electrostatic links with the purine adenine base at two locations, while the (somewhat stronger) guanine-cytosine pair relies on three possible hydrogen bonds. The base pairs code for the construction of the organism. Since there are only bases in the DNA molecule, and there are about 20 different amino acids, the minimum number of bases that can uniquely code for an amino acid is three. This is called the triplet codon.

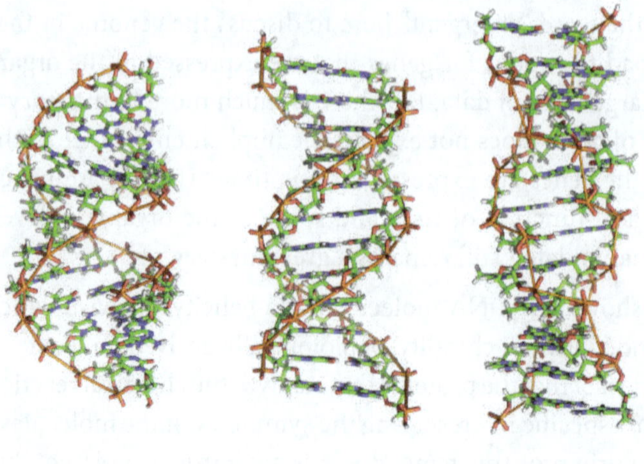

Figure 7: The DNA double helix, in three of its possible configurations.

Source: *Wikimedia Commons, GNU Free Documentation License, Version 1.2. Author: Zephyris (Richard Wheeler), 4 February 2007.*

The remarkable specificity of molecular interactions in biology is actually a common and all-important theme. It is also a physics problem: How well do we have to understand the potentials of molecular interactions before we can begin to predict the structures that form? We will discuss this vexing problem a bit more in the protein section, but it remains a huge problem in biological physics. At present, we really cannot predict three-dimensional structures for biological structures, and it isn't clear if we ever will be able to given how sensitive the structures are to interaction energies and how complex they are.

An example of this extreme sensitivity to the potential functions and the composition of the polymer can be found in the difference between ribonucleic acids (RNA) and deoxyribonucleic acids (DNA).

Structurally, the only difference between RNA and DNA is that at the 2' position of the ribose sugar, RNA has a hydroxyl (OH) molecule—a molecule with one hydrogen and one oxygen atom—while DNA just has a hydrogen atom. Figure 8shows what looks like the completely innocuous difference between the two fundamental units. From a physicist's bottom-up approach and lacking much knowledge of physical chemistry, how much difference can that lone oxygen atom matter?

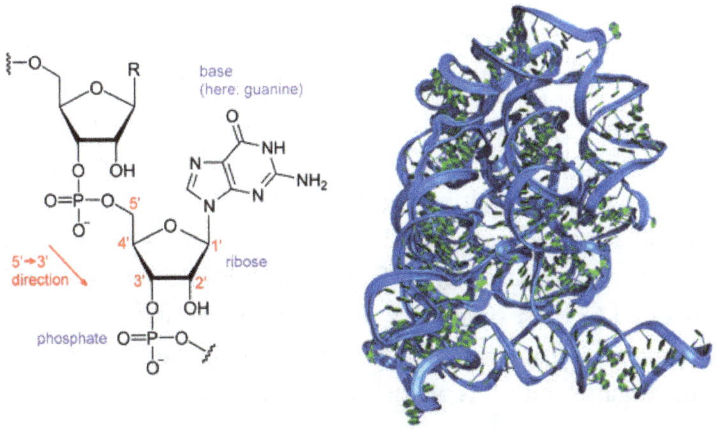

Figure 8: The chemical structure of RNA (left), and the form the folded molecule takes (right).

Source: *Left: Wikimedia Commons, Public Domain. Author: Narayanese, 27 December 2007.*

Commons, Public Domain. Author: MirankerAD, 18 December 2009.

Unfortunately for the bottom-up physicist, the news is very bad. RNA molecules fold into a far more complex structure than DNA molecules do, even through the "alphabet," for the structures are just four letters: A,C, G, and bizarrely U, a uracil group that Nature for some reason has favored over the thymine group of DNA. An example of the complex structures that RNA molecules can form is shown in Figure 11. Although the folding rules for RNA are vastly simpler than those for DNA, we still cannot predict with certainty the three-dimensional structure an RNA molecule will form if we are given the sequence of bases as a starting point.

1.7 THE PUZZLE OF PACKING DNA: CHROMOSOMES

1.7.1 Mapping, Sequencing, and Controversy

Since the DNA molecules code for the proteins that are so critical to life, knowing the sequence of the base pairs is vital to knowing what proteins will be produced. This kind of single base pair resolution fractionation is key to "sequencing." At a much more coarse level, you might want to know the basic ordering of various proteins on a strand of DNA; we call this kind of low resolution "mapping" the DNA. The National Institutes of Health have established a National Center for Biotechnology Information with the express purpose of trying to centralize all the information pouring in from sequencing and mapping projects. The sequencing and mapping of the human genome has been a huge national and private effort, and a very contentious one based upon the raft of ethical, legal, and social implications. The Human Genome Initiative was an astonishing success. However, one school of thought posits that the effort was partly (or maybe mostly) successful because of the efforts of a rival private program headed by entrepreneur Craig Venter, which used a different but complementary approach.

Let's consider a simpler problem than RNA folding: packaging DNA in the cell. A gene is the section of DNA that codes for a particular protein. Since an organism like the bacterium *Escherichia coli* contains roughly 4,000 different proteins and each protein is roughly 100 amino acids long, we would estimate that the length of DNA in *E. coli* must be about 2 million base pairs long. In fact, sequencing shows that the *E. coli* genome actually consists of 4,639,221 base pairs, so we are off by about a factor of two, not too bad. Still, this is an extraordinarily long molecule. If stretched out, it would be 1.2 mm in length, while the organism itself is only about 1 micron long.

The mathematics of how DNA actually gets packaged into small places, and how this highly packaged polymer gets read by proteins such as RNA polymerases or copied by DNA polymerases, is a fascinating exercise in topology. Those of you who are fishermen and have ever confronted a highly tangled fishing line can appreciate that the packaging of DNA in the cell is a very nontrivial problem.

The physics aspect to this problem is the stiffness of the double helix, and how the topology of the twisted and folded molecule affects its biological function. How much energy does it take to bend or twist the polymer into the complex shapes necessary for efficient packaging of DNA in a cell? And how does the intrinsic twist of the double helix translate into the necessity to break the double helix and reconnect it when the code is read by proteins? In other words, biological physics is concerned with the energetics of bending DNA and the topological issues of how the DNA wraps around in space.

The incredible length of a DNA molecule, already bad enough for bacteria, gets more outrageous for higher organisms. Most mammals have roughly 3 x 109 base pairs wrapped up into chromosomes, which are very complex structures consisting of proteins and nucleic acids. However, although we view ourselves as being at the peak of the evolutionary ladder, there seems to be much more DNA in organisms we view as our intellectual inferiors: Some plants and amphibians have up to 1011 base pairs! If we laid out the DNA from our chromosomes in a line, it would have a length of approximately 1 meter; that of amphibians would stretch over 30 meters!

1.7.2 Dark matter in the Genome

Why is the human DNA genome so long, and other genomes even longer still? We don't know exactly how many genes the human genome contains, but a reasonable guess seems to indicate about 30,000. If we imagine that each gene codes for a protein that has about 100 amino acids, and that three base pairs are required to specify each amino acid, the minimal size of the human genome would be about 107 base pairs. It would seem that we have at least 1,000 times as much DNA as is necessary for coding the genome. Clearly, the amount of "excess" DNA must be much higher for plants and amphibians.

Apparently, the DNA is not efficiently coded in the cell, in the sense that lots of so-called "junk" DNA floats around in a chromosome. In fact, a large amount of noncoding DNA has a repeating motif. Despite some guesses about what role this DNA plays, its function remains a substantial puzzle. Perhaps the information content of the genome is not just the number of base pairs, but that there is much "hidden" information contained in this dark genome.

We have succeeded in sequencing the coding part of the human genome, but not the dark part. Are we done now that we know the coding sequence of one given individual? Hardly. We don't know how to extract the information content of the genome at many levels, or even how to define the genome's information quantitatively. The concept of "information" is not only a tricky concept, but also of immense importance in biological physics. Information is itself an emergent property in biology, and it is contextual:

The environment gives meaning to the information, and the information itself means little without the context of the environment.

One problem is that we don't know how to measure information in the genome. Paradoxically, information to a physicist is related to entropy, which is a quantitative measure of disorder. The lower the entropy, the higher the information content. We do, however, need to be careful how we define entropy, because the standard equation in undergraduate physics courses does not apply to a string of base pairs.

1.7.3 Different Meanings of Information

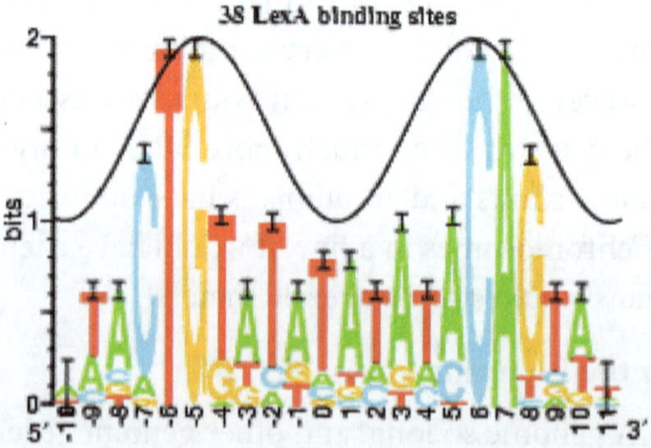

Fig. 9: This sequence logo is a compact way of displaying information contained in a piece of genetic material.

Source: *P. P. Papp, D. K. Chattoraj, and T. D. Schneider,Information Analysis of Sequences that Bind the Replication Initiator RepA, J. Mol. Biol., 233, 219-230, 1993.*

The introduction of entropy emphasizes a critical point: Information has a different meaning to a biologist than it does to a physicist. Suppose you look at some stretch of the genome, and you find that all four of the bases are present in roughly equal numbers—that is, a given base pair has a 25 percent chance to be present in the chain. To a biologist, this implies the sequence is coding for a protein and is informationrich. But a physicist would say it has high entropy and low information, somewhat like saying that it may or may not rain tomorrow. If you say it will rain tomorrow, you convey a lot of information and very little obvious entropy. The opposite is true in gene sequences. To a biologist, a long string of adenines, a is useless and conveys very little information; but in the physics definition of entropy, this is a very low entropy state. Obviously, the statistical concepts of entropy and the biological concepts of information density are rather different.

The dark matter that makes up a huge fraction of the total genome is still very much *terra incognito*. Entropy density maps indicate that it has a lower information density (in the biologist's use of the word) than the "bright matter" coding DNA. Unraveling the mysteries of the dark matter in the genome will challenge biologists just as much as exploring the cosmological variety challenges astrophysicists.

1.8 PROTEINS

Having explored the emergent genome in the form of DNA from a structural and informational perspective, we now move on to the globular polymers called "proteins," the real molecular machines that make things tick. These proteins are the polymers that the DNA codes and are the business end of life. They regulate the highly specific chemical reactions that allow living organisms to live. At 300 K (80°F), the approximate temperature of most living organisms, life processes are characterized by tightly controlled, highly specific chemical reactions that take place at a very high rate. In nonliving matter, highly specific reactions tend to proceed extremely slowly. This slow reaction rate is another result of entropy, since going to a highly specific reaction out of many possible reactions is extremely unlikely. In living systems, these reactions proceed much faster because they are catalyzed by biological proteins called enzymes. It is the catalysis of very unlikely chemical reactions that is the hallmark of living systems.

The mystery of how these protein polymers do their magical chemical catalysis is basically the domain of chemistry, and we won't pursue it further here. As physicists, we will turn our attention to the emergent structure of biological molecules. We saw in the previous section how DNA, and its cousin RNA, have a relatively simple structure that leads, ultimately, to the most complex phenomena around. In this section, we will ask whether we can use the principles of physics to understand anything about how the folded structure of proteins, which is incredibly detailed and specific to biological processes, arises from their relatively simple chemical composition.

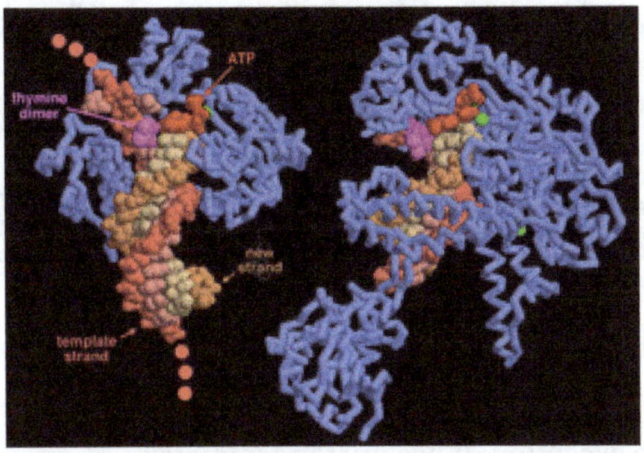

Fig. 10: The enzyme on the left has a much easier time reading DNA than the enzyme on the right due to structural details that are difficult to predict from first principles.

Source: © RCSB Protein Data Bank.

Proteins: The Emergence of Order from Sequence

As polymers go, most proteins are relatively small but much bigger than you might expect is necessary. A typical protein consists of about 100 to 200 monomer links; larger polymers are typically constructed of subunits consisting of smaller balls of single chains. For example, the protein RNA polymerase, which binds to DNA and creates the single-strand polymer RNA, consists (in *E. coli*) of a huge protein with about 500,000 times the mass of a hydrogen atom, divided into five subunits. Despite their small size, folded proteins form exceedingly complex structures. This complexity originates from the large number of monomer units from which the polymers are formed: There are 21 different amino acids. We saw that RNA could form quite complex structures from a choice of four different bases. Imagine the complexity of the structures that can be formed in a protein if you are working with a choice of 21 of them.

Note that you can assign a handedness to the bonding pattern within the protein: Some proteins are lefthanded, and others are right-handed. Experimentally, it was observed that naturally occurring biological molecules (as opposed to molecules synthesized in the laboratory) could rotate the plane of polarization of light when a beam of light is passed through a solution of the molecule. It is easy to see this by getting some maple syrup from a store and observing what happens when a polarized laser beam passes through it. First, orient an "analyzing" polarizer so that no laser light passes through it. Then put the syrup in the laser's path before the analyzing polarizer. You will notice that some light now passes through the polarizer. The beam polarization (as you look at it propagating toward you) has rotated counterclockwise, or in a right-handed sense using the right-hand rule. The notation is that the sugar in the syrup is dextrorotary (D-rotary), or right-handed. In the case of the amino acids, all but one are left-handed, or L-rotary.

Fig. 11: As polarized light passes through corn syrup, which is full of right-handed sugar molecules, its plane of polarization is rotated.

Source: *Technical Services Group, MIT Department of Physics.*

Glycine is the one exception. It has mirror symmetry.

We know how to denote the three-dimensional structure of a protein in a rather concise graphical form. But when you actually see the space-filling picture of a protein—what it would look like if you could see something that small—your physicist's heart must stop in horror. It looks like an ungodly tangled ball. Who in their right mind could possibly be interested in this unkempt beast?

We can say some general things about protein structure. First, the nearest-neighbor interactions are not totally random; they often show a fair amount of order. Experiments have revealed that nature uses several "motifs" in forming a globular protein, roughly specified by the choice of amino acids which naturally combine to form a structure of interest. These structures, determined primarily by nearestneighbor interactions, are called "secondary structures." We commonly see three basic secondary structures: the -helix, the -strand (these combine into sheets), and the polyproline helix.

We can now begin to roughly build up protein structures, using the secondary structures as building blocks. For example, one of my favorite proteins is myoglobin because it is supposed to be simple. It is not. We can view it as basically a construction of several alpha helices which surround a "prosthetic group," the highly conjugated heme structure used extensively in biology. Biologists often regard myoglobin as a simple protein. One possible function is to bind oxygen tightly as a storage reservoir in muscle cells. There may be much more to this molecule than meets the eye, however.

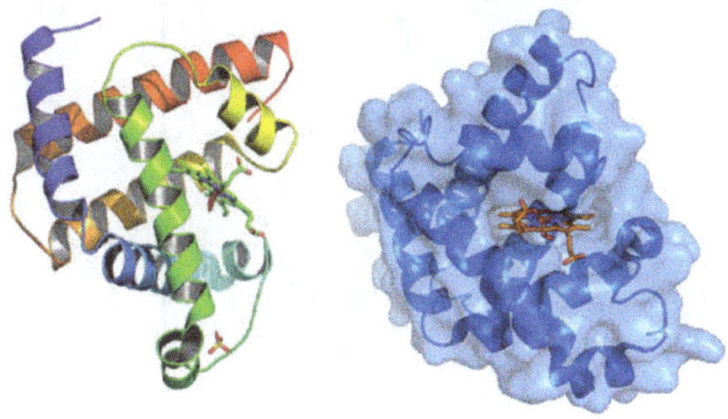

Fig. 12: The structure of myoglobin (left) and the form it actually takes in space (right).

Source: *Left: Wikimedia Commons, Public Domain. Author: AzaToth, 27 February 2008; right: Wikimedia Commons Creative Commons Attribution-Share Alike 3.0 Unported License. Author: Thomas Splettstoesser, 10 July 2006.*

As their name indicates, globular proteins are rather spherical in shape. Also, the polarizability of the various amino acids covers quite a large range, and the protein is designed (unless it is membranebound) to exist in water, which is highly polar. As biologists see it, the polarizable amino acids are predominantly found in the outer layer of the globular protein, while the non-polar amino acids reside deep in the interior. This arrangement is not because the non-polar amino acids have a strong attraction for one another, but rather because the polar amino acids have strong interactions with water (the socalled hydrophilic effect) and because introducing non-polar residues into water gives rise to a large negative entropy change (the so-called hydrophobic effect). So, physics gives us some insight into structure, through electrostatic interactions and entropy.

One kind of emergence we wish to stress here is that, although you would think that a polymer consisting of potentially 21 different amino acids for each position would form some sort of a glue-ball, it doesn't. Many proteins in solution seem to fold into rather well-defined three-dimensional shapes. But can we predict these shapes from the amino acid sequence? This question is known as the "protein-folding problem," and has occupied many physicists over the past 30 some years as they attempt with everincreasingly powerful computers to solve it. While Peter Wolynes and Jose Onuchich have been able to sketch out some powerful ideas about the general path of the protein folding that make use of the physics concept of free energy, it could well be that solving the puzzle to a precise answer may be impossible.

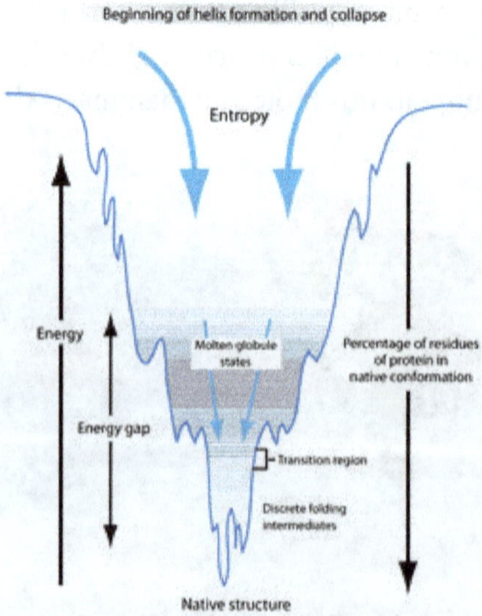

Fig. 13: A schematic of how minimizing the free energy of a molecule could lead to protein folding.

There may well be a fundamental reason why a precise answer to the folding problem is impossible: because in fact there may be no precise answer! Experiments by Hans Frauenfelder have shown that even for a relatively simple protein like the myoglobin presented in Figure 13, there is not a unique ground state representing a single free energy minimum but rather a distribution of ground states with the same energy, also known as a conformation distribution, which are thermally accessible at 300 K. It is becoming clear that this distribution of states is of supreme importance in protein function, and that the distribution of conformations can be quite extreme; the "landscape" of conformations can be extremely rugged; and within a given local valley, the protein cannot easily move over the landscape to another state. Because of this rugged landscape, a protein might often be found in metastable states: trapped in a low-lying state that is low, but not the lowest, unable to reach the true ground state without climbing over a large energy barrier.

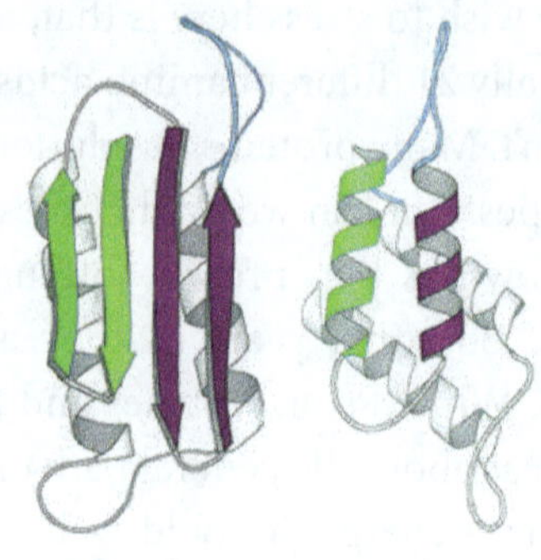

Fig. 14:Two possible conformations of a prion protein: on the left as a beta sheet; on the right as an alpha helix.

Source: © *Flickr, Creative Commons License. Author: AJC1, 18 April 2007.*

An extreme example of this inherent metastability of many protein structures, and the implication to biology, is the class of proteins called "prions." These proteins can fold into two different deep valleys of free energy: as an alpha-helix protein rather like myoglobin, or as a beta-sheet protein. In the alpha-helix conformation, the prion is highly soluble in water; but in the beta-sheet conformation, it tends to aggregate and drop out of solution, forming what are called "amyloid plaques," which are involved with certain forms of dementia. One energy valley leads to a structure that leads to untreatable disease; the other is mostly harmless.

The apparent extreme roughness of biological landscapes, and the problems of ascertaining dynamics on such landscapes, will be one of the fundamental challenges for biological physics and the subject of the next section.

1.9 FREE ENERGY LANDSCAPES

A certified unsolved problem in physics is why the fundamental physical constants have the values they do. One of the more radical ideas that has been put forward is that there is no deeper meaning.

The numbers are what they are because an unaccountable number of alternate universes are forming a landscape of physical constants. We just happen to be in a particular universe where the physical constants have values conducive to form life and eventually evolve organisms who ask such a question.

This idea of a landscape of different universes actually came from biology, evolution theory, in fact, and was first applied to physics by Lee Smolin. Biology inherently deals with landscapes because the biological entities, whether they are molecules, cells, organisms, or ecologies, are inherently heterogeneous and complex. Trying to organize this complexity in a systematic way is beyond challenging. As you saw in our earlier discussion of the protein folding problem, it is easiest to view the folding process as movement on a free energy surface, a landscape of conformations.

1.9.1 Glasses, Spin Glasses, Landscapes

There is a physical system in condensed matter physics that might provide a simpler example of the kind of landscape complexity that is characteristic of biological systems. Glasses are surprisingly interesting physical systems that do not go directly to the lowest free energy state as they cool. Instead, they remain frozen in a very high entropy state. For a physical glass like the windows of your house, the hand-waving explanation for this refusal to crystallize is that the viscosity becomes so large as the system cools, that there is not enough time in the history of the universe to reach the true ground state.

Fig. 15: As a glass cools, the viscosity increases so rapidly that the atoms get frozen in a disordered state.

Source: *OHM Equipment, LLC.*

A more interesting glass, and one more directly connected to biology, is the spin glass. It actually has no single ground state, which may be true for many proteins as well. The study of spin glasses in condensed matter physics naturally brings in the concepts of rough energy landscapes, similar to those we discussed in the previous section. The energy landscape of a spin glass is modified by interactions within the material. These interactions can be both random and frustrated, an important concept that we will introduce shortly. By drawing an analogy between spin glasses and biological systems, we can establish some overriding principles to help us understand these complex biological structures.

A spin glass is nothing more than a set of spins that interact with each other in a certain way. At the simplest level, a given spin can be pointing either up or down, as we saw in Unit 6; the interaction between two spins depends on their relative orientation. The interaction term Jij specifies how spin interacts with spin j. Magically, it is possible to arrange the interaction terms between the spins so that the system has a large set of almost equal energy levels, rather than one unique ground state. This phenomenon is called "frustration."

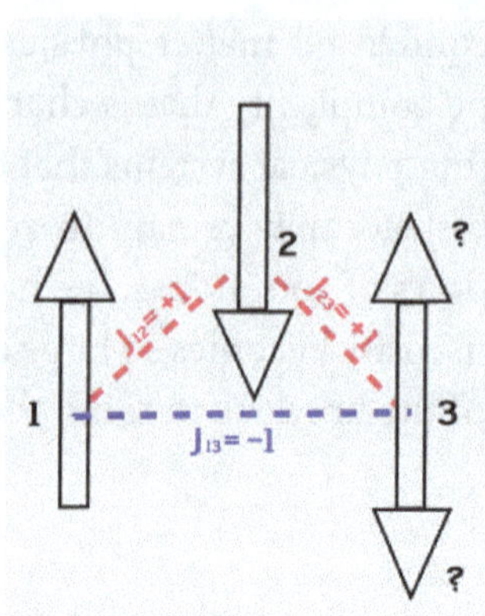

Fig. 16:This simple system of three spins is frustrated, and has no clear ground state.

For a model spin glass, the rule that leads to frustration is very simple. We simply set the interaction term to be +1 if the two spins point in the same direction, and -1 if they point in different directions. If you go around a closed path in a given arrangement of spins and multiply all the interaction terms together, you will find that if the number is +1, the spins have a unique ground state; and if it is -1, they do not. Figure 16 shows an example of a simple three-spin system that is frustrated. The third spin has contradictory commands to point up and point down. What to do? Note that this kind of a glass is different from the glass in your windows, which would find the true ground state if it just had the time. The spin glass has no ground state, and this is an emergent property.

Frustration arises when there are competing interactions of opposite signs at a site, and implies that there is no global ground energy state but rather a large number of states with nearly the same energy separated by large energy barriers. As an aside, we should note that this is not the first time we've encountered a system with no unique ground state. Systems with spontaneously broken symmetry also had many possible ground states. The difference here is that the ground states of the system with broken symmetry were all connected in field space—on the energy landscape, they are all in the same valley—whereas the nearly equal energy levels in a frustrated system are all isolated in separate valleys with big mountains in between them. The central concept of frustration is extremely important in understanding why a spin glass forms a disordered state at low temperatures, and must play a crucial role in the protein problem as well.

1.9.2 Hierarchical States

Fig. 17: A Rubik's Cube is a familiar example of a hierarchical distribution of states.

Source: *Wikimedia Commons, GNU Free Documentation License 1.2. Author: Lars Karlsson (Keqs), 5 January 2007.*

Take a look at a Rubik's cube. Suppose you have some random color distribution, and you'd like to go back to the ordered color state. If you could arbitrarily turn any of the colored squares, going back to the desired state would be trivial and exponentially quick. However, the construction of the cube creates large energy barriers between states that are not "close" to the one you are in; you must pass through many of the allowed states in some very slow process in order to arrive where you want to be. This distribution of allowed states that are close in "distance" and forbidden states separated by a large distance is called a hierarchical distribution of states. In biology, this distance can mean many things: how close two configurations of a protein are to each other, or in evolution how far two species are apart on the evolutionary tree. It is a powerful idea, and it came from physics.To learn anything useful about a hierarchy, you must have some quantitative way to characterize the difference between states in the hierarchy. In a spin glass, we can do this by calculating the overlap between two states, counting up the number of spins that are pointing the same way, and dividing by the total number of spins. States that are similar to one another will have an overlap close to one, while those that are very different will have a value near zero. We can then define the "distance" between two states as one divided by the overlap; so states that are identical are separated by one unit of distance, and states that are completely different are infinitely far apart.\

Knowing that the states of a spin glass form a hierarchy, we can ask what mathematical and biological consequences this hierarchy has. Suppose we ask how to pass from one spin state to another. Since the spins interact with one another, with attendant frustration "clashes" occurring between certain configurations, the process of randomly flipping the spins hoping to blunder into the desired final state is likely to be stymied by the high-energy barriers between some of the possible intermediate states.

Fig. 18: Here, we see two possible paths across an energy landscape strewn with local minima.

A consistent and logical approach would be to work through the hierarchical tree of states from one state to another. In this way, one always goes through states that are closely related to one another and hence presumably travels over minimum energy routes. This travel over the space is movement over a landscape. In Figure 18, we show a simulated landscape, two different ways that system might pick its way down the landscape, and the local traps which can serve as metastable sticking points.

In some respects, this landscape picture of system dynamics is more descriptive than useful to the central problems in biological physics that we are discussing in this course. For example, in the protein section, we showed the staggering complexity of the multiple-component molecular machines that facilitate the chemical reactions taking place within our bodies, keeping us alive. The landscape movement we have described so far is driven by pre-existing gradients in free energy, not the time-dependent movement of large components. We believe that what we observe there is the result of billions of years of evolution and the output of complex biological networks.

1.10 EVOLUTION

Biological evolution remains one of the most contentious fields to the general public, and a dramatic example of emergent phenomena. We have been discussing the remarkable complexity of biology, and it is now natural to ask: How did this incredible complexity emerge on our planet? Perhaps a quote from French Nobel Laureate biologist Jacques Monod can put things into perspective: "Darwin's theory of evolution was the most important theory ever formulated because of its tremendous philosophical, ideological, and political implications." Today, over 150 years after the publication of "On the Origin of the Species," evolution remains hotly debated around the world, but not by most scientists. Even amongst the educated lay audience, except for some cranks, few have doubt about Newton's laws of motion or Einstein's theories of special and general relativity, but about half of the American public don't agree with Darwin's theory of evolution. Surely, physics should be able to clear this up to everybody's satisfaction.

Fig. 19: The variation in Galapagos finches inspired Charles Darwin's thinking on evolution, but may evolve too fast for his theory.

Source: © Public Domain from *The Zoology of the Voyage of H.M.S. Beagle, by John Gould. Edited and superintended by Charles Darwin. London: Smith Elder and Co., 1838.*

Or maybe not. The problem is that simple theories of Darwinian evolution via random mutations and natural selection give rise to very slow change. Under laboratory conditions, mutations appear at the low rate of one mutated base pair per billion base pairs per generation. Given this low observed rate of mutations, it becomes somewhat problematic to envision evolution via natural selection moving forward to complex organisms such as humans. This became clear as evolution theories tried to move from Darwin's vague and descriptive anecdotes to a firmer mathematical foundation.

Recent work on the Galapagos Islands by the Princeton University biologists Peter and Rosemary Grant revealed something far more startling than the slow evolution of beak sizes. The Grants caught and banded thousands of finches and traced their elaborate lineage, enabling them to document the changes that individual species make in reaction to the environment. During prolonged drought, for instance, beaks may become longer and sharper, to reach the tiniest of seeds. Here is the problem: We are talking about thousands of birds, not millions. We are talking about beaks that change over periods of years, not thousands of years. How can evolution proceed so quickly?

1.10.1 Fitness Landscapes and Evolution

In our protein section, we discussed the concept of a free energy landscape. This indicates that proteins do not sit quietly in a single free energy minimum, but instead bounce around on a rough landscape of multiple local minima of different biological functional forms. But this idea of a complex energy landscape did not originate from proteins or spin glasses. It actually came from an American mathematical biologist named Sewall Wright who was trying to understand quantitatively how Darwinian evolution could give rise to higher complexity—exactly the problem that has vexed so many people.

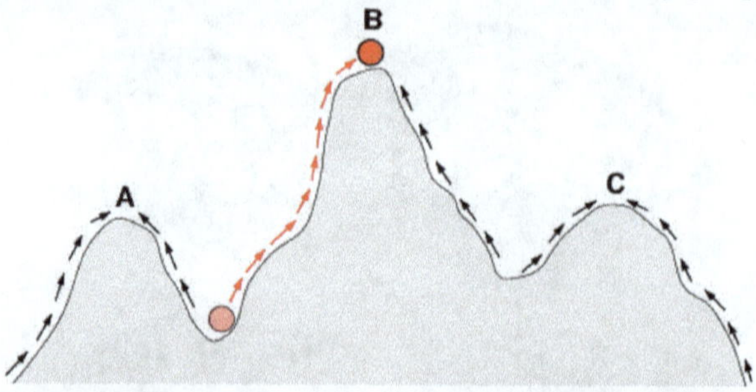

Fig. 20: Natural selection can be viewed as movement on a fitness landscape.

Source: *Wikimedia Commons, Public Domain. Author: Wilke, 18 July 2004.*

We can put the problem into simple mathematical form. Darwinian evolution is typically believed to be due to the random mutation of genes, which occurs at some very small rate of approximately 109 mutations/base pair-generation under laboratory conditions. At this rate, a given base pair would undergo a random mutation every billion generations or so. We also believe that the selection pressure—a quantitative measure of the environmental conditions driving evolution—is very small if we are dealing with a highly optimized genome. The number of mutations that "fix," or are selected to enter the genome, is proportional to the mutation rate times the selection pressure. Thus, the number of "fixed" mutations is very small. A Galapagos finch, a highly evolved creature with a genome optimized for its environment, should not be evolving nearly as rapidly as it does by this formulation.

There is nothing wrong with Darwin's original idea of natural selection. What is wrong is our assumption that the mutation rate is fixed at 10-9 mutations/base-pair generation, and more controversially perhaps that the mutations occur at random on the genome, or that evolution proceeds by the accumulation of single base-pair mutations: Perhaps genomic rearrangements and basepair chemical modifications (a process called "epigenetics") are just as important. Further, we are beginning to understand the role of ecological complexity and the size of the populations. The simple fitness landscape of Figure 20 is a vast and misleading simplification. Even in the 1930s, Seawall Wright realized that the dynamics of evolution had to take into account rough fitness landscapes and multiple populations weakly interbreeding across a rough landscape. Figure 24 dating all the way back to 1932, is a remarkably prescient view of where evolution biological physics is heading in the 21st century.

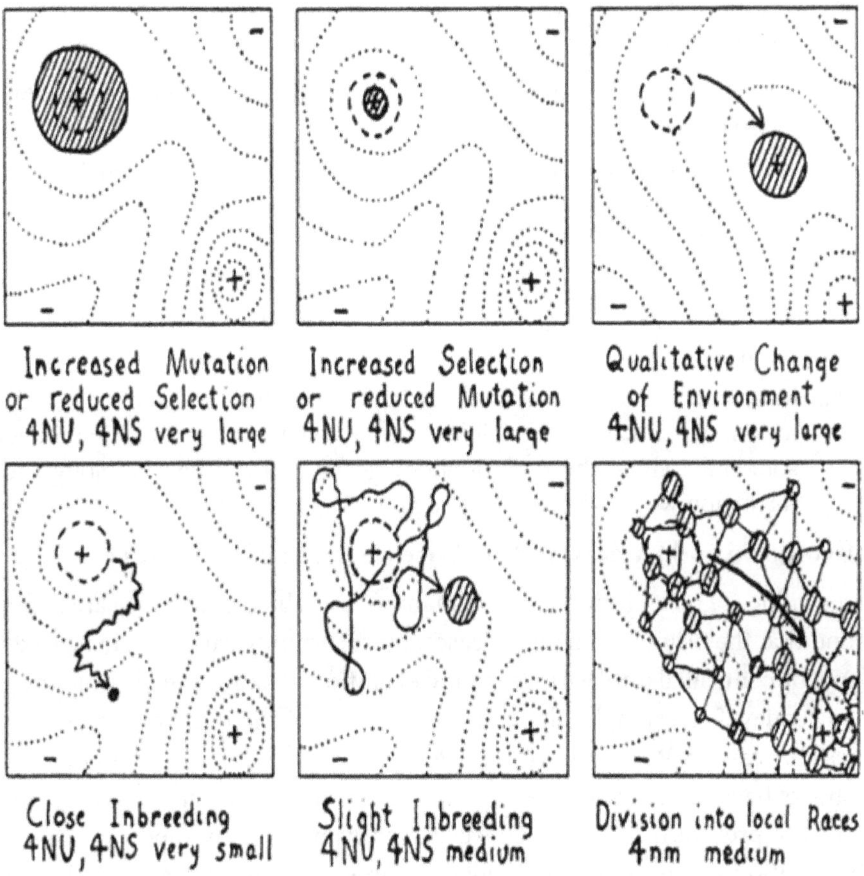

Fig. 21: Sewall Wright sketched the path different populations might take on the fitness landscape.

Source: *Sewall Wright, "The Role of Mutation, Inbreeding, Crossbreeding, and Selection in Evolution," Sixth International Congress of Genetics, Brooklyn, NY: Brooklyn Botanical Garden, 1932.*

Darwinian evolution in a broader sense is also changing the face of physics as the fundamental concepts flow from biology to physics. Darwinian evolution as modified by recent theories teaches us that it is possible to come to local maxima in fitness in relatively short time frames through the use of deliberate error production and then natural selection amongst the errors (mutants) created. This seems somewhat counterintuitive, but the emergence of complexity from a few simple rules and the deliberate generation of mistakes can be powerfully applied to seemingly intractable problems in computational physics.

Applications of Darwinian evolution in computational physics have given rise to the field of evolutionary computing. In evolutionary computing, principles taken from biology are explicitly used. Evolutionary computation uses the same iterative progress that occurs in biology as generations proceed, mutant individuals in the population compete with other members of the population in a guided random search using parallel processing to achieve the increase in net fitness. To be more specific, the steps required for the digital realization of a genetic algorithm are:

i. A population of digital strings encode candidate solutions (for example, a long, sharp beak) to an
ii. optimization problem (needing to adapt to drought conditions).
iii. In each generation, the fitness of every string is evaluated, and multiple strings are selected based
iv. on their fitness.
v. The strings are recombined and possibly randomly mutated to form a new population.
vi. Re-iterate the next generation. It is possible that by exploring artificial evolution, which came from biology and moved into physics, that we will learn something about the evolutionary algorithms running in biology and the information will flow back to biology.

1.10.2 Evolution and Understanding Disease in the 21st Century

The power influence of evolution is felt in many areas of biology, and we are beginning to understand that the origins of many diseases, most certainly cancer, may lie in evolution and will not be controlled until we understand evolution dynamics and history much better than we do today. For example, shark cartilage is one of the more common "alternative medicines" for cancer. Why? An urban legend suggests that sharks do not get cancers. Even if sharks have lower incidence rates of cancer than *Homo sapiens*, they possess no magic bullet to prevent the disease. However, sharks possess an important characteristic from an evolution perspective: They represent an evolutionary dead-end. Judging from the fossil record, they have evolved very little in 300 million years, and have not attempted to scale the fitness landscape peaks that the mammals eventually conquered.

We can ask two questions based on what we have developed here: Is cancer an inevitable consequence of rapid evolution, and in that sense not a disease at all but a necessary outlier tail of rapid evolution? And is cancer, then, inevitably connected with high evolution rates and high stress conditions and thus impossible to "cure"?

Fig. 22: Cartilage from the fin of the Mako shark.

Source: www.OrangeBeach.ws.

Stress no doubt drives evolution forward, changing the fitness landscapes we have discussed from a basically smooth, flat, and boring plane into a rugged landscape of deep valleys and high peaks. Let us assume that in any local habitat or ecology is a distribution of genomes that includes some high-fitness genomes and some low-fitness genomes. The low-fitness genomes are under stress, but contain the seeds for evolution. We define stress here as something that either directly generates genomic damage, such as ionizing radiation and chemicals that directly attack DNA, viruses, or something that prevents replication of the genome, such as blockage of DNA polymerases or of the topological enzymes required for chromosome replication. Left unchallenged, all these stress inducers will result in the extinction of the quasi-species.

This is the business end of the grand experiment in exploring local fitness peaks and ultimately in generating resistance to stress. The system must evolve in response to the stress, and it must do this by deliberately generating genetic diversity to explore the fitness landscape—or not. Viewed in the perspective of game theory's prisoner's dilemma (see sidebar), the silent option under stress is not to evolve—to go down the senescent pathway and thus not attempt to propagate. Turning on mutational mechanisms, in contrast, is a defection, in the sense that it leads potentially to genomes which can propagate even in what should be lethal conditions and could, in principle, lead to the destruction of the organism: disease followed by death, which would seem to be very counterproductive. But it may well be a risk that the system is willing to make. If ignition of mutator genes and evolution to a new local maximum of fitness increases the average fitness of the group, then the inevitable loss of some individuals whose genome is mutated into a fitness valley is an acceptable cost.

1.11 NETWORKS

The complex biological molecules we have trying to understand are the building blocks of life, but it is far from obvious to put these building blocks together into a coherent whole.

Biological molecules, as well as cells and complete organisms, are organized in complex networks. A network is defined as a system in which information flows into nodes, is processed, and then flows back out. The network's output is a function of both the inputs and a series of edges that are the bidirectional paths of information flow between the nodes. The theory and practice of networks is a vast subject, and with one ultimate goal of understanding that greatest of mysteries, the human brain. We will return to the brain and its neural networks in the next section. For now, we will discuss the more prosaic networks in living organisms, which are still complex enough to be very intimidating.

It isn't obvious when you look at a cell that a network exists there. The cytoplasm of a living cell is a very dynamic entity, but at least at first glance seems to basically be a bag of biological molecules mixed chaotically together. It is somewhat of a shock to realize that this bag of molecules actually contains a huge number of highly specific biological networks all operating under tight control. For example, when an epithelial cell moving across a substrate, patterns of specific molecules drive the motion of the cell's internal skeleton. When these molecules are tagged with a protein that glows red, displaying the collective molecular motion under a microscope, a very complex and interactive set of networks appears.

- The emergent network of the cytoplasm is a system of interconnected units. Each unit has at least an input and an output, and some sort of a control input which can modulate the relationship between the input and the output. Networks can be analog, which means that in principle the inputs and outputs are continuous functions of some variable; or they can be digital, which means that they have finite values, typically 1 or 0 for a binary system. The computer on which this text was typed is a digital network consisting of binary logic gates, while the person who typed the text is an analog network. There are many different kinds of biological networks, and they cover a huge range of length scales, from the submicron (a micron is a millionth of a meter) to the scales spanning the Earth. Outside of neural networks, the most important ones are (roughly in order of increasing abstraction): Metabolite networks: These networks control how a cell turns food (in the form of sugar) into energy that it can use to function. Enzymes (proteins) are the nodes, and the smaller molecules that represent the flow of chemical energy in the cell are the edges.
- Signal transduction networks: These networks transfer information from outside the cell into its interior, and translate that information into a set of instructions for activity within the cell. Proteins are the nodes, typically proteins called kineases, and diffusible small signaling molecules which have been chemically modified are the edges. This is a huge class of networks, ranging from networks that process sensory stimuli to chemical inputs such as hormones.
- Transcriptional regulatory networks: These networks determine how genes are turned on and off (or modulated).
- Interorganism networks: This is a very broad term that encompasses everything from the coordinated behavior of a group of bacteria to complex ecologies. The

nodes are individual cells, and the edges are the many different physical ways that cells can interact with each other.

- There are probably fundamental network design principles that must be obeyed independent of their biological or manmade (which is still biological) origin if the network is to be stable to perturbations.

Instability in a network is not generally viewed as a good thing, although there are exceptions to this rule. For example, the aerodynamics of most modern fighter jets makes the plane inherently unstable. This sounds like a very bad thing, except that it makes the fighter extremely adaptive to direction changes.

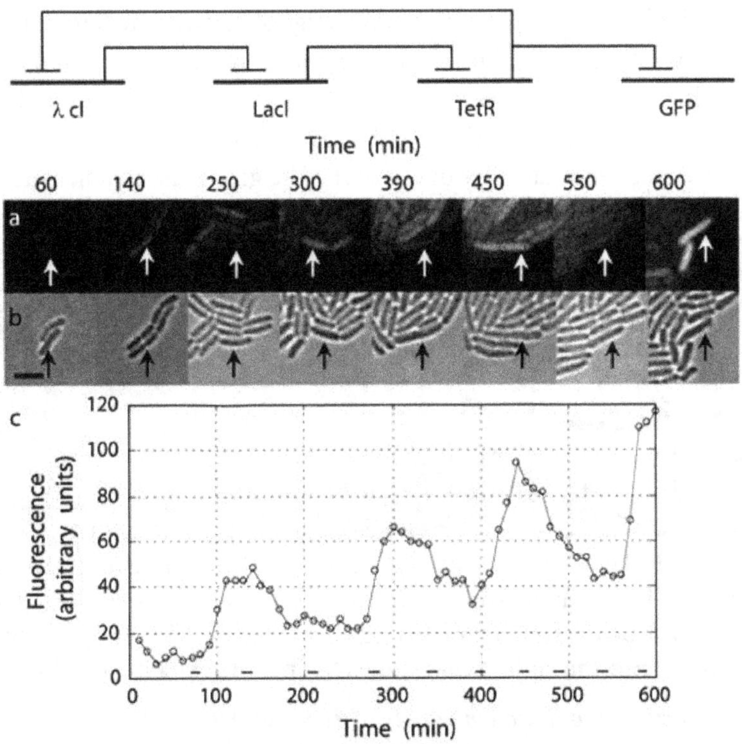

Fig. 23: The circuit diagram (top), bacterial population (center), and plot of the dynamics (bottom) of the repressilator, an example of a simple synthetic biological network.

Source: *Top: Wikimedia Commons, Public Domain. Author: Timreid, 26 February 2007. Center and bottom: Reprinted by permission from Macmillan Publishers Ltd: Nature 403, 335-338 (20 January 2000).*

Modern computers can constantly monitor and instantaneously correct the instability, so we end up with aircraft that are far more maneuverable—and thus more effective fighters—than the ordinary, stable variety.

The kind of stability issues the fighter jet and other similar networks face are deterministic, and can be modeled by ordinary differential equations that are straightforward to write down. One might then imagine designing a network based on a set of these equations. One of the pioneering exercises in designing "from scratch" was the work of Elowitz and Leibler of an oscillating gene expression pattern. It is sobering to understand the depth of understanding that was necessary to have made this simple oscillator work.

For an electrical engineer, it is straightforward to design an oscillator following some basic rules of electromagnetism. However, in a biological network, the parameters are much less cleanly defined.

Despite the inherent challenges, we now have a basic set of biological modules that is being developed in the new field of "synthetic biology," which is a mix of physics, engineering, and biology that exploits our knowledge of networks to design new functional biological "circuits." Figure 27 shows an example of a biological "film" consisting of bacteria. To do this, a gene was inserted into *E. coli* that coded for a protein that causes the bacteria to make a black pigment. The pigment production was coupled to a light sensor, so that pigment would be made only in the dark. The group used stencils to pattern light exposure and produce bacterial photography.

Fig. 24: Portrait of the University of Texas at Austin (UT Austin)/University of California, San Francisco (UCSF) Synthetic Biology team. The left panel shows the projected image of the students and professors from UT Austin and UCSF who participated in the project, and the right panel shows the resulting bacterial photo.

Source: *Aaron Chevalier and the University of Texas at Austin.*

In addition to deterministic stability issues in biological networks, there is also the issue of stability in the presence of noise. For example, at the macro-scale, the sensory network of the dog's nose is about a million times more sensitive than a human noise. Despite this extreme sensitivity, the dog nose is not overwhelmed by the presence of an enormous background of other molecules. That is, the dog nose sensory network is extremely good at noise rejection. At the micro-scale of the single cell, the very finite number of molecules actually involved in the network node edges leads to statistical noise that can either confound the network's stability or increase its sensitivity. The dog has clearly resolved this issue to its benefit, but it remains a problem in the design of synthetic biological networks.

The obvious end goal of synthetic biology could be something truly astonishing: synthetic life. The key to synthetic life, if it is indeed achieved, will be our ability to harness biological networks. So far, scientists have synthesized genetic material and other important biological molecules from scratch, but have not put the pieces together into a complete living organism. The feat hailed by the media as the so-called creation of a "new form of life" by Craig Venter is something of a misnomer. While Venter's synthesis of a functional bacterial chromosome of one million base pairs was a fantastic technical achievement, it is very far from synthetic life, as the new chromosome was inserted into an already functioning cell consisting of a enormous collection of interacting networks, which we neither can understand nor can reproduce.

Until we can understand the emergence of life from the networks of biology, we will remain very far from achieving synthetic life.

The Emergence of the Mind

So far, we have wrestled with the structural diversity of proteins and its relationship to the free energy landscape, and we have tried to find some of the unifying and emergent properties of evolution that might explain the diversity of life and the increase in complexity. We have also taken a look at how the biological networks necessary to bind a collection of inanimate objects into a living system emerge. At the highest level lies the greatest mystery of biological physics: the emergence of the mind from a collection of communicating cells.

Fig. 25: Ruby-throated hummingbird.

Source: *Steve Maslowski, U.S. Fish and Wildlife Service.*

We started our discussion of biological physics by considering a chicken egg. Birds lay eggs, so let's consider a bird: the ruby-throated hummingbird presented in all its glory in Figure 28. About 7 cm long and weighing about 5 grams, this bird is capable of some rather amazing biophysical things. Its wings beat about 50 times each second, and they rotate around their central axis through almost 180 degrees, allowing the bird to fly backwards and forwards and hover. A lot of these fascinating mechanical properties can be considered the subject of biological physics.

But there is far more to these hummingbirds than just flying ability. They live for about nine years, spend their summers in the northern parts of North America and their winters in tropical Central America. So, this small animal can navigate in the fall over thousands of miles, including over hundreds of miles of open water, to certain locations and then return in the spring to the region it was born. How is it that a tiny hummingbird can do all this remarkable navigation? The advancement in capabilities of the digital computer over the past 30 years has been truly staggering, yet it pales against what the hummingbird's brain can do. The human brain is far more impressive. Why can a couple of pounds of neurons drawing a few watts of chemical power with an apparent clock speed of maybe a kilohertz at best do certain tasks far better than a machine the size of a large truck running megawatts of power? And at a much more troubling level, why do we speak of the soul of a person when no one at this point would seriously ascribe any sense of self-recognition to one of our biggest computers? We seem to be missing something very fundamental.

1.12 TRADITIONAL COMPUTERS VS. BIOLOGY

We have moved into the computer age via the pathway pioneered by British mathematician Alan Turing, whom we first met in the introduction to this unit. Our modern-day computers all basically use the model described in Figure 26, coupled with the idea that any number is to be presented by bits in a binary representation. We have made things much faster than those early computers, but the basic idea has not changed.

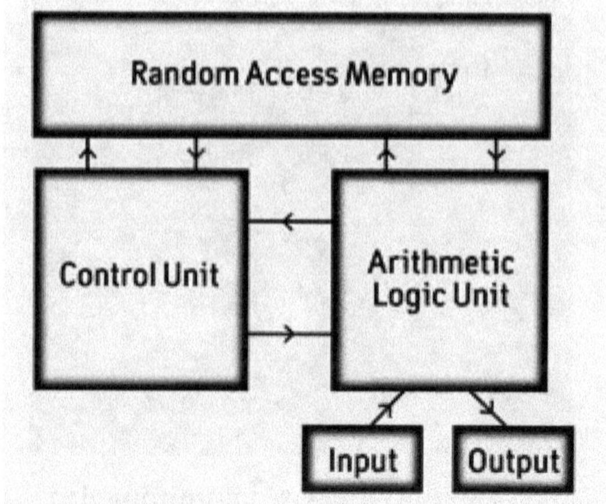

Fig. 26: Schematic of a modern digital computer.

But this is not how biology has developed its own computers. The basic design has four major flaws as far as biology is concerned:

- The machine must be told in advance, in great error-free detail, the steps needed to perform the algorithm.

- Data must be clean; the potential loss of a single bit can crash the code.
- The hardware must be protected and robust; one broken lead and the machine can crash.
- There is an exact correspondence between a bit of data and a hardware location: The information in the machine is localized.

None of this is any good for a biological system. As far as biology is concerned, our computers are evolutionary dead-ends. We started this unit by considering the fragility of the egg in a large fall. Yet, as the example of Phineas Gage in the side bar shows, our brain can take enormous abuse and remain basically functional. I challenged you initially to drop the possibly cooked egg and see what happens.

Now I challenge you to take a 2 cm diameter steel rod and thrust it through your laptop with great force, then try to surf the web.

1.13 THE BRAIN OF A NEMATODE

The human brain is probably the most complex structure in the universe that we know, but not only humans have brains. The adult hermaphrodite of the "lowly" nematode *C. elegans* consists of only 959 cells; yet when you watch it navigating around on an agar plate, it certainly seems to be computing something based on its sensory input. The creature displays an astonishingly wide range of behavior: locomotion, foraging, feeding, defecation, egg laying, larva formation, and sensory responses to touch, smell, taste, and temperature, as well as some complex behaviors like mating, social behavior, and learning and memory. It would be quite hard to build a digital computer that could do all that, and certainly impossible to pack it into a tube about 1 mm long and a 100 microns in diameter that can reproduce itself.

The *C. elegans* doesn't have a brain per se, but it does have about 302 information-carrying neuronsthat form approximately 7,000 synapses. We believe that any real brain capable of making some sort of a computation, as opposed to the collective behavior seen in single-celled organisms, must consist of neurons that transfer information. That information is not transferred to some sort of a central processing unit. Biological computers are systems of interconnected cells that transfer and process information. The network of neurons in *C. elegans* displays the common feature in interconnectivity: the synaptic connections formed by the neurons.

1.13.1 The Brain Versus the Computer

I want to concentrate on one thing here: how differently the brain, even the pseudo-brain of *C. elegans*, is "wired" from the computer that you're using to read this web page. Your computer has well-defined regions where critical functions take place: a section of random access memory (RAM) and a central processing unit (CPU). Each part is quite distinct, and buses transfer binary data between the different sections. Take out a single bus line or damage one of the RAM chips, and the system shuts down.

Brains in biology seem to have evolved in a different way. First, they are spatially diffuse. The computer is basically a two-dimensional device. Brains at every level seem to be basically three-dimensional. The interconnection takes place not via a bus, but rather through a vast network of input-output synaptic connections. For example, *C. elegans* has roughly 20 interconnects per neuron. In the human brain, we believe that the number is on the order of 103. Since the human brain has around 1012 neurons, the number of interconnects is on the order of 1015—a huge number.

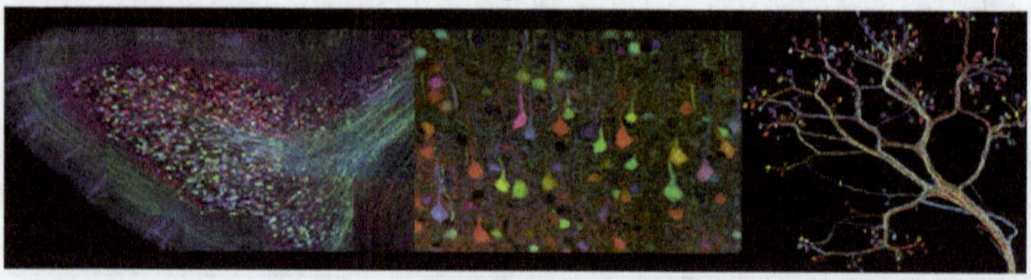

Fig. 27:Rainbow images showing individual neurons fluorescing in different colors. By tracking the neurons through stacks of slices, we can follow each neuron's complex branching structure to create the treelike structures in the image on the right.

Source: *Jeff Lichtman, Center for Brain Science, Harvard University.*

iIt would be a mistake to think that the 1012 neurons in the brain correspond to about 1012 bits of information, or about 100 Gigabytes. The number is much higher, because of the three-dimensional interconnections linking each neuron with about 103 other neurons. Returning to our theme of spin glasses, we can estimate the information capacity by making the simple assumption that each neuron can be like a spin which is either up or down depending on its storage of a bit of information. This means that the total number of differing configurations of the brain is on the order of , an absurdly huge number, far greater than even the number of atoms in the universe. We can only assume that the brains of living organisms emerged as they exploited this immense 3-D information capacity owing to the ability of communities of cells to form neuronal interconnections throughout space.

1.13.2 How Does the Brain Reason?

Given the large information capacity of even a small network of neurons and the fact that the human brain's capacity exceeds our ability to comprehend it, the next question is: How does a brain reason? As usual, we need to start by defining what we're talking about. According to the *Oxford English Dictionary*, "reasoning" is "find[ing] a solution to a problem by considering possible options." I suppose this dodges the question of the emergent property of consciousness, but I don't see this problem being solved any time soon, although I hope I am wrong.

The hummingbird has a big problem, essentially asking itself: How shall I fly back to a place I was at six months ago that is thousands of miles away from where I am now? Presumably, the bird uses different physics than that of a traditional computer, because the information content that the bird has to sort out would cause it to fail catastrophically. So, we finally have the problem that perhaps physics can attack and clarify in the 21st century: How can a set of interacting neurons with a deep level of interconnects take previously stored information and determine an optimal solution to a problem it has not yet seen?

The hummingbird faces a problem rather reminiscent of the traveling salesman problem, explained in the sidebar. To choose the correct locations to pass through on its springtime journey north, it must consider a number of combinations far beyond the power of any computer system to resolve. How does the hummingbird do it? Is it magic?

Physics shows that it isn't magic. As we have previously discussed, while a protein may fold or a species play with its genome in an almost uncountable number of ways, basic free energy minima schemes lead quite efficiently to a vastly smaller set of combinations that are roughly optimal. Nature doesn't necessarily find the "best" solution, but it seems able to efficiently find a subset of solutions that works well enough. In the case of the traveling salesman problem, the vast combinatorics interconnects of a neural network of many neurons provides exactly the kind of search over a free energy surface that we need.

1.13.3 The "Reasoning" Ability of Neural Networks

We have discussed how landscapes—either fitness landscapes or free energy landscapes—can give rise to vastly complex surfaces with local minima representing some particular desired state. John Hopfield,a theoretical physicist at Princeton University, has explored ways for a system to find these minima. The three basic ideas below highlight how biological computers differ from their electronic counterparts:

- Neural networks are highly interconnected. This interaction network can be characterized by a matrix, which tabulates the interaction between each pair of neurons.
- Neurons interact in a nonlinear analog way. That is, the interconnection interaction is not an "all or nothing" matter, but a graded interaction where the firing rate of neurons varies smoothly with the input potential.
- An "energy function" can be constructed that allows us to understand the collective (or emergent)
- Dynamics of the neuron network as it moves over the information landscapes and finds local minima that represent effective solutions to the problem.

Fig. 28: An optimal traveling salesman problem (TSP) tour through Germany's 15 largest cities. It is the shortest among 43,589,145,600 possible tours visiting each city exactly once.

Source: *Wikipedia, Creative Commons Attribution-Share Alike License.*

Hopfield and molecular biologist David Tank set out to make an analogy between neural networks and the energy network of a glassy system characterized by a large number of degrees of freedom. Following the three principles outlined above, they used this analogy to write an equation for the free energy of a neural network in terms of the interaction between each pair of neurons, the threshold for each neuron to self fire, and the potential for each of the neurons in the network. They also recognized that the interaction between pairs of neurons can change with time as the neural network learns.

The solution to a problem such as the traveling salesman problem emerges in the neural network as interaction strengths between the neutrons are adjusted to minimize the free energy equation. The flow of the neuron states during the computation can be mapped onto a flow on a free energy surface, similar to the flow of a spin glass toward its ground state or natural selection on a fitness landscape (but in the opposite direction). Clearly, quite complex and emergent neuronal dynamics can evolve with even the simple system we are considering here.

Hopfield and Tank showed that this neuronal map has quite impressive "reasoning" ability. A set of 900 neurons encoded to solve a 30-city traveling salesman problem was able to find 107 "best" solutions out of the 1030 possible solutions, a rejection ratio of 1023 in just a few clock cycles of the neural network.

Although we clearly are a long way from understanding the emergent nature of consciousness, this example reveals the immense computational power of neural networks. Surely, one of the grand challenges in 21st century physics will be to move from these simple physical models derived from very concrete physics concepts to the vastly more complex terrain of the human brain.

2

BIOPHYSICS IN DRUG DISCOVERY

2.1 BIOPHYSICAL TECHNIQUES

Biophysical methods are techniques to study the structure, properties, dynamics or function of biomolecules at an atomic or molecular level. They encompass a range of techniques including microscopy, spectroscopy, electrophysiology, single-molecule methods and molecular modelling.

A range of biophysical techniques are currently used to analyse the binding of a compound to a target. The available technologies differ in the physical principle that underlies the detection of binding, throughput, infor- mation content, sample requirements, sensitivity and robustness of the resulting data. TABLE 1 summarizes the techniques used most frequently in drug discovery, together with the characteristics of typical experiments, and FIG. 1 shows typical data obtained from the different methods described in TABLE 1. Given the large number of biophysical techniques, this Review is inevitably not able to comprehensively cover all existing or developing techniques, and so some techniques are not discussed in detail or not mentioned.

Two main biophysical techniques are currently used to investigate the structure of protein–ligand complexes. The most common technique is X-ray crystallogra- phy, which can be used for proteins of any size. X-ray diffraction patterns of either protein–ligand co-crystals or apoprotein crystals soaked with a ligand are used to determine the structure of the complex at atomic resolu- tion and derive the geometry and other details of protein–ligand interactions.

Table 1: Selection of established biophysical methods for analysis of protein–ligand interactions

Technique or method	Strengths	Limitations	Typical material Affinity requirements range		Throughput day*
		X-ray crystallography			
Protein–compound co-crystallization or apoprotein crystal soaking	Identification of binding site for establishment of modes of action direct visualization at atomic resolution of target–ligand interactions for structural optimization of lead compounds	Need for diffraction quality crystals suitable for ligand soaking or co-crystallization often requires access to large research infrastructures (synchrotrons) No quantitative affinity information	~2.5–5 nmol crystallizable protein (for example, 15 μl of protein solution at 5–10 mg per ml with a MM of~30 kDa) per 96-well nanodrop crystallization plate	No lower limit to 1 M	100s
		NMR			
Ligand-observed NMR	Confirmation of ligand binding to unlabelled proteins of any size Integrity of ligand and protein in every experiment	Considerably large amounts of protein required (typically tens of milligrams for screening) Limited derived structural information	~3–6 nmol protein per compound or cocktail tested for 1H experiments (for example, 0.1–0.2 mg protein with a MM of ~30 kDa) amount of protein needed is ~10 times less for 19F experiments competitor ligand to confirm specific binding	100 nM–10 mM	100s
Protein-observed NMR	Monitor protein integrity upon ligand binding Titration can reliably determine Kd Binding epitope resolved from pattern of perturbations can determine structure if NMR spectrum is assigned	A large amount of isotopically labelled protein is required Screening of large libraries is only possible in multiplexing mode (see REFS 142–144) Only suitable for small proteins (MM <40 kDa)	Protein with a MM of <40 kDa ~30 nmol of an isotopically labelled protein per compound tested (for example, ~1 mg protein with a MM of ~30 kDa)	100 nM–1 mM	100s
		SPR			
Microfluidic surface-based biosensor (immobilized protein)	Direct time-resolved determination of interactions over a broad range of conditions high sensitivity allows analysis of fragments	Requires immobilization of functional target with high stability over time and towards regeneration procedures signals affected by solvent effects	~15 nmol protein (for example, ~0.5 mg with a MM of ~30 kDa) for assay development and screening of 2,000 compounds reference compounds for evaluation of protein functionality	1 nM–500 μM	100s

Technique or method	Strengths	Limitations	Typical material Affinity requirements range		Throughput day*
Microarray-based biosensor with CCD camera detection (immobilized compound library)	Very high sensitivity ultra-high throughput (12,000 fragment interactions in triplicates per day)	Requires chemical modification of compounds to immobilize ligands through a covalent linker to a surface, preferably in different orientations detergents are not tolerated	~100–150 nmol protein (for example, 3–5 mg with a MM of ~30 kDa) to screen 120,000 compounds reference compounds to assess binding specificity or to identify allosteric binders	No lower limit to 500 µM	10,000s
		TSA			
Dsf for	• fast and robust assay	• requires a fluorescent	• 80 pmol protein per	1 Nm–100µm	1,000S
Monitoring of	Development	Dye	Analysis (40 µl at 2 µm)		
Protein thermal	• Functional knowledge	• Artefacts occur owing to	• Ideally use a protein with		
Unfolding with	of target not necessary	fluorescence quenching	Intermediate tm (<50 °c)		
A fluorescent		or aggregation			
Reporter ligand		Not suitable for			
(Typically using		Disordered or			
Qpcr)		Hydrophobic proteins			

Technique or method	Strengths	Limitations	Typical material requirements	Affinity range	Throughput per day*
Microcalorimetry					
Itc	Direct determination of thermodynamic parameters for a binary system	Only useful for binding events with an enthalpic componentvery high protein consumption, requires high solubility of titrated component	~6–60 Nmol protein per titration (for example, 0.2–2 Mg with a mm of~30 kda)concentrations needed are ~10 times the estimated kd (when kd is unknown, a 10–100 μm concentration range is used)	1 Nm–100 μm	10S
Dsc	Determination of the effect of a ligand on the thermal stability of a protein can be used to study ultra-tight binding that cannot be measured by other methods	Very high protein consumption	~1.5–30 Nmol protein per scan (for example, ~500 Ml at 0.1–2 Mg per ml with a mm of~30 kda) A kd measurement requires a series of scans at different concentrations	Dissociation constants down to 10–20 M can be estimated	10S
Ms					
Nc-ms (also called native ms; usually esi-ms)	Direct visualization of complex formation high sensitivity very low protein consumption accurate mass measurements high-content information no labelling required	Requires sample desalting protein has to be stable in an es-ms suitable buffer (usually a low-concentration ammonium buffer) detergents are not tolerated unspecific binding makes the detection of low-affinity binders more difficult	10–100 Pmol protein per compound tested (2–5 μl at 5–20 μm)	No lower limit to 500 μm	100S
As-ms (also called sec-ms)	Ultra-high throughput when compound cocktails used can be applied to solubilized membrane proteins ligand mass detection enables verification of compound structure	Low-affinity binders (including fragments) are hard to detect because they tend to dissociate from the protein during the sec step owing to high off-rates	10 Pmol protein (~2 μl at 5 μm) per single-binding reaction with a 2,500-compound library	No lower limit to 10 μm	1,000,000S (Compound cocktails)
Hdx-ms	Direct detection of	Spatial resolution is	~30 Nmol protein (for	No lower limit	10S
	Protein binding site	Limited and depends	Example, ~1 mg with a	To 20 μm	
	And/or changes in	On peptide lengths	Mm of ~30 kda), mostly		
	Protein conformation	And coverage of target	For assay development		
	Or protein dynamics	Sequence after protease			
	Upon binding	Cleavage			
Mst					

Technique or method	Strengths	Limitations	Typical material requirements	Affinity range	Throughput per day*
Monitoring of	In-solution	Requires labelling	Protein can either	1 Pm–1 mm	100S
Changes in the	Measurements	Or strong intrinsic	Be fluorescently		
Thermophoretic	Applicable to	Fluorescence	Labelled or have		
Motion of a	Solubilized membrane		Detectable intrinsic		
Fluorescently	Proteins		Fluorescence		
Labelled or			~7 Pmol protein per		
Intrinsically			Compound tested		
Fluorescent			At 12 serial dilutions		
Protein upon			(Twelve 10 µl aliquots		
Ligand binding			Of protein solution at		
			60 Nm)		
Technique or method	**Strengths**	**Limitations**	**Typical material requirements**	**Affinity range**	**Through-put per day***
Saxs					
Low-angle elastic scattering of x-rays, bringing information about the shape and size of macromolecules	Larger-scale structural changes of proteins upon ligand binding can be visualized at low resolution with high throughput	Low resolution (1–2nm) does not allow for elucidation of atomic details of ligand interactions	Monodisperse protein material ~1 Nmol protein per sample (for example, ~0.03 Mg with a mm of ~30 Kda) ~100 Nmol protein for an entire study (for example, ~3 Mg with a mm of ~30 Kda) Typically, a concentration series is measured for each sample (for example, 1, 2, 5 and 10 mg per ml)	No lower limit to 30 µm	100S
Qcm					
Monitoring of	Direct time-resolved	Requires immobilization	~15 Nmol protein (for	1 Nm–500 µm	100S
The vibration	Determination of	Of functional target with	Example, ~0.5 Mg with		
Frequency of a	Interactions with	High stability over time	A mm of ~30 kda) for		
Quartz crystal	Proteins and cells	And robustness towards	Assay development and		
Upon interaction		Regeneration procedures	Screening campaign		

Technique or method	Strengths	Limitations	Typical material requirements	Affinity range	Throughput per day*
Between an					
Analyte and an					
Immobilized					
Protein or cell					
Rotating cell-based ligand binding assay using radioactivity or fluorescence					
Monitoring	Direct time-resolved	Laboratory requirement	~105 Cells per assay	1 Pm–1 μm	10S
Of molecular	Determination of	To work with radioactivity	Fluorescent or		
Interactions	Interactions between	(Not with fluorescent	Radioactively		
At the surface	Labelled ligands with	Compounds)	Labelled analytes (in		
Of cells using	Surface-associated		Approximately picomolar		
Radioactivity or	Proteins		Amounts)		
Fluorescence					

AS-MS, affinity selection mass spectrometry; CCD, charge-coupled device; DSC, differential scanning alorimetry; DSF, differential scanning fluorimetry; ESI-MS, electrospray ionization mass spectrometry; HDX-MS, hydrogen–deuterium exchange mass spectrometry; ITC, isothermal titration calorimetry; MM, molecular mass; MS, mass spectrometry; MST, microscale thermophoresis; NC-MS, non-covalent mass spectrometry; NMR, nuclear magnetic resonance; qPCR, quantitative PCR; QCM, quartz crystal microbalance; SAXS, small-angle X-ray scattering; SEC-MS, size-exclusion chromatography mass spectrometry; SPR, surface plasmon resonance; Tm, transition midpoint for thermal unfolding; TSA, thermal shift analysis. *Number of protein-ligand experiments per day for a well-behaved system with optimal facilities.

Nuclear magnetic resonance (NMR) spectroscopy can also be used to solve structures of protein–ligand complexes for targets with a molecular mass of less than ~40 kDa. NMR also has a much wider application in probing ligand binding, through two alternative experimental setups: ligand-observed NMR and protein-observed NMR. In ligand-observed NMR, changes in the NMR parameters (mostly chemical shifts, relaxation rates and diffusion rates) of molecules or cocktails of molecules are measured in the presence of a target protein. Typical experiments performed include magnetization transfer experiments (saturation transfer difference (STD) NMR, water-ligand observed via gradient spectroscopy (waterLOGSY), transferred nuclear Overhauser effect (NOE), NOE pumping and other NOE-based methods), relaxation editing (longitudinal, transverse and double- quantum relaxation) and diffusion editing. The infor- mation that can be obtained can range from simply determining whether the ligand interacts with a protein or not, to binding specificity and information on the binding mode (such as epitope mapping, pharmacophore mapping and mapping of ligand orientation). With some techniques it is possible to estimate binding affinities. In protein-observed NMR, the chemical shift pertur- bations in the 2D NMR spectra of isotopically labelled proteins are monitored in the presence of ligands. Such measurements are limited to small proteins (prac- tically, those with a molecular mass <40 kDa), but when resonances have been assigned, the binding epitope and the dissociation constant (K) can be inferred from the data. Both methods require relatively large amounts of a protein, which needs to be stable and of high purity.

Surface plasmon resonance (SPR) is a spectroscopic technique that monitors changes in refractive index at the interface of a liquid sample and a surface with an immobilized sensor molecule, typically the target pro- tein. The signal is shifted as a result of analyte binding or induced conformational changes. When using micro- fluidic systems with continuous registration of the signal, it is possible to characterize the binding mechanism and determine the corresponding kinetic parameters (the association rate constant (k d the dissociation rate constant (koff)) and the affinity (Kd). Thermodynamic parameters can also be estimated by performing analyses at multiple temperatures. An alternative configuration of SPR involves imaging of a microarray with immobilized small-molecule sensors using a charge-coupled device (CCD) camera, providing information on binding (yes or no) and binding specificity through on-array competition experiments. Both of these SPR techniques require immobilization of one of the binding partners, either the protein in the microfluidic system or a ligand library in the microarray technology. In the latter case, the protein is in the solution over the microarray, and thus, as the signal depends on the mass of the analyte (in this case, the protein), the detection of binding in this setup is extremely sensitive.

Thermal shift analysis (TSA), also known as differential scanning fluorimetry (DSF), measures the tem- perature at which a protein unfolds through binding of a fluorescent probe to the exposed hydrophobic sur- faces. If a ligand stabilizes or (less frequently) desta- bilizes the fold upon binding, then there will be a change in the transition midpoint for thermal unfolding (Tm) (positive or negative, respectively). In some cases, it is also possible to determine K values. The major advan- tages of the method are its fast set up, its inexpensiveness and the requirement of only small amounts of protein material.

Two major microcalorimetry techniques are used for studying protein–ligand interactions. Isothermal titration calorimetry (ITC) measures the generation or consumption of heat following titration of a ligand on to a protein (or the reverse). The data obtained include stoichiometry, K d, change in enthalpy (ΔH), change in entropy (ΔS) and the heat capacity change (ΔCP). Differential scanning calorimetry (DSC) measures heat changes (and corresponding thermodynamics) resulting from the thermal unfolding of a protein. An advantage of these techniques is that experiments are performed in solution, but large amounts of protein are typically required.

Mass spectrometry (MS) is also used in different modes. Non-covalent MS (NC-MS), also called native MS, is usually carried out using electrospray ionization MS (ESI-MS), generating data that can confirm binding, assess specificity through competition experiments, define stoichiometry, measure K d and give an estimate of the enthalpic component of binding.

Affinity-selection MS (AS-MS), also called size-exclusion chromatography MS (SEC-MS), can be used to study protein–ligand interactions, but with decreased information content. Typically, a protein is incubated with a library of compounds and passed through a SEC column to isolate any ompounds that are bound to the protein a d therefore coelute with the protein. The protein complexes are then dissociated and the ligands are identified by LC-MS or ESI-MS. In this case, confirmation of binding and K d values can be obtained. Finally, hydrogen– deuterium exchange MS (HDX-MS) can be used to measure the altered accessibility to deuterium exchange of amino acids involved in ligand binding. This method has successfully been used for the characterization of protein–protein interactions, such as those between epitope binding regions of therapeutic antibodies and their antigens. Furthermore, small-molecule binding and alterations in protein dynamics can be investigated. The method works for soluble and membrane-bound proteins and relies on an appropriate protease cleavage pattern to get accurate sequence coverage.

Microscale thermophoresis (MST) is a relatively new methodology that monitors fluorescence in an infrared laser-heated spot. It is an equilibrium-based method that can detect ligand binding-induced changes in thermophoretic mobility (the motion of protein molecules along a microscopic temperature gradient), which depend on size, charge and hydration shell. These changes in thermophoretic mobility can be used to estimate K d values.

Small-angle X-ray scattering (SAXS) can be used to study protein–protein, protein–DNA, protein–RNA and protein–small molecule interactions in solution, providing information on the folding, oligomerization state and intrinsic flexibility of a protein and its complexes, as well as the shape of an assembly and an envelope structure to a 1–2-nm resolution. Although we are not aware of any extensive ligand screening efforts, SAXS can be used to monitor the effect of ligands on the modulation of protein–protein interactions and of the conformation and oligomeric state of intrinsically disordered proteins.

Quartz crystal microbalance is a continuous flow surface biosensor that measures changes in the vibration frequency of a quartz crystal following an interaction between an analyte and an immobilized target protein or cell, quantified as K d, k on, k off or thermodynamic parameters.

Finally, the rotating cell-based ligand binding assay is a relatively new method for real-time monitoring of molecular interactions on cells using radioactivity or fluorescence. This method provides data that can be used to characterize the binding mechanism, as well as K d, k on and k off.

2.1.1 What do these Techniques Require?

All of these techniques have certain requirements for the compound to be tested. Ideally, the compound must be soluble well above its binding affinity constant, stable and non-aggregating under the given experimental conditions. However, even under the same experimental conditions, different techniques demand different levels of solubility relative to affinity. For instance, in ligand-observed NMR, a low occupancy of the protein binding site by the ligand (<20%) suffices to observe binding, whereas in X-ray crystallography a substantial percentage (>50%) of binding sites must be occupied to unambiguously identify binding and resolve the protein–ligand co-structure.

Similarly, for a robust measurement, all of the techniques require the protein target to be homogenous and well behaved at relatively high concentrations. However, each of the techniques has differing requirements for the amount, characteristics and preparation of the target.

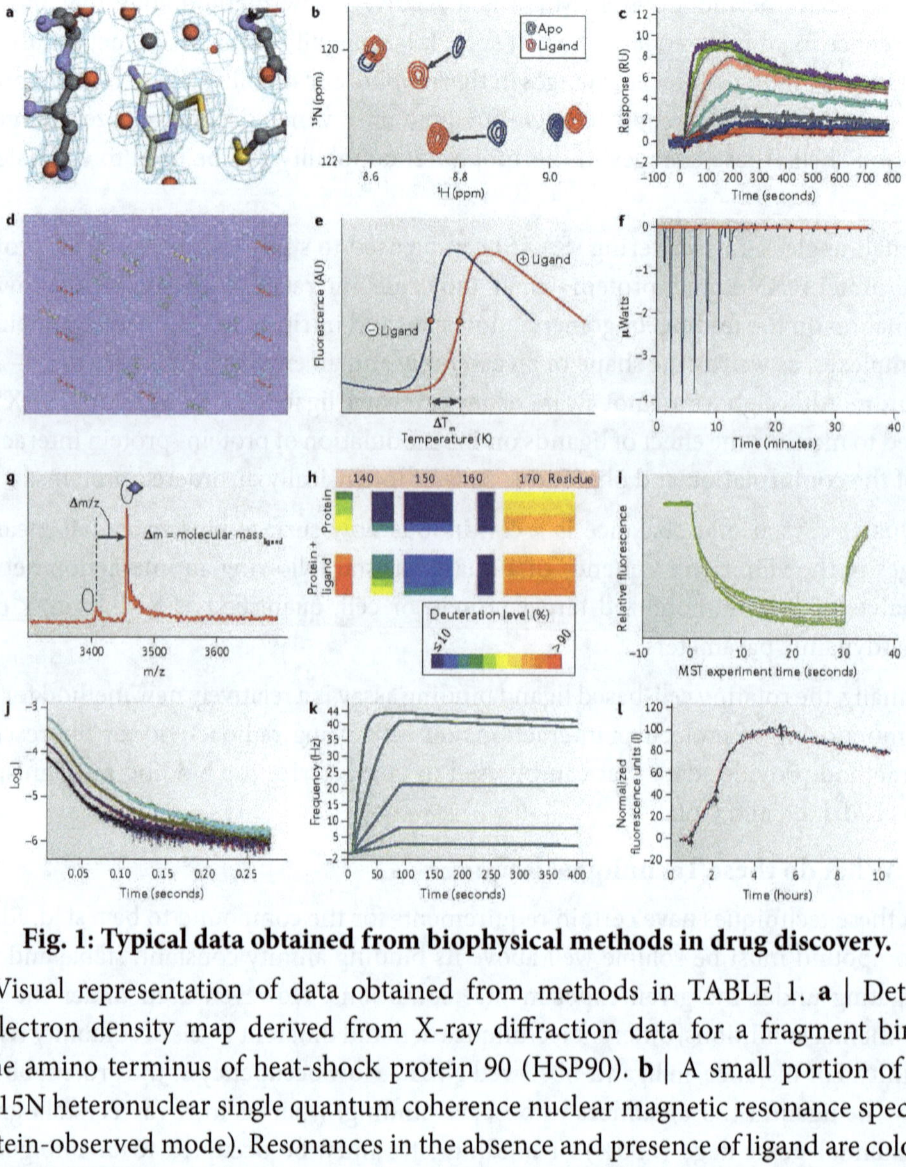

Fig. 1: Typical data obtained from biophysical methods in drug discovery.

Visual representation of data obtained from methods in TABLE 1. **a** | Detail of an electron density map derived from X-ray diffraction data for a fragment binding to the amino terminus of heat-shock protein 90 (HSP90). **b** | A small portion of a 2D 1H–15N heteronuclear single quantum coherence nuclear magnetic resonance spectrum (protein-observed mode). Resonances in the absence and presence of ligand are coloured blue and red, respectively. **c** | A typical channel-based surface plasmon resonance (SPR) sensorgram at different analyte concentrations. **d** | A fingerprint from a microarray SPR screening, showing the SPR shift on a relative colour scale. The horizontal and vertical axes correspond to the *x* and *y* coordinates on the microarray. Fragment hits are easily identified from triplicate patterns. **e** | A typical recording of relative fluorescence intensity versus temperature for the unfolding of a protein target in the presence of a fluorescent dye. The blue and red curves are the recordings in the absence and presence of test compound, respectively. In the depicted case, the test compound stabilizes the protein (change in transition midpoint for thermal unfolding (ΔTm) >0). **f** | A thermogram showing heat absorption rates (microjoules per second) upon sequential ligand additions to a target solution (isothermal titration calorimetry; ITC) (raw data). **g** | A portion of a native mass spectrum. Upon ligand addition, the peaks corresponding to the apoprotein decrease and the peaks corresponding to the protein–ligand complex increase. The mass difference can be used to evaluate the binding stoichiometry. **h** | The deuteration level (observed deuterium uptake over the maximum theoretical uptake) for each peptide at increasing time points (from top to bottom) in the absence and presence of a ligand. **i** | Fluorescence monitoring of the thermophoretic motion of a target at different ligand concentrations. **j** | Small-angle X-ray scattering curves for a protein–protein complex at different concentrations. **k** | Typical quartz crystal microbalance (QCM) sensorgram at different analyte concentrations. **l** | Relative intensity versus time showing the interaction of a fluorescent antibody with receptors at the surface of living cells. The red and blue curves depict two independent experiments. Δm/z, change in mass-to-charge ratio; MST, microscale thermophoresis; ppm, parts per million.

Clearly, X-ray crystallography requires crystals, which usually means at least a few milligrams of protein at more than 10 mg per ml concentration. Advances such as cryocooling, microfocus beamlines and new detector technologies have dramatically reduced the number of crystals required to determine 3D structures. New developments — such as serial crystallography done at room temperature in combination with the ultra-high brilliance of X-rays from free-electron lasers — have the potential to work with crystals of sub-micrometre dimension. For NMR, improved electronics and the advent of cryogenic probes have increased sensitivity by more than tenfold over the past 10 years, reducing either the amount of sample required or the measurement time; nevertheless, a protein solubility of around 10 μM or greater is still required. Ligand-observed NMR measurements can be made on unlabelled targets of any molecular mass, whereas protein-observed NMR is mostly limited to proteins of less than 40 kDa and requires isotope labelling (one or more of 15N, 13C and 2H).

Routine labelling is only possible by overproduction in engineered bacterial cell lines; the recent innovation of selective labelling of specific amino acid types has mad it possible to exceed the 40 kDa limit of protein-observed NMR. Improvements in label-free native ESI-MS have led to greater throughput and the ability to screen hundreds of compounds while still allowing the detailed characterization of the interaction.

There is a growing collection of methods in which either the target or the compound is 'tethered' to some aspect of the instrument such that the binding of the other 'free' component can be characterized. The different methods, such as SPR, bio-layer interferometry (BLI) and weak affinity chromatography (WAC), have the common advantage of allowing the binding of many compounds to be measured if a small quantity (less than 1 mg) of the target can be tethered without disturbing target folding, ligand binding properties or function. Such methods are also prone to similar artefacts, such as nonspecific interaction of compounds with the immobilization medium and the possible influence of additives or co-solvents (for example, DMSO) on the binding signals. The assay sensitivity of larger target proteins can be compromised if steric constraints limit the amount that can be tethered, and the binding of small 'free' ligands can lead to small signals when the technique relies on changes in mass or refractive index close to the sensor surface. This is a class of techniques for which increased experience and education has greatly improved application36,37, and there is now widespread understanding and acceptance of the stringent controls needed to validate the experimental protocol.

The thermal techniques (TSA, ITC and DSC) are all solution-based and do not require immobilization of either the target or ligand. The artefacts in TSA measurements are well documented38, but TSA benefits from ease of use and requires only micrograms of material.

ITC is the gold standard for direct binding measurements. Unfortunately, despite substantial improvements in calorimeter instrument sensitivity, milligrams of protein and high compound solubility are still required to characterize the binding of each compound. Finally, there are a number of techniques that rely on fluorescence (either intrinsic or extrinsic) of either the target or the compound, such as MST and fluorescence polarization (FP). These solution-based methods bring additional considerations and a need for validation steps if extrinsic labelling is used, as the fluorophore itself may affect some aspect of the binding.

2.1.2 What Information can Currently be Obtained?

One of the challenges for drug discovery researchers who are not familiar with biophysical techniques is to understand what type of questions can be addressed by such techniques and which techniques are the most suitable. The following discussion provides some examples of 'typical' questions posed and the techniques that are currently routinely and robustly used in our laboratories.

The discussion uses the phrase 'tool compound' to represent a compound that is known to bind with a particular affinity to a specific site on the target, and is soluble, stable and non-aggregated under the conditions used to make measurements.

Is the target folded?

A 1D proton NMR spectrum of an unlabelled protein is relatively quick to obtain. A folded protein will have narrow peaks spread over a large chemical shift range (–0.5 to +10.5 ppm), often with distinct peaks at the extremes of the distribution. An unfolded or partially folded protein will have broader and less well-dispersed peaks, often close to their random chemical shift values. Similarly, a circular dichroism spectrum will reflect folding, particularly for proteins containing substantial numbers of α-helices. A more rapid way of characterizing protein samples is by TSA, through which a clear thermal melt curve is a quick way to characterize folding and homogeneity across samples, particularly if a tool compound is available.

Does a compound bind to the target?

All of the techniques in TABLE 1 can be configured to detect the binding of a compound to the target, with the dynamic ranges indicated. Some techniques provide more information than others (as discussed here), but all provide the essential first assessment of whether the compound binds or not — with real confidence if a tool compound is available for assay validation. Measurements at different ligand concentrations and evaluation of dependence on concentration provide additional information on the binding behaviour and an estimation of the affinity.

2.2 STOICHIOMETRY OF BINDING

Accurate stoichiometry measurements are difficult, as they require accurate knowledge of both the total ligand and macromolecule concentrations and the fraction that is able to form a complex. The absolute errors need to be less than 20% on each of these to be able to confidently distinguish a 1:1 from a 2:1 complex. ITC is probably the most routinely used method to determine stoichiometry, but it consumes large quantities of protein. SPR is a good alternative when a suitable tool compound is available.

Single-molecule methods would avoid the need for precise concentration determination and may overcome sample consumption issues. Currently, the closest technique that matches this is NC-MS, which has many attributes that make it a valuable biophysical technique for drug discovery. However, further development of the method is required for more routine use.

2.2.1 Where does the Compound Bind?

In the absence of an X-ray structure of the ligand complex, the most robust technique for identifying where a compound binds is heteronuclear single quantum coherence (HSQC) NMR spectroscopy (usually 15N but sometimes 13C). Each peak in the spectrum reports on the local chemical environment of the resonant nuclei, which can be altered by compound binding. If the spectrum is assigned, such that peaks can be linked to individual protein residues, then the actual site of binding can be identified; if not, then a fingerprint of peak movements can determine whether compounds are likely to bind to the same site or different sites. Once a tool compound for a site has been identified, many types of competition displacement assay can be configured using other techniques to more rapidly test and characterize new compounds. However, it is important to remember that these displacement techniques do not directly report the site of binding, unlike NMR and X-ray crystallography, and allosteric site binders may displace probe compounds owing to long-range conformational effects. In any case, the use of site-specific mutants of the target can help to provide information about binding site location.

2.3 ATOMIC STRUCTURE OF THE TARGET–COMPOUND COMPLEX

X-ray crystallography is the most powerful, robust and routine method for providing a detailed atomic picture of a compound binding to its target. There are rare examples of discrepant observations between soaked and co-crystallized complexes (for example, when targets require conformational change or when crystal packing interferes with ligand binding) or when the crystallization conditions themselves affect the ligand binding mode (for example, pH). Therefore, as with other experimental methodologies, corroborating data from X-ray structures obtained by alternative crystallization protocols or from orthogonal solution studies such as NMR provide a more complete picture.

2.3.1 Thermodynamics of Compound Binding

ITC is the most direct method used to measure the thermodynamic parameters of binding (K d, ΔH and ΔS) and, with variation of temperature, ΔCP. The most straightforward and routinely used experiment is a direct binding titration, in which the heat change upon binding (ΔH) is monitored as small aliquots of a concentrated compound solution are injected into a solution containing the target at constant temperature. This configuration can be reversed when the compound is less soluble and the protein target is highly soluble. Affinity measurements, most often using SPR, at different temperatures can provide an alternative way of determining thermodynamic parameters using a van't Hoff analysis (see below). However, this method involves additional assumptions and has not been widely used so far.

2.3.2 Kinetics of Compound Binding

SPR biosensors with a microfluidic flow system and dextran surfaces are the most extensively used instruments for measuring ligand binding kinetics. With an appropriate configuration, k on and k off can be determined by simply fitting an equation corresponding to a suitable model to the sensorgram that results from the SPR experiment. K d can also be derived as the ratio of k off to k on. For more complex interaction mechanisms, other kinetic parameters can be estimated. Many compounds have a relatively high k on (limited by the rate of diffusion) with an increasingly low k off as the affinity for the target increases. There is growing evidence from the literature demonstrating that ligand binding kinetics can influence efficacy and safety and therefore therapeutic success for some targets5. Attempts to increase our understanding of how to modulate these kinetics parameters by correlating them with X-ray ligand complexes to give structure–kinetic relationships (SKRs) are an area of great interest.

2.4 WHERE ARE BIOPHYSICAL TECHNIQUES APPLIED?

Two major developments in drug discovery have been synergistic with the deployment of a wide range of sensitive biophysical methods: the increased focus on drug-like properties during compound optimization and the rise of FBDD. We begin with a brief overview of biophysical methods in FBDD, as this has driven and challenged the development of methods more than any other application in recent years. We then outline how these advances have been exploited in areas such as high-throughput screening (HTS), hit confirmation and more detailed characterization of compound binding.

2.4.1 Fragment Screening.

The first published example of FBDD was 20 years ago, with the seminal work of the Abbott group, which demonstrated that nanomolar affinity drug candidates could be derived from small fragments that bound with a millimolar K d . This group not only pioneered the use of protein-observed NMR experiments to detect weak interactors but also were the first to conduct fragment screening by crystallography. Since then, many companies have been founded using fragment-based screening (FBS) as their primary hit identification strategy. Several of these are renowned for advancing methodologies for their chosen biophysical approach: for example, X-ray crystallography ligand-observed NMR spectroscopy46, SPR47, small-molecule microarray SPR18,19 and NC-MS19,25. In both large and small companies, multiple established methods and increasingly emerging biophysical techniques (for example, MST and WAC)48 are applied concurrently during FBS hit validation. Comparisons of the range of FBS methods, including illustrative successful fragment-to-lead case studies, are well documented in literature reviews40. However, faced with a growing array of biophysical methods capable of FBS, two related questions are often posed and debated: why do different techniques result in different hit lists, and which is the 'best' FBS technique to use?

Although many studies report different hits depending on the screening method, much of this discrepancy can be explained by variations in assay protocol and detection method. Techniques have differential sensitivity to the properties of individual compounds within the fragment library (for example, solubility, aggregation potential, stability and ability to interfere with the assay) and assay conditions used (for example, the effect of buffer conditions, including temperature and pH, on both the fragments and the targets). Consequently, weak-affinity fragments may lie just above or below an assay detection threshold as the balance of factors changes.

The goal of FBS is to detect the binding of lowmolecular- mass compounds (typically <300 Da) with low affinity to a protein target (in the micromolar to millimolar range). It is widely acknowledged that there is no single 'best' technique; instead, the choice of technique, or often techniques, for primary FBS is frequently dictated by logistical and practical considerations such as equipment and protein availability. More importantly, we believe that success in FBS depends strongly on users' experience and expertise, specifically the ability to establish robust assays, stringent data interpretation protocols and a well-considered screening cascade. In some instances, an inclusive strategy is adopted, progressing all hits from all of the applied methods; in others, only overlapping hits validated by more than one technique are included. Both strategies can be appropriate depending on the circumstances and when they are chosen in light of an understanding of the acquired FBS knowledge.

The adoption of FBDD in some organizations has required changes in the processes and gate-keeping criteria for progression through early drug discovery. Fragments straight out of FBS often have no measurable target function (either agonism or antagonism). Therefore, it is important to develop robust models of how fragments bind and to understand the opportunities and probability for successful affinity optimization. After the initial investment to improve affinity independently of activity, for fragments that register detectable functional activity, conventional activity coupled with ligand efficacy criteria can be used to prioritize further chemistry efforts towards generating lead molecules. Another barrier is that many larger companies use multiple hit-finding strategies, so it can be difficult to persuade a medicinal chemistry team to undertake the optimization of a 500 μM fragment if a 100 nM hit is available from HTS. For these reasons, structural information on fragment–target complexes, usually obtained by X-ray crystallography, is considered essential in FBDD to provide insight into chemical opportunities for efficient fragment evolution.

Most fragment screens are configured to identify compounds that bind to known binding sites (usually the active site of an enzyme). It is often straightforward to find many fragment hits for such 'conventional' targets, provided that the target protein can be readily generated and, in particular, that tool compounds are available to validate the various assays. For other target classes, innovative experimental design may be required to identify and successfully develop these hits. The first example of a difficult target class is represented by proteins with druggable allosteric sites (often initially unknown and without a natural biological function). For this class, there are some striking examples in which careful design of NMR experiments has enabled: detection of hits that inhibit BCR–ABL kinase at a myristoyl binding site; generation of non-bisphosphonate farnesyl yrophosphate synthase inhibitors; and discovery of compounds that activate enzymes such as glycoside hydrolase. The second example of a difficult target class includes proteins that transiently interact with other proteins or that form multi-protein complexes — two different types of important protein–protein interactions. For such systems, thorough experimental design and assessment with many orthogonal biophysical techniques is needed to validate fragment binding.

2.4.2 HTS

Here, HTS is defined as screening a collection of ~1 million compounds or more against a protein target to identify a compound subset that shows activity. Such screens are typically based on biochemical or cellular assays, often with optical detection. Few biophysical methods satisfy the throughput requirements of HTS55, although many are used at the secondary hit-validation stage. TSA58 and, to some extent, MS59 are techniques that can be configured to provide the throughput, ease of access and low sample consumption needed for HTS screening. Cocktails of compounds are often used to reduce the number of measurements, material consumption and costs. TSA was initially used for proteins of unknown function or when obtaining reagents required to establish a traditional biochemical displacement or enzymatic assay proved challenging. The simple format of TSA has facilitated its application to other protein classes, especially within academic laboratories.

2.4.3 Hit Confirmation and Validation

Biophysical methods are now routinely incorporated as an integral component of the HTS hit validation and characterization process of most pharmaceutical ompanies. In the past, many HTS triage strategies relied solely on hits progressing through a series of negative selection assays designed to filter out compounds acting through unwanted effects, such as pan-assay interference compounds (PAINS)60 — for example, compounds that cause target oxidation through their intrinsic redox activity, unstable or reactive compounds and compounds that induce protein aggregation. Unfortunately, this filtering approach is imperfect as not all nuisance mechanisms are known, and problematic compounds for some targets may be true high-quality ligands for others, as evidenced by the finding that 10% of a set of known drugs can exhibit 'promiscuous' activity61. This proves especially difficult for challenging targets such as protein interactions, for which the number of authentic hits is often very low.

The use of positive selection assays provides an effective and efficient alternative approach, and many biophysical methodologies have been adopted to serve this purpose. The major roles and benefits of biophysical approaches within hit validation fall into three categories.

First, some approaches can allow the orthogonal confirmation of functional activity. For example, many screening assays use optical readouts such as fluorescence and luminescence to confirm activity. For enzyme assays, non-optical detection using RapidFire MS or NMR provides alternative ways to monitor substrate consumption or product build-up.

The second major type of positive selection assays is those that verify direct target engagement. For example, a simple biophysical binding experiment using SPR, NMR, NC-MS or MST can confirm that hits bind directly to the protein target of interest rather than to any other biochemical assay component, such as capturing antibodies or coupling enzymes.

Third, positive selection assays can be used to determine whether the mode of action of a compound on target protein is aligned with the biological understanding of the disease mechanism. This relies on the proein used in the assays being in a functionally relevant state (for example, with regard to post-translational modification and binding partners). Briefly, biophysical assays can yield information on binding specificity and stoichiometry, and ITC and SPR provide additional quantitative data on binding thermodynamics and kinetics that implicitly carry information on the mode of action of the hits (see below). X-ray crystallography (and NMR) can identify the ligand binding site on the target and elucidate the atomic details of the protein–ligand interactions. Taken together, these methods provide a comprehensive data package that adds confidence and optimizes success in the prioritization and selection of HTS hits for the following chemical optimization phase.

2.4.4 In-Depth Characterization of Compound Binding

The advent of sensitive biophysical technologies for timeresolved and thermodynamic analysis of protein–ligand interactions that do not require reporter groups such as enzyme substrates has had a tremendous impact on the ability to characterize details of molecular interactions.

The kinetics of interactions are typically determined by SPR biosensor technology and the value of kinetic data is well recognized; however, the technology can provide additional data for understanding interactions and optimizing lead compounds. Interpretation of the SPR data is frequently done assuming a simple one-step reversible 1:1 interaction model, but exceptions occur and additional information can be obtained for some systems. For example, interactions between targets and optimized leads are often very tight, resulting in essentially irreversible dissociation rates that are challenging to quantify owing to measurement errors and limited stability of the assay system. Qualitative analyses and ranking of compounds can still provide useful information for prioritization of analogues and further optimization63. Another example is dynamic proteins and protein interactions whose function and regulation is dependent on complex interaction mechanisms involving conformational changes of free or bound protein. In particular, the determination of kinetic parameters requires the use of appropriate mathematical models that take all parameters and the potential mechanism of binding into consideration. Qualitative details of interactions can also be obtained by analysis under different conditions, termed chemodynamic analysis. Chemodynamic analysis involves a series of identical experiments that only differ in buffer pH, ionic strength or the presence of other substances that influence the interaction. This is valuable information as it provides an improved understanding of the dominating interaction forces, which is important for structural optimization of leads. It also provides a measure of the robustness of the interaction throughout different experi mental conditions as well as the relevance of the data for cellular and in vivo experiments.

In addition, Eyring analysis of the variation of kinetic rate constants (for example, from SPR) with temperature allows the entropy and enthalpy of the transition state for binding to be explored, supplementing the information on entropy and enthalpy of binding that can be determined from calorimetry. Although it is difficult to predict these thermodynamic parameters with computational methods, the changes that occur during compound optimization can sometimes be rationalized and provide insight into the key features required for high-affinity binding.

2.4.5 Assisting Lead Optimization

In late-stage hit-to-lead or lead optimization, the focus of drug discovery projects typically moves away from binding affinity and selectivity optimization to include a multitude of additional optimization parameters. These include cellular activity, in vitro drug metabolism and pharmacokinetic (DMPK) properties, physicochemical properties, in vivo harmacokinetics and efficacy in disease-relevant animal models. More recently, biophysical parameters have been used as additional optimization criteria. For instance, tailoring the residence time of a compound by extending the dissociation rates has been reported to be a useful complement to affinity optimization and is in some cases the most important parameter4. In some projects, the potency of very active compounds is prohibitive for accurate quantification of binding in biochemical assays, whereas biophysical assays are still able to measure activities in the pM range. For in vivo experiments with inconclusive pharmacokinetic–pharmacodynamic (PK–PD) relationships, knowledge of the drug residence time on the target derived from SPR may be extremely helpful to explain the observed data.

BOX 2 shows examples of the successful use of biophysics in drug discovery for different purposes: hit identification, hit-to-lead optimization, improvement of absorption, distribution, metabolism and excretion (ADME) properties, attainment of longer residence times and identification of allosteric inhibitors. Emerging technologies and wider applications TABLE 2 highlights a selection of emerging biophysical techniques used in drug discovery. These emerging techniques can either provide new information (for example, conformational changes) or the same information (k on, k off, K d and stoichiometry) in a way that is faster or simpler than established techniques. Many of these techniques use completely new types of physical principles for detection. In addition to the dvantages related to the information they can provide, crucial factors for their establishment include material requirements, ease of use and implementation in current drug discovery settings, and cost.

2.5 ENABLING DRUG DISCOVERY FOR CHALLENGING TARGETS

Numerous targets prove difficult in small-molecule drug discovery, including protein–protein interactions, multi-protein complexes and intrinsically isordered proteins. In particular, protein–protein interactions represent a highly challenging area. The binding sites are large and shallow, with binding hotspots separated by long distances. Further challenges include protein conformational flexibility and competition with the endogenous ligand for the binding site. Biochemical assays as used in HTS are frequently not successful in detecting weakly binding compounds that could provide a starting point for a medicinal chemistry programme.

In such cases, biophysical methods are instrumental in providing essential ligand binding information. FBDD has made some progress for this class of target, with the selective B cell lymphoma 2 (BCL-2) inhibitor venetoclax (also known as ABT-199)66 being approved by the US Food and Drug Administration (FDA) in April 2016 and clinical candidates identified for other targets (for example, hepatitis C virus nonstructural protein 3 (HCV NS3))67. In addition, early inhibitors have been identified for the first time for another member of the BCL-2 family, induced myeloid leukaemia cell differentiation protein MCL-1 (REF. 68) and for one of the most highly validated (but challenging) drug targets in cancer, KRAS69–71. NMR was used to identify fragment hits in most of these examples, exploiting the ability of this technique to robustly identify weak binding compounds. Small-molecule microarray SPR has enabled some hits to be identified for protein–protein interactions (by differential screening against the isolated partners and the protein–protein complex) and for intrinsically disordered proteins, such as the Tau protein.

2.6 EVALUATION OF TARGET ENGAGEMENT IN COMPLEX SYSTEMS

Lack of target engagement at the in vivo site of action is believed to be a substantial source of attrition in phase II proof-of-concept studies. This emphasizes the need for novel technologies to assess target engagement in more complex cellular and in vivo systems, preclinically as well as in clinical studies, so that the best compounds are chosen and optimized and more reductionist approaches can be validated at an early stage to avoid wasted resources. The recently reported cellular thermal shift assay (CETSA) is one of a few currently vailable biophysical methods for this purpose. Although its strengths and weaknesses have not been fully elucidated, the possibility to apply the technology directly in live unmodified cells and with unmodified compounds within ex vivo tissue is tantalizing. In a typical CETSA experiment, vehicle and ligand-treated live cells are heated to different temperatures and the remaining soluble target protein is quantified by affinity reagents; for example, Western gels or quantitative MS technologies. In its simplest interpretation, ligand-induced thermal stabilization indicates permeability to a relevant cellular compartment and target binding. Coupling CETSA to a quantitative MS proteomic analysis is a recent development, and the fingerprint of protein changes highlights both the power and complexity of the method, as proteins can be destabilized as well as stabilized, and there can be pathway-dependent as well as direct effects. Nonetheless, CETSA has already demonstrated insights within more physiological systems, especially for selectivity profiling and target identification studies.

2.7 STUDY OF MEMBRANE PROTEINS IN A MORE NATIVE-LIKE ENVIRONMENT

Studying membrane proteins by biophysical techniques is challenging. This is mainly due to low gene expression, the low overall yield of protein after purification, the typically poor protein stability and the need to work with detergents and lipids. More than 10 years ago, the application of SPR to detergentsolubilized membrane proteins was demonstrated on two chemokine receptors, C-C chemokine receptor 5 (CCR5) and C-X-C chemokine receptor 4 (CXCR4). Both receptors were solubilised directly from cells with a mixture containing detergents and lipids, and captured on a SPR sensor using a specific antibody without prior purification. Further studies on isolated membrane receptors included fragment-screening applications on wildtype β-2adrenergic receptor that was solubilized and purified in detergent micelles, and finally immobilized with a carboxy-terminal polyhistidine-tag.

Furthermore, specific antibodies targeting fusion proteins or tags have enabled biophysical analysis of G protein-coupled receptors (GPCRs). Most current studies on purified GPCRs use stabilized variants of receptors, and substantial progress in biophysical screening (SPR and target-immobilized NMR screening (TINS)) has been demonstrated by Heptares and ZoBio. After careful assessment of ligand binding properties compared to wild-type protein, the engineered receptors can be used in biophysical studies. Examples include SPR screening of lowmolecular- mass ligands for stabilized β1-adrenergic and adenosine A2A receptors that were captured by a polyhistidine tag (His-tag) on the sensor surface. Recently, the discovery of dual inhibitors for orexin receptors (OX_1 and OX_2) was reported85. To date, stabilized GPCRs applied in SPR-based fragment screening have been developed by iterative single-point mutagenesis and alanine scanning. Directed molecular evolution represents an alternative approach to stabilize membrane proteins, as shown with the neurotensin 1 receptor. This receptor was used for a fragment screening effort with 6,369 compounds by SPR followed by hit validation by NMR, resulting in 4 confirmed hits. This example represents the first successful fragment screening for a GPCR with a peptidic endogenous ligand. The ability of NMR to detect target–ligand interactions from nanomolar to millimolar affinity with high sensitivity is a fundamental advantage for membrane proteins. For the screening of larger libraries, the TINS approach has been successfully applied to wild-type and stabilized membrane GPCRs.

Recently, lipid bilayer nanodiscs were developed to enable detergent-free membrane protein preparation. For example, stabilized human A2A receptor with a combination of affinity tag and green fluorescent protein was used for SPR measurements to characterize smallmolecule binding in a more native-like environment89. Although the off-rates measured this way are similar to k off values derived from detergent-solubilized receptors, one order of magnitude faster on-rates lead to lower observed K d values. The substantially increased stability of the nanodisc sample holds great promise regarding the efficient use of less-stabilized receptors for biophysical methods.

Table 2: Selection of emerging biophysical techniques used in drug discovery

Name	Type	What is measured?	Derived data
Biolayer interferometry	Glass fibre-based biosensor	Interaction between white light sent down a glass fibre and the light reflected back up to the instrument. The reflected light comes from interfaces represented by a sensor surface exposed to an analyte and a reference surface with a proprietary bio-compatible layer	Kd, kon and koff
Backscattering interferometry	Microscale interferometry of HeNe laser light and CCD array detection	Change in specific refractive index signature of target and liga nd measured as the mean polarizability of the sample as probed in a microfluidic channel	Kd
Surface acoustic wave	Surface-based fluidic biosensor	Real-time detection of binding-induced conformational changes through monitoring the shift in the phase of surface acoustic waves that travel along the biosensor	Conformational changes, stoichiometry, Kd, kon and koff
Second-harmonic generation	Optical prism microplate based biosensor	Proteins labelled with a second harmonic generation active dye are tethered to supported lipid bilayers by a His-tag. An incident femtosecond laser light creates an evanescent wave through total internal reflection when it strikes the sensor surface. Conformational changes that alter the orientational distribution of the label in space or time result in a signal change	Conformational changes
Waveguide-based grating-coupled interferometry	Integrated microfluidic surface biosensor	Changes in total mass on the biosensor result in a shift of the wave's phase; this phase shift provides information about the on and off rates, as well as the stoichiometry of the interaction	Kd, kon and koff

Besides GPCRs, SPR methods have been developed to characterize ligand binding to ion channels, and binding affinity as well as binding kinetics were investigated for example, acid-sensing ion channel 1a90, γ-aminobutyric acid type A91, 5-hydroxytryptamine (5-HT3) 92 and nicotinic receptors93. ITC measurements with membrane proteins are challenging because of the special care required for handling of detergents in measurements and analysis of the samples. Reports are limited to monotopic membrane proteins such as carnitine palmitoyltransferase 2 and high-affinity small-molecule characterization in the lead-optimization phase.

2.7.1 Single-Molecule Studies

Single-molecule measurements make use of a diverse range of optical analysis methods, ranging from atomic force microscopy to various optical microscopies with a broad spectrum of light combined with ultra-high-resolution imaging techniques. With the invention of novel fluorescent methods in combination with near-field microscopes in the mid-1990s, the application of single-molecule techniques to probe properties at the molecular level started to increase96–99. Single-molecule methods represent potential option for identifying rare as well as transient structural states of a protein target or measuring quantitative kinetic parameters, or monitoring protein dynamics over time trajectories from seconds to minutes while consuming only small amounts of sample material.

One advantage of single-molecule methods is the low sample consumption, as probe volumes only require a femtolitre amount of sample with picomolar protein concentrations. Despite several limitations (such as the long data collection time, high background to noise level and difficulties in analysing low-affinity compounds), the popularity of single-molecule spectroscopy has increased in the past 10–20 years, especially in combination with Forster resonance energy transfer (FRET) to study the complex conformational landscape of proteins, including the determination of kinetic parameters in each conformational state97,100. Although the well-known single-molecule patch clamp technique has been used in drug discovery since the 1970s, the application of other singlemolecule experiments in drug discovery is scarce owing to the complexity of experiments and the limitations of the methods101. Further developments — including higher-throughput detection on microfluidic chips102, progress in fluorophore chemistry, and technical improvement of instrumentation, software for data collection and analysis — are still needed to enable wider application of single-molecule technologies in an industrial environment.

2.7.2 Trends in structure-determination technologies.

Knowledge about the 3D structures of biomolecules of pharmaceutical relevance and their drug complexes has proven to be of great value for an efficient and successful drug discovery effort. Structural biology provides a detailed view on the mode of binding of functionally active ligands and enables their structure-guided optimization. The generation of structural information in drug discovery has been dominated by X-ray crystallography.

In contrast to NMR-based methods, there is no target size limitation, and the determination of multiple structures of target–ligand complexes (an essential step for probing the protein-binding pharmacophore with structurally distinct ligands) is easily facilitated by either co-crystallization or soaking protocols. Serial femtosecond crystallography (SFX) using ultrashort pulses of coherent and extremely intense radiation generated by X-ray free electron lasers (XFELs) 103,104 and high-resolution single-particle cryo electron microscopy (cryo-EM)105,106 are two new promising technologies in structural biology. With XFELs, it is now possible to obtain structural information for notoriously difficult targets for which either only very small crystals can be obtained (which are not suitable for diffraction experiments on regular synchrotron beamlines) or radiation damage is a serious problem (for example, see Kern et al.107). In the context of drug discovery, the enhanced access to novel structures through SFX — in particular, structures of challenging membrane protein targets — s of great value. SFX does not require work at cryogenic temperatures, which helps to derive information on molecular dynamics. Furthermore, SFX enables full automation of sample handling from crystallization to X-ray data collection, as no manual crystal manipulation is required. The use of this method in high-throughput ligand–complex crystallography will need further technological developments. Currently, the high level of effort involved in determining a structure and the relatively poor data quality limit applications to non-routine, low-throughput and high-value projects. Cryo-EM is used to study large macromolecular complexes (~200–4,000 kDa). Some of these complexes reside in membranes, can adopt multiple conformational states and/or have been refractive to crystallization105. Recent examples of targets studied by cryo-EM include transient receptor potential cation channel subfamily V member 1 (TRPV1)108, γ-secretase109 and β-galactosidase, for which a 2.2 A-resolution structure was obtained. For some systems, cryo-EM is able to achieve the resolution needed to elucidate details of protein– ligand interactions, thus enabling structure-based design. Today, the minimal molecular size for high-resolution cryo-EM is about 200 kDa106, leaving a large set of relevant drug targets out of scope. This limitation can probably be reduced by future developments in methodology and instrumentation or by studying molecular targets in the context of larger, functional protein complexes.

2.8 PERSPECTIVES

The fundamental science of molecular recognition and computational methods for data interpretation have evolved in parallel with the implementation of bio physical methods for drug discovery. The ability to accurately determine kinetic rate constants and thermodynamic parameters for protein–ligand interactions provides the possibility of new insights. For example, for inhibitors of HIV1 protease, it was demonstrated that k on and k off change from one generation of inhibitors to the next. In fact, the fully optimized, latest-generation drug molecules have a particularly long residence time.

Furthermore, the widespread assumption that similar ligands have similar association rates was clearly not valid, as small differences in compound structures have a large effect on binding kinetics. Another example is the confirmation that the tight binding of ligands containing moieties such as hydroxamates and sulfonamides is dominated by the interaction with metal ions (for example, Zn^{2+} ions), and the modulating effects of the other parts of the molecule on the kinetics and affinities are not very significant. It may thus be difficult to achieve high selectivity with compounds that include such moieties. As with any in vitro assay, it is important to carefully explore how the experimental conditions could lead to false-positive and false-negative results. Ad hoc selection of pH, ionic strength or type of buffer driven by convenience or preliminary protein handling considerations needs to be replaced by more attention to the detailed physiological conditions for the protein of interest. Questions such as 'what is the tissue location of the active drug target?' or 'what are the conditions in a particular cellular compartment?' need to be considered before the study, and an informed decision made for the selection of the experimental conditions to improve the translation of the results to cellular and in vivo assays. As an example, a pH shift of two units for the aspartic protease β-secretase 1 (BACE1) substantially reduced the inding of all ligands from some chemotypes with micromolar affinity at physiological pH. The pH dependency of inhibitor interactions with BACE1 has been demonstrated to be a unique characteristic of the compounds, influenced not only by the active site aspartic residues, but also by changes in the ligand's pK a upon binding. The implications for drug design and translation of *in vitro* data to in vivo effects also requires the development of modelling methods.

2.9 USE OF KINETIC AND THERMODYNAMIC DATA: EXPECTATIONS AND REALIZATION

Expectations were raised during the past decade that the application of binding kinetics and thermodynamics in the selection and optimization of leads would improve the success rate of preclinical drug discovery. After the pioneering publication of Copeland et al. in 2006, suggesting a correlation of drug–target residence time with drug efficacy and safety, many examples have been published in which binding kinetics and residence time were incorporated into the design strategies of drug candidates. A popular conclusion therefore has been to advocate the use of dissociation rates in addition to or instead of affinity as a key parameter in compound optimization. However, this approach has a number of caveats. First, for many novel targets, the desired kinetic drug profile is not known at the outset of a drug discovery programme, so it is impossible to define a clear goal for off-rate optimization. Second, in the context of drug efficacy, binding kinetics cannot be viewed as being independent from pharmacokinetics; in fact, in many cases, favourable elimination properties of drugs may largely compensate for faster dissociation kinetics. Furthermore, target turnover rates can be faster than dissociation rates, such that beneficial effects of long residence times will not materialize. Finally, as the molecular determinants of binding kinetics are still only incompletely understood, the prospective design of kinetic profiles is currently not possible except for rare cases, in particular those in which long residence times are tuned through a covalent mode of inhibition.

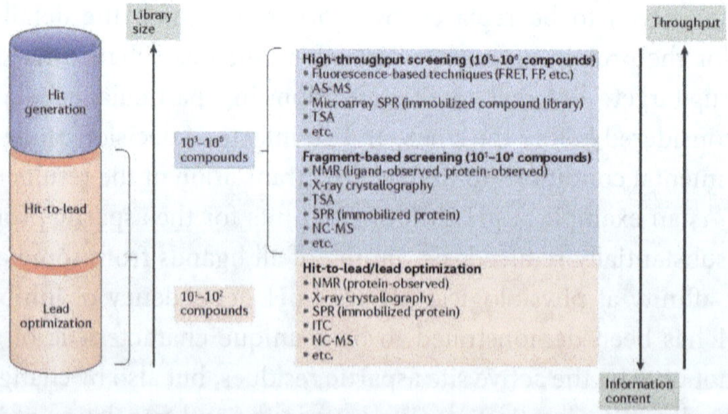

Fig. 2: Biophysical techniques in drug discovery: throughput versus content.

The typical size of the compound collections involved at the different stages of drug discovery decreases from 105–106 for high-throughput screening (HTS) to 103–104 for fragment-based screening (FBS) and then to 101–102 for the hit-to-lead stage followed by lead optimization. The throughput of the biophysical techniques used at different stages should be adapted to the size of the corresponding compound collections. In general, the information content of the techniques is inversely proportional to their throughput. For instance, HTS provides a high to ultra-high throughput but a low information content (usually just 'yes or no' binding information). Conversely, lead optimization requires the highest information content to guide the choice of the best compounds for the next step (drug candidate selection) and the methods used generally have lower throughput. AS-MS, affinity-selection mass spectrometry; ITC, isothermal calorimetry; FP, fluorescence polarization; FRET, Förster resonance energy transfer; NC-MS, non-covalent mass spectrometry; NMR, nuclear magnetic resonance; SPR, surface Plasmon resonance; TSA, thermal shift analysis.

Therefore, using binding kinetics in compound optimization is currently mainly applied in a monitoring mode, which, for example, can enable retrospective analyses of SKRs or resolve potential cases of PK–PD disconnects. Similarly, there have been proposals that detailed analysis of the contributions from enthalpy and entropy to the free energy of binding could aid compound selection before lead optimization. In these proposals there were simplified assumptions that a large negative binding enthalpy is advantageous for target selectivity and ADME properties, because it reflects predominantly specific, direct and polar interactions between the ligand and protein; conversely, a large negative binding entropy is disadvantageous, because it is built mainly on nonspecific van der Waals interactions. However, in recent years, wider analysis has demonstrated that the global parameters ΔH and ΔS: integrate multiple contributions of the complex protein–ligand binding process (most importantly the dominant role of the water structure); are very sensitive to influences of the experimental conditions applied in generating the experimental data9; and are subject to the phenomenon of enthalpy–entropy compensation125. Retrospective interpretation of differentiated thermodynamic repertoires of ligands is only possible with additional detailed information at hand — most importantly, highresolution crystal structures of the protein–ligand complex containing information on the water network within the binding site9,126. Taken together, these factors currently preclude the prospective use of thermodynamics in compound prioritization. Therefore, current applications focus on deriving mechanistic information on binding modes and protein conformational changes when ITC data are used in concert with structural and computational analyses.

2.10 SELECTION OF A BIOPHYSICAL METHOD TO FIT THE PURPOSE

The choice of a biophysical method will first depend on the drug discovery stage — hit generation (HTS or FBS), hit-to-lead or lead optimization — in order to find the best compromise between the throughput, the size of the library to be screened, the sample availability and the expected information content (FIG. 2). For instance, ITC will clearly not be the best choice for a primary screening, but it will bring a wealth of valuable information for compound classification and prioritization at a later stage. In FBS, different screening methods retrieve distinct sets of binders49. The choice of which methods to use for which targets is based on a number of factors, some pertaining to the availability and type of target protein. Moreover, skill in the application and choice of a method is as important as the method itself. An often unrecognized factor is that some biophysical technology can be deceptively user-friendly and easily generates large amounts of data. However, without skilled experimentalists the data are not meaningful. The reasons for this vary, but include inappropriate experimental design and conditions, poorly performed experiments, lack of suitable controls and incorrect or inadequate data analy sis. For example, in SPR biosensor analysis, the basic art of fast and accurate pipetting and performing appropriate controls is crucial. Moreover, the visualization, presentation and interpretation of data also require experience. Although there is no excuse for poor quality work, each method shows limits and artefacts but, by recognizing and addressing them, a suitable combination of methods can provide the required information.

All in all, the best way to obtain a high rate of valuable hits from a screening campaign seems to be to use a robust, high-throughput, low-sample-consuming technique for primary screening and to confirm hits using lower-throughput but higher-content orthogonal techniques. Contrary to the belief that all drug discovery challenges are best solved through the introduction of new technologies, substantial advances can also be driven by innovative application. The kinetic probe competition assay (kPCA) for the discovery and characterization of GPCR ligands provides such an example of creative experimental design rather than novel instrumentation to extend the toolbox of technologies to progress drug discovery.

2.11 APPLICATION OF BIOPHYSICAL METHODS: AT A GLANCE

- The increasing application of biophysical methods, especially over the past 10 years, has led to an improved process for compound prioritization and decision-making in early-stage drug discovery, based on valuable additional experimental information. For conventional soluble target classes, the use of biophysical measurements is well established, as the preceding steps of protein production and enablement of target-specific assays are normally well understood. This enables biophysical Primary screening, secondary screening and detailed characterization assays to be developed on timelines that match other screening activities, and thus enables their integration into hit identification approaches.
- One major benefit of such integration is the ability to reliably and quickly identify true positive hits, thereby reducing wasted resources and increasing return on investment. Furthermore, biophysical methods are crucial to the success of FBDD. The delivery of clinical candidates for difficult drug targets, such as protein–protein interactions, and an increasing number of marketed drugs derived from FBDD are testimony to the integral role it now has in drug discovery. Finally, biophysical methods are extensively used by medicinal chemists working on the challenge of multi-parameter optimization of compounds.
- The future holds considerable promise for the extension of applications of biophysical methods. Improved abilities to produce large amounts of high-quality proteins, including membrane proteins, will increase the percentage of projects in the drug discovery portfolio that can benefit from these methods. Translation of biophysical methods to cellular environments will strengthen the physiological relevance of measurements, with techniques such as isothermal microcalorimetry and in-cell NMR131 now possible. Single-molecule and cell biophysical measurements are also advancing rapidly and enabling a degree of characterization that was previously inconceivable.
- All of these developments are substantially enhancing our ability to look at more relevant and complex biological systems. However, for traditional and emerging biophysical techniques, care needs to be taken in the design of the experiments, the evaluation of the experimental conditions and the critical analysis of the results, and qualified personnel are important in all steps of the work.
- Overall, methodological improvements in detection sensitivity (better signal-to-noise ratios), protein consumption and throughput, the development of novel technologies investigating new facets of functional behaviour of biomolecules, and an improved understanding of how to interpret results and relate them to existing functional data, will ensure that biophysical methods have a great future in pharmaceutical research.

3

BIOPHYSICS AND CHROMATOGRAPHY

3.1 INTRODUCTION

The International Union of Pure and Applied Chemistry (IUPAC) has defined chromatogra- phy as: A method, used primarily for separation of the components of a sample, in which the components are distributed between two phases, one of which is stationary while the other moves. The stationary phase may be a solid, or a liquid supported on a solid, or a gel. The stationary phase may be packed in a column, spread as a layer, or distributed as a film, etc.; in these defini- tions chromatographic bed is used as a general term to denote any of the different forms in which the stationary phase may be used. The mobile phase may be gas or liquid.

3.2 CHROMATOGRAPHY

Chromatography is a physical method for separation of compounds. Tswett, Russian botanist (referred as the father of chromatography) is credited for the development of chromatogra- phy. He employed the technique to separate various plant pigments such as chlorophylls and xanthophylls by passing solutions of these compounds through a glass column packed with finely divided calcium carbonate. The separated species appeared as colored bands on the column, which accounts for the name he chose for the method (Greek chroma meaning 'color' and graphein meaning 'writing').

Chromatography is based on a very simple principle. The sample to be examined (called the solute or analyte) is allowed to interact with two immiscible phases – a mobile phase and a stationary phase. The two immiscible phases could be a solid and a liquid, or a gas and a liquid or a liquid and another liquid. The stationary phase (i.e. sorbent), which may be a solid or a liquid supported on a solid, does not move. The mobile phase moves the sample through the stationary phase. The mobile phase may be a liquid (liquid chromatography) or a gas (gas chromatography). All chromatographic methods involve passing a mobile phase through a stationary (immobile) phase. The two phases are chosen so that the components of the sample distribute themselves between the mobile and stationary phases to varying degrees.

3.2.1 Classification of Chromatographic Methods

Chromatographic methods can be classified in three fundamental ways:

- Based on the shape of the chromatographic bed.
- Based on the physical state of the mobile and stationary phases.
- Based on the mechanism of separation.

3.2.1.1 Based on the Shape of the Chromatographic Bed

Based on the shape of the chromatographic bed, there are two types of chromatography —*planar* chromatography and *column* chromatography. In **planar** chromatography, the stationary phase is spread on a flat, planar surface. The plane can be a paper impregnated by a substance acting as a stationary phase (paper chroma- tography, PC) or a thin layer of a substance acting as a stationary phase spread on a glass, metal or plastic plate (thin layer chromatography, TLC). Planar chromatography is also termed open-bed chromatography. In thin layer chromatography, the stationary phase is a thin layer of silica gel or alumina on a glass, metal or plastic plate. Most commonly, silica gel is used It is the partition coefficient, σ, which describes how much of the internal volume is avail- able for the solute ($0 < \sigma < 1$). When σ is compared with the values measured for solutes of known size, it provides information about the molecular size of an unknown solute. If a series of solutes of known size is subjected to size exclusion chromatography, a linear relationship between partition coefficient and size is observed.

Ion exchange chromatography

Ion exchange chromatography is applicable for the separation of charged molecules. In this chromatographic technique, ionic solutes display reversible electrostatic interactions with a charged stationary phase. The stationary solid phase commonly consists of an insoluble matrix with covalently attached anions or cations (called ion exchangers). Solutes entering the column may be negatively charged, positively charged, or neutral under the experimental conditions. Solute ions of the opposite charge in the mobile liquid phase bind reversibly to the ion exchanger by electrostatic interactions. The strength of interactions depends on the size of the charge and the charge density (amount of charge per unit volume) of the solute. The greater the charge or the charge density, the stronger the interaction. Neutral solutes show little or no affinity for the stationary phase and move with the eluting buffer. The bound solutes can be released by eluting the column with a buffer of increased ionic strength or pH. An increase in buffer ionic strength releases bound solutes by displacement. Increasing the buffer pH decreases the strength of the interaction by reducing the charge on the solute or on the resin.

Ion exchangers are made up of two parts – an insoluble matrix and chemically bonded charged groups within and on the surface of the matrix. An ion exchanger is classified as cationic or anionic depending on whether it exchanges cations or anions.

Cation exchanger (also called acidic ion exchanger): It is used for cation separation. Anion exchanger (also called basic ion exchanger): It is used for anion separation.

Each type of exchanger is also classified as strong or weak according to the ionizing strength of the functional group. An exchanger with a quaternary amino group is, therefore, a strongly basic anion exchanger, whereas primary or secondary aromatic or aliphatic amino groups would lead to a weakly basic anion exchanger. A strongly acidic cation exchanger contains the sulfonic acid group.

Table 1 Commonly used ion exchangers

Name	Type	Functional group
Anion exchanger DEAE-cellulose	Weakly basic	Diethylaminoethyl (DEAE)
QAE-Sephadex	Strongly basic	Quaternary aminoethyl (QAE)
Q-Sepharose	Strongly basic	Quaternary ammonium (Q)
Cation exchanger CM-cellulose	Weakly acidic	Carboxymethyl (CM)
SP-Sepharose	Strongly acidic	Sulfopropyl (SP)
SOURCE S	Strongly acidic	Methylsulphate (S)

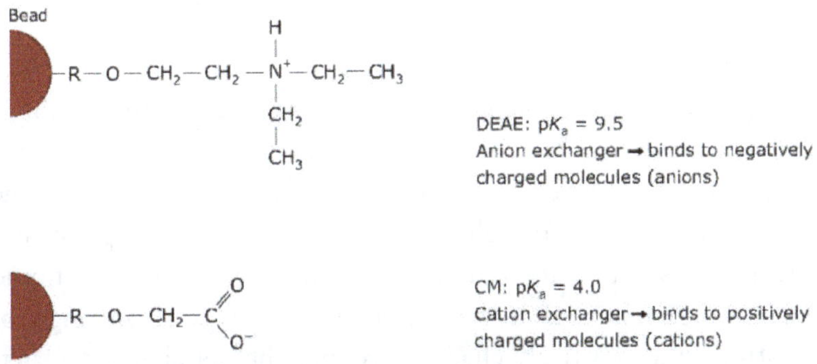

Fig. 1: Ion exchangers–diethylaminoethyl (DEAE) and carboxymethyl (CM). The positive charge of DEAE attracts negatively charged molecules. CM is suitable for binding with positively charged molecules.

A solute in a given sample, which has the opposite charge to that of the charged group of the ion exchanger binds to the column. Separation of charged solutes occurs because different solutes have different degree of interaction with the ion-exchanger due to differences in their charges, charge density and distribution of charge on their surfaces. If a solute has a net positive charge at pH 7, it will usually bind to ion exchanger (cation exchanger) containing carboxylate groups, whereas a negatively charged protein will not.

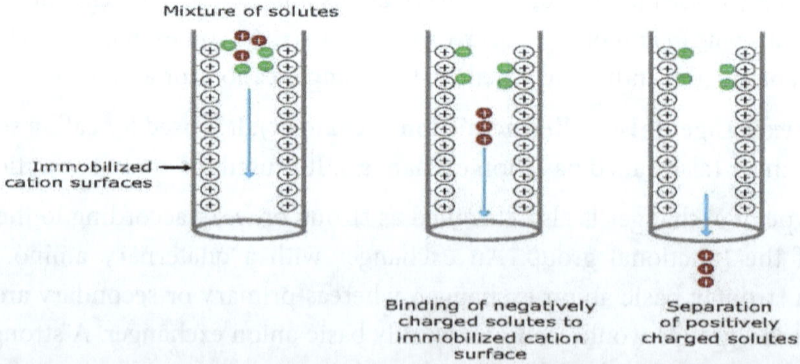

Fig. 2: Column materials used for ion exchange chromatography contain charged groups covalently linked to the surface of an insoluble matrix. The charged groups of the matrix can be positively or negatively charged. Anion exchangers have positively charged groups that will attract negatively charged anions. When a mixture of solutes is loaded into the anion exchanger, negatively charged solutes bind to the exchanger.

The bound molecules can be eluted by altering the pH of the eluting buffer or by increasing the salt concentration of the eluting buffer. A positively charged protein bound to cation exchanger can be eluted by increasing the salt concentration in the eluting buffer because cations present in the buffer compete with positively charged groups on the protein for binding to the ion exchanger. Proteins that have a low density of net positive charge will tend to emerge first, followed by those having a higher charge density.

Selection of the ion exchanger

Before a proper choice of ion exchanger can be made, the nature of the solutes to be separated must be considered. The choice of ion exchanger (whether to use a cationic or anionic exchanger) for the purification of a biomolecule largely depends on the isoelectric point, pI, of the biomolecule. If the solute molecule has only one type of charged group, the choice is simple. A solute that has a positive charge will bind to a cationic exchanger and vice versa. However, many solutes have more than one type of ionizing group and may have both negatively and positively charged groups. The net charge on such molecules depends on pH. At the isoelectric point, the solute has no net charge and would not bind to any type of ion exchanger. At a pH value above the pI of a solute, it will have a net negative charge and adsorb to an anion exchanger. Below the pI, the solute has a net positive charge and will adsorb to a cation exchanger.

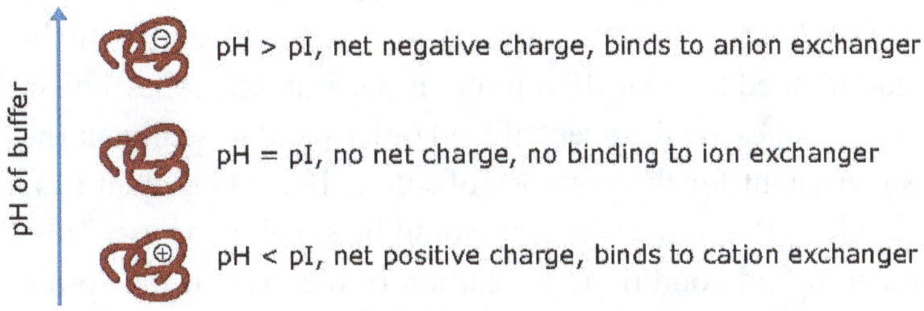

Fig. 3: The net charge on a protein is influenced by the pH of its solvent. At pH=pI, the protein has zero net charge and, therefore, will not bind to a cation exchanger or an anion exchanger. Adjusting the pH above or below the pI of the protein will lead to a net charge and protein binding to either an anion exchanger (pH > pI) or a cation exchanger (pH < pI).

In principle, solutes with both negatively and positively charged groups should bind to both anionic and cationic exchangers. However, when one is dealing with large biomolecules, the pH range of stability must also be evaluated. The range of stability refers to the pH range in which the biomolecule is not denatured. For example, if the pI of a protein is 4 then, in most cases, it is advisable to choose an ion exchanger which binds to the protein at a pH > 4. Since at pH > 4 this protein is negatively charged, the ion exchanger has to be an anion exchanger, e.g. DEAE. One could also use a pH < 4 and a cation exchanger, but many proteins are not stable or aggregate under these conditions. If, in contrast, the protein we want to purify has a pI = 10, it is positively charged at a pH around 7. Thus, in general for this protein type we have to choose a cation exchanger, which is negatively charged at neutral pH.

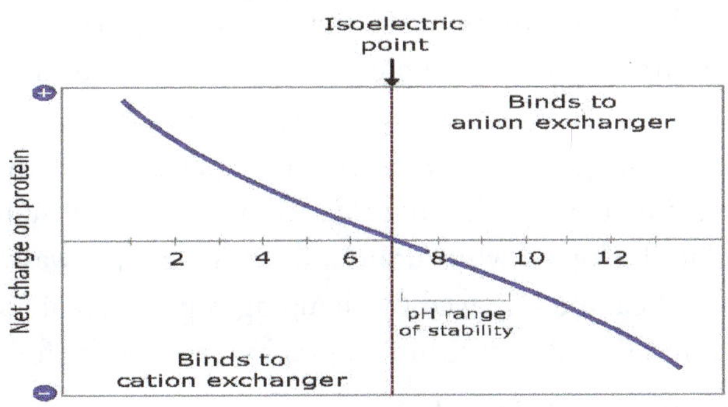

Fig. 4: Diagram shows how the net charge of a hypothetical protein changes as a function of pH. Below the isoelectric point, the molecule has a net positive charge and would be bound to a cation exchanger. Above the isoelectric point, the net charge is negative, and the protein would bind to an anion exchanger. Superimposed on this graph is the pH range of stability for the hypothetical protein. The range of stability refers to the pH range in which the biomolecule is not denatured. Because it is stable in the range of pH 7.0-9.0, the ion exchanger of choice is an anionic exchanger. In most cases, the isoelectric point of the protein is not known. The type of ion exchanger must be chosen by trial and error.

In most cases, the isoelectric point of the protein is not known. The type of ion exchanger must be chosen by trial and error as follows. Small samples of the solute mixture in buffer are equilibrated for 10 to 15 minutes in separate test tubes, one with each type of ion exchanger. The tubes are then centrifuged or let stand to sediment the ion exchanger. Check each supernatant for the presence of solute. If a supernatant has a relatively low level of added solute, that ion exchanger would be suitable for use. This simple test can also be extended to find conditions for elution of the desired macromolecule from the ion exchanger. The ion exchanger charged with the macromolecule is treated with buffers of increasing ionic strength or changing pH. The supernatant after each treatment is analyzed as before for release of the macromolecule.

Choice of buffer

This decision includes not just the buffer substance but also the pH and the ionic strength. Buffer ions will, of course, interact with ion-exchange resins. Buffer ions with a charge opposite to that on the ion exchanger compete with solute for binding sites and greatly reduce the capacity of the column. Cationic buffers should be used with anionic exchangers; anionic buffers should be used with cationic exchangers. The pH chosen for the buffer depends first of all on the range of stability of the macromolecule to be separated. Second, the buffer pH should be chosen so that the desired macromolecule will bind to the ion exchanger. In addition, the ionic strength should be relatively low to avoid damping of the interaction between solute and ion exchanger. Buffer concentrations in the range 0.05 to 0.1 M are recommended.

3.3 AFFINITY CHROMATOGRAPHY

Affinity chromatography is a technique enabling purification of a biomolecule with respect to biological function or individual chemical structure. The substance to be purified is specifically and reversibly adsorbed to a ligand (binding substance), immobilized by a covalent bond to a chromatographic bed material (i.e. matrix). Samples are applied under favourable conditions for their specific binding to the ligand. Substances of interest are consequently bound to the ligand while unbound substances are washed away. Recovery of substances of interest can be achieved by changing experimental conditions to favour desorption. The operation of affinity chromatography involves the following steps:

- Choice of an appropriate ligand.
- Immobilization of the ligand onto a support matrix.
- Binding of the molecules of interest with the ligand.
- Removal of non-specifically bound molecules.
- Elution of the molecules of interest in a purified form.

A biospecific ligand that can be attached to a chromatography matrix covalently is one of the requirements for successful affinity purification. The binding between the ligand

and molecules of interest must be reversible to allow the molecules to be removed in an active form. After washing away the contaminants, the coupled ligand must retain its specific binding affinity for the molecules of interest. Some examples of types of ligands that are usually used in affinity chromatography are listed in table 2.

Table 2 Typical biological interactions used in affinity chromatography

Types of ligand	Target molecules or molecules of interest
Enzyme	Substrate analogue, inhibitor, cofactor
Antibody	Antigen
Lectin	Polysaccharide, glycoprotein, cell surface receptor, cell
Nucleic acid	Complementary base sequence, nucleic acid binding protein
Avidin	Biotin
Calmodulin	Calmodulin-binding molecule
Poly(A)	RNA containing poly(U) sequences
Glutathione	Glutathione-S-transferase or GST fusion proteins
Proteins A and G	Immunoglobulins

4

BIOPHYSICS AND ELECTROPHORESIS

4.1 ELECTROPHORESIS: AN INTRODUCTION

Electrophoresis (Electro refers to the energy of electricity and Phoresis, from the Greek verb phoros, means to carry across) is a technique for separating or resolving charged molecules (such as amino acids, peptides, proteins, nucleotides, and nucleic acids) in a mixture under the influence of an applied electric field. Charged molecules in an electric field move or migrate, at a speed determined by their charge to mass ratio. According to the laws of electrostatics, an ion with charge 'Q' in an electric field of strength 'E' will experience an electric force, F electrical

$$F \text{ electrical} = Q.E$$

The resulting migration of the charged molecule through the solution is opposed by a frictional force, Ffrictional

$$F \text{ frictional} = V.f$$

where, V is the rate of migration of charged molecule and f is its frictional coefficient The frictional coefficient measures the resistance encountered by the molecule in moving through the solvent. It depends on the size and shape of the migrating molecule and the viscosity of the medium. For a spherical molecule, it is given by the Stokes' law;

$$f = 6\pi\eta r$$

where, η is the viscosity of the solvent and r is the radius of the molecule In the constant electric field, the force on charged molecule balances each other;

$$Q.E = V.f$$

so that each charged molecule moves with a constant characteristic velocity. The migration of the charged molecule in the electric field is generally expressed in terms of electrophoretic mobility (μ), which is the ratio of the migration rate of a charged molecule to the applied electric field:

$$\mu = \frac{V}{E} = \frac{Q}{f}$$

So according to the above equation, electrophoresis mobility is directly proportional to the charge and inversely proportional to the viscosity of the medium, size and shape of the molecule.

Electrophoresis is of two types – *moving boundary* (or free boundary) *electrophoresis* and *zone electrophoresis.*

Moving boundary electrophoresis is the electrophoresis in a free solution, without a supporting media. It was developed by Tiselius in 1937. To separate the different charged molecules present in a mixture, the sample (dissolved in a buffer solution that serves as an electrolyte and maintains the desired pH) is placed in a glass tube connected to electrodes. When an electrical potential is applied across the tube, the charged molecules migrate toward one or the other electrode. Because different charged molecules migrate at different rates, a number of interfaces or boundaries are formed between the leading edge of each charged molecules and the remaining mixture.

In zone electrophoresis, a sample is constrained to move in some kind of inert matrix such as filter paper moistened with buffer (paper electrophoresis) or a gel (gel electrophoresis). A matrix is required because the electric current passing through the electrophoresis solution generates heat, which causes diffusion and convective mixing of the bands in the absence of a supporting matrix. The kind of supporting matrix used depends on the type of molecules to be separated and on the desired basis for separation.

4.2 GEL ELECTROPHORESIS

Gel electrophoresis is a simple, rapid and sensitive analytical tool for separating charged molecules. Under the influence of an electrical field charged molecules migrate in the direction of the electrode bearing the opposite charge. Most biomolecules carry a net charge at any pH other than at their isoelectric point. These charged biomolecules have different electrophoretic mobilities due to their differences in charge, shape and size. In gel electrophoresis, gel serves as porous matrix and behave like a molecular sieve. Typically, a gel matrix provides resistance to the migration of the charged molecules that are being subjected to the electrical force. There are two basic types of materials used to make gels: agarose and polyacrylamide.

Agarose is a natural colloid extracted from seaweed. It is a linear polysaccharide made up of the basic repeating unit agarobiose, which comprises alternating units of galactose and 3,6-anhydrogalactose. Agarose gels have a relatively large pore size and are used primarily to separate larger molecules with a molecular mass greater than 200 kDa. Agarose is usually used at concentrations between 1% and 3%. Agarose gels are used for the electrophoresis of both proteins and nucleic acids. Because the pores of an agarose gel are large, agarose is used to separate macromolecules such as nucleic acids, large proteins and protein complexes.

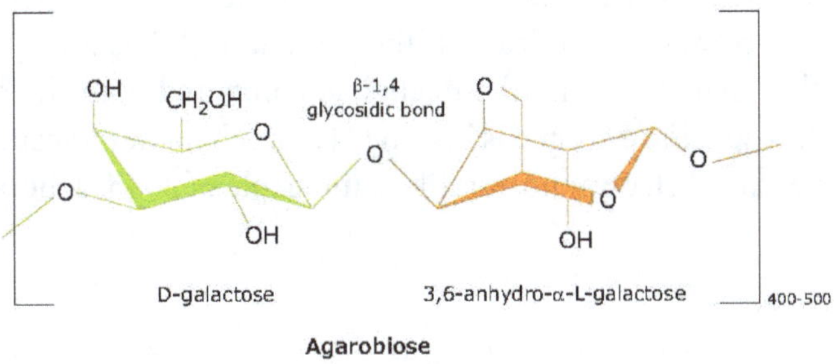

Fig. 1: Agarose is a linear polysaccharide made up of the basic repeat unit agarobiose, which comprises alternating units of galactose and 3,6-anhydro-α-L-galactose.

In discontinuous PAGE, discontinuity is based on four parameters:

- The gel structure
- The ph value of the buffer
- The ionic strength of the buffer
- The nature of the ions in the gel and in the electrode buffer

Discontinuous PAGE can be native or denaturing. In discontinuous native PAGE, proteins are prepared in nonreducing, nondenaturing sample buffer, and electrophoresis is also performed in the absence of denaturing and reducing agents. In discontinuous denaturing PAGE, electrophoresis is performed in the presence of denaturing detergent sodium dodecyl sulfate (SDS), called SDS-PAGE.

4.2.1 SDS-PAGE

Electrophoretic separation of proteins is most commonly performed in polyacrylamide gels. The relative movement of proteins through a polyacrylamide gel depends on the charge density (charge per unit of mass), mass (or size) and shape of the molecules. If two proteins have the same mass and shape, the one with the greater charge density will move faster through the gel. Similarly, if two proteins having the same charge density and shape, the one of smaller mass or size will migrate faster than the large size protein. The shape is also a factor because compact globular proteins move more rapidly than elongated fibrous proteins of comparable mass.

In SDS-PAGE, proteins are treated with the negatively charged anionic detergent sodium dodecyl sulfate (also called sodium lauryl sulfate) before and during gel electrophoresis. SDS imparts a large net negative charge on protein. Proteins treated with SDS have similar charge-tomass ratio i.e. binds to proteins to make them uniformly negatively charged. This is because the amount of SDS bound per unit weight of the protein is constant ~1.4 g of SDS per gram of protein (a stoichiometry of about one SDS molecule per two amino acids). The amount of SDS in the complex depends only on the size of the protein, not on charge or sequence. The negative charge acquired by protein due to binding of SDS is usually much greater than the charge on native protein; this native charge thus becomes insignificant. If the protein itself has a very large positive or negative charge, this charge may not be negligible compared with the charge produced by the bound SDS. A specific example is histone H1 which has a molecular mass of about 21 kDa, but behaves in gel electrophoresis like a molecule of 30 kDa, because of its strong positive charge.

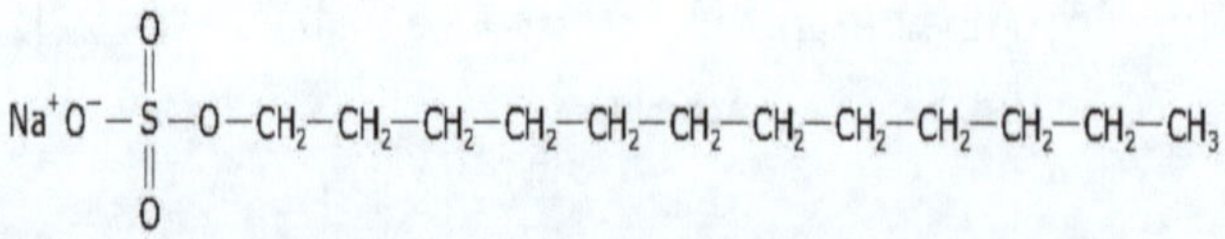

Fig. 2: Structure of sodium dodecyl sulfate (SDS).

SDS bound polypeptides are flexible rod-shaped with a uniform negative charge per unit length. In SDS-PAGE, migration is, therefore, determined not by the intrinsic electric charge of polypeptides but by molecular weight because SDS treatment eliminates the effect of differences in charge density and shape. Thus, when a current is applied, all SDS-bound proteins in a sample will migrate through the gel toward the positively charged electrode. Proteins with less mass travel more quickly through the gel than those with greater mass because of the sieving effect of the gel matrix.

4.2.2 Reducing and Non-Reducing SDS-PAGE:

SDS denatures proteins, causing multimeric proteins to dissociate into their subunits, and all polypeptide chains are forced into extended conformations. In reducing SDS-PAGE, along with SDS, protein is also treated with reducing agents such as β-mercaptoethanol, dithiothreitol to break intrachain or interchain disulfide bonds between cysteine residues. In a multimeric protein, subunits are dissociated when treated with SDS and a reducing agent. Thus, reducing SDS-PAGE gives monomeric mass. A protein shows a band corresponding to 100kDa in native-PAGE but 25kDa in reducing SDS-PAGE suggest the protein is an oligomeric protein with four identical subunits. In nonreducing SDS-PAGE, reducing agents are not used.

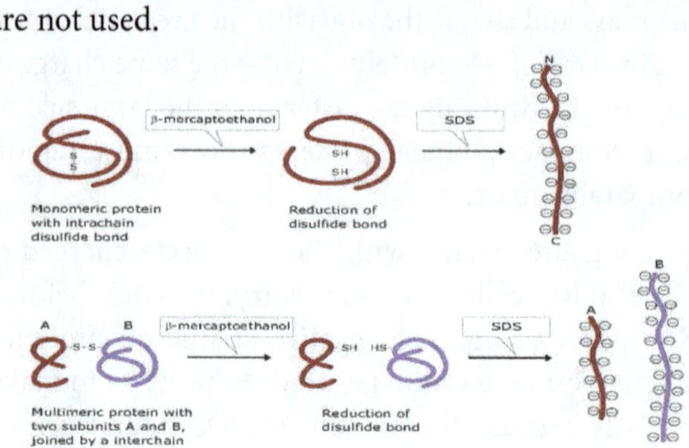

Fig. 3: Effect of SDS on the conformation and charge of a protein. The protein is denatured and coated with a molecule with a uniform charge per unit length. The intrinsic charge of the protein is neutralized. Addition of reducing agents such as β-mercaptoethanol reduces any disulfide bonds.

4.2.3 Native Page

In SDS-PAGE, the gel is cast in a buffer containing SDS, an anionic detergent. The SDS denatures proteins in the sample. Usually, a *reducing agent* such as β-mercaptoethanol, dithiothreitol is also added to break intrachain and interchain disulfide bonds and ensure that no tertiary or quaternary protein structure remains. Consequently, when these samples are electrophoresed, proteins separate according to mass alone.

In native PAGE, proteins are separated according to the net charge, size and shape of their native structure. Most protein molecules carry a net charge at any pH other than their isoelectric point and hence electrophoretic migration occurs because most proteins carry a net negative charge in alkaline running buffers. The higher the negative charge density, the faster a protein will migrate. At the same time, the frictional force of the gel matrix creates a sieving effect, toxicity than EtBr, but they are considerably more expensive. Moreover, all of the alternative dyes either cannot be or do not work well when added directly to the gel, therefore the gel will have to be post stained after electrophoresis.

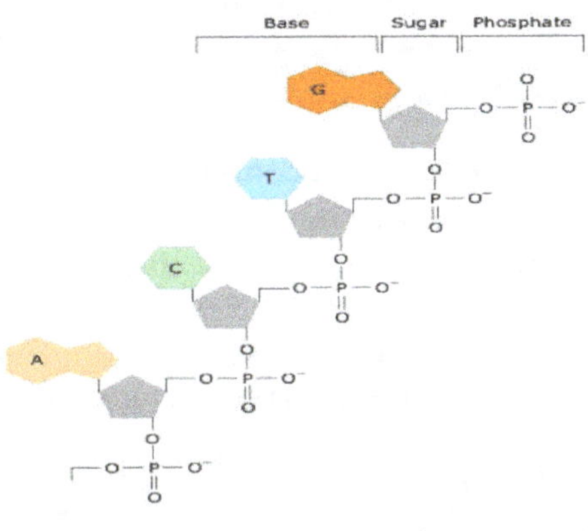

Figure 4 The sugar-phosphate backbone of nucleic acid composed of alternating sugar and phosphate groups is negatively charged at neutral to basic pH and hydrophilic.

4.3 SEQUENCE AND CONFORMATION OF NUCLEIC ACID MOLECULES

Nucleic acid molecules have different electrophoretic mobility when they are of different sizes. The mobility of nucleic acid molecules in gel electrophoresis is determined mainly on the basis of size. However, nucleic acids with the same number of nucleotides i.e. same size but different sequence composition and conformation (shape) may have different mobilities during electrophoresis.

Sequence: AT-rich nucleic acid may migrate more slowly than GC-rich nucleic acid of the same size and shape. The anomalous migration is likely due to sequence composition affecting their molecular conformation.

Conformation: Electrophoresis separates nucleic acid molecules, not only according to their size, but also according to their conformation (*or* shape). The conformations of nucleic acids can be circular and supercoiled, circular and relaxed, and linear. The migration of DNA molecules of the same sequence but differing conformations is affected by the compactness of each conformation. The more compact the conformation of DNA, the more easily it is able to migrate through the gel. Thus, a relaxed circular DNA migrates more slowly than a highly supercoiled form of same circular DNA.

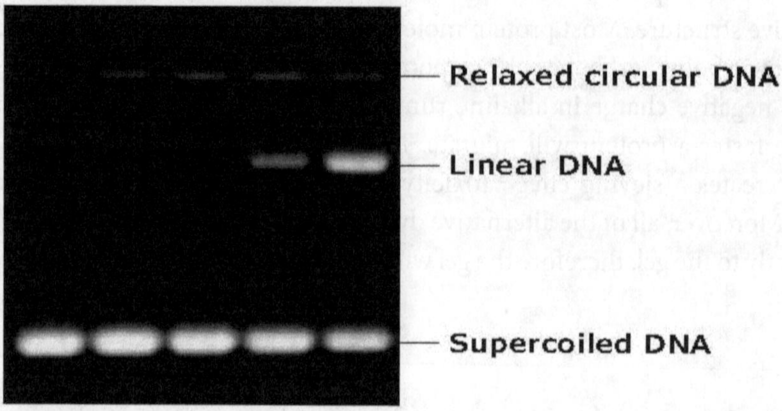

Fig. 5: DNA molecules having different conformations move through the gel at different rates. Supercoiled covalently closed circular DNA, because of its compact conformation, moves through the gel fastest, followed by a linear DNA fragment of the same size, with the relaxed circular form traveling the slowest. The basis of this separation is that the greater the supercoiling, the more compact the conformation of a covalently closed circular DNA (cccDNA).

4.4 UV SHADOWING

In place of staining with a dye, nucleic acids may be indirectly visualized by a method call UV shadowing, taking advantage of UV absorption by nucleic acids. For detection by UV shadowing, nanograms to micrograms of samples are needed, and a thin and transparent gel-like polyacrylamide should be used to ensure UV absorption and transmission. In a UV shadowing protocol, the gel is removed from the cassette after electrophoresis to maximize detection, wrapped in clear plastic film for protection, and then placed on a UV-fluorescent thin layer chromatography (TLC) plate. When the gel is exposed to UV radiation, absorption by the nucleic acid bands casts shadows on the TLC plate. The shadowy areas of the gel of desired sizes are cut out for further processing.

4.5 QUANTIFICATION

Qualitative analysis of gels for the presence or absence of a band or relative mobilities of two bands can easily be performed by visual inspection. Answering 'How much?' and 'What size?' questions require additional work.

Amount: The amount of material in a band can be determined by a number of methods. The simplest is to visually compare the intensity of a band, either stained or autoradiographic, to standards of known quantity on the same gel. More accurate answers can be determined by using a densitometer to scan the stained gel or photograph/autoradiograph of the gel.

Size: The migration distances of nucleic acids in gel electrophoresis generally display a predictable correlation with their sizes, enabling calculation of the size of nucleic acids in a given sample. For linear double-stranded DNA fragments, migration distance is inversely proportional to the log of the molecular weight (expressed in base pairs), within a certain range. For approximate sizing, migration distances are compared to molecules of known sizes (called ' molecular weight standards'). For a given gel concentration, usually, there will be some molecular weight range in which log of the molecular weight and migration distance are approximately linearly related. The DNA standard contains a mixture of DNA fragments of known sizes that can be compared against the unknown **DNA samples:** The exact sizes of separated DNA fragments can be determined by plotting the log of the molecular weight for the different bands of a DNA standard against the distance traveled by each band. Because shape affects the mobility of a molecule through a sieving gel, all the molecules in one gel must have similar shapes for valid comparisons. This does not present a problem for double-stranded DNA, because the shape of the molecules is virtually sequence-independent. Single-stranded nucleic acids, however, must be denatured to ensure similar shapes. For RNA or single-stranded DNA, denaturants added to the buffer may include formamide, urea, formaldehyde, or methylmercury hydroxide. Nucleic acids can also be denatured by treatment with glyoxal before electrophoresis.

4.6 PULSED-FIELD GEL ELECTROPHORESIS

The sizes of the DNAs that can be separated by conventional gel electrophoresis are limited to ~50 kb in size. Very large DNA fragments are unable to penetrate the pores in an agarose gel and thus cannot readily be resolved. However, larger DNA fragments can be resolved.

5

BIOPHYSICS AND SPECTROSCOPY

5.1 INTRODUCTION: AN OVERVIEW

Spectroscopy is the study of the interaction between electromagnetic radiation and matter as a function of the wavelength or frequency of the radiation. The matter can be atoms, molecules or ions. The nature of the interaction between radiation and matter may include – absorption, emission or scattering. It is the absorption, emission or scattering of radiation by matter that is used to quantitatively or qualitatively study the matter or a physical process.

A study of the radiation absorbed or emitted by an atom or a molecule will give information about its identity and this technique is known as qualitative spectroscopy. Measurement of the total amount of radiation will give information about the number of absorbing or emitting atoms or molecules and is called quantitative spectroscopy.

5.2 ELECTROMAGNETIC RADIATION

Electromagnetic radiation is a form of energy and has both electrical and magnetic characteristics. A representation of electromagnetic radiation with *electric field* (E) and the *magnetic field* (B) – at right angle to the direction of the wave – is depicted in the figure 1. The electric and magnetic fields in an electromagnetic wave oscillate along directions perpendicular to the propagation direction of the wave.

Electromagnetic spectrum ranges from very short wavelengths (such as gamma rays) to very long wavelengths (radio waves). The visible region of the spectrum extends approximately over the wavelength range 400–700 nm, the shorter wavelengths being the blue end of the spectrum and the longer wavelengths the red. The wavelengths between 400 and 200 nm make up the near ultraviolet region of the spectrum and wavelengths above 700 nm to approximately 2000 nm (2 μm) the infrared region. The boundaries describing the electromagnetic spectrum are not rigid, and an overlap between spectral regions is possible.

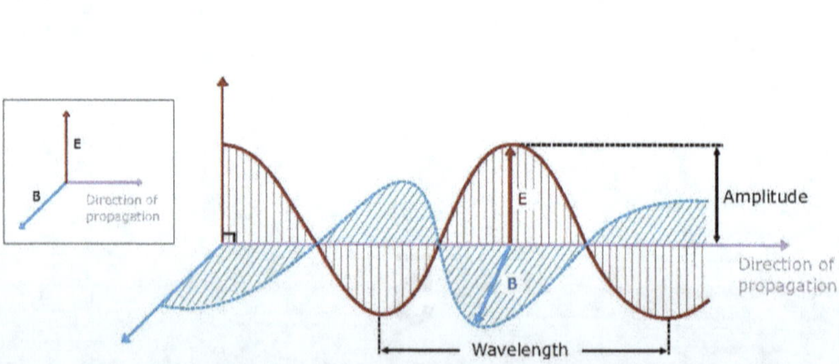

Fig. 1: A representation of electromagnetic radiation with the electric field (E) and the magnetic field (B) at right angles to the direction of the wave movement. Both fields oscillate at the same frequency.

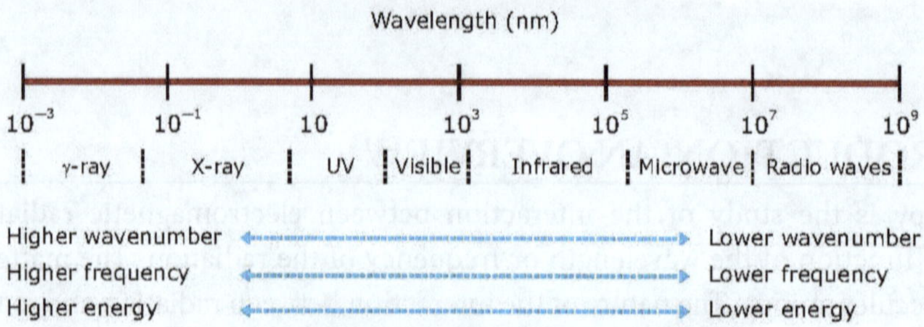

Fig. 2: Range of electromagnetic radiation.

An electromagnetic radiation is characterized by several fundamental properties such as its velocity, amplitude, frequency, energy and polarization. The energy associated with a given segment of the spectrum is related to its frequency and wavelength. Frequency (ν) is the number of wave cycles that pass through a point in one second. It is measured in Hz, where 1 Hz = 1 cycle/sec. The wavelength (λ) is the length of one complete wave cycle. It is often measured in centimeters. It is inversely proportional to the frequency (ν) and is governed by the relationship;

$\nu = \frac{c}{\lambda}$ where, c = speed o light.

The energy is directly proportional to frequency and inversely proportional to wavelength. It is related to wavelength and frequency by the following equation:

$E = h\nu = \frac{hc}{\lambda}$ where, h is the planck's constant (6.6×10^{-34} joules-sec).

The radiation in the infrared region of the electromagnetic spectrum is also expressed in terms of wavenumber, rather than wavelength. It is the reciprocal of the wavelength. It is denoted by $\bar{\nu}$ Wavenumber is expressed in units of per centimeter (cm^{-1}).

$\bar{\nu} = \frac{c}{\lambda}$ where, $\bar{\nu}$ is in units of cm^{-1} and λ is in units of cm.

The main reason chemists prefer to use wavenumbers as units is that they are directly proportional to energy (*a higher wavenumber corresponds to a higher energy*).

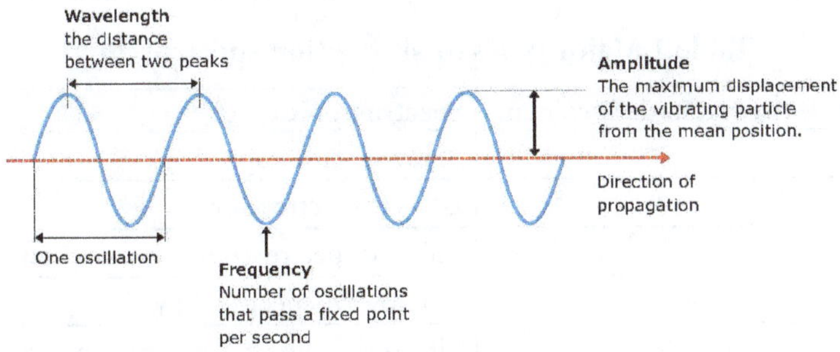

Fig. 3:Higher Wavenumber Corresponds to a Higher Energy

5.3 TYPES OF SPECTROSCOPY

Spectroscopy is the collective term for a diverse group of techniques in which the interactions of electromagnetic radiation with a matter of interest are studied. When electromagnetic radiation meets matter, the radiation is either reflected off, scattered, emitted, transmitted or absorbed. This gives rise to three principal branches of spectroscopy: *absorption spectroscopy*, *emission spectroscopy* and *scattering spectroscopy*.

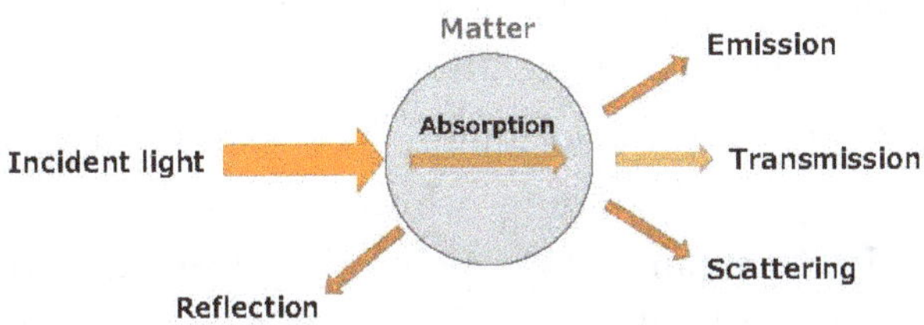

Fig. 4: Principal branches of spectroscopy

5.3.1 Absorption Spectroscopy

Absorption spectroscopy studies radiation *absorbed* at various wavelengths. When a beam of electromagnetic radiation passes through a sample, much of the radiation passes through the sample without a loss in intensity. However, at selected wavelengths, the radiation's intensity is attenuated (decrease in the number of photons). This process of attenuation is called absorption. A plot of the amount of light absorbed by a sample versus the wavelength of the light is called an absorption spectrum. Through absorption spectroscopy, one can find both *qualitative* and *quantitative* information about the sample. In absorption spectroscopy, electromagnetic radiation is absorbed by an atom or molecule, which undergoes a transition from a lower-energy state to a higher energy or excited state. Absorption occurs only when the energy of radiation matches the difference in energy between two energy levels. The type of transition depends on the energy of electromagnetic radiation. Absorption spectroscopy can be categorized into different spectroscopic types (X-ray spectroscopy, UV-Vis spectroscopy, infrared spectroscopy) based on the region of the electromagnetic spectrum absorbed by anatom or molecule.

Table 1 Major types of absorption spectroscopy

Region of the electromagnetic spectrum	Spectroscopic type
X-ray	X-ray spectroscopy
UV–Vis	UV-Vis spectroscopy
Infrared (IR)	Infrared spectroscopy Raman spectroscopy
Microwave	Microwave spectroscopy
Radio wave	Electron spin resonance spectroscopy Nuclear magnetic resonance spectroscopy

Effect of interaction between electromagnetic radiation and matter

Atomic and molecular spectroscopies are mainly related to the absorption, emission, or scattering of electromagnetic radiation and the changes taking place in those systems due to the energy of the radiation. In absorption spectroscopy, the effect of electromagnetic radiation depends on the energy associated with the radiation. The absorption of very energetic radiations to an unoccupied orbital of greater potential energy. Generally, the transition of electrons occurs from the *highest occupied molecular orbital* (HOMO) to the *lowest unoccupied molecular orbital* (LUMO).

Molecular orbitals with lowest energy are the σ-orbitals. The π-orbitals lie at high energy levels and non-bonding orbitals (*n*) lie at even higher energies. The non-bonding orbitals contain a lone pair of electrons and they are stable, filled orbitals. Anti-bonding orbitals (π^* and σ^*), are normally empty and have higher energy than bonding or non-bonding orbitals.

Jablonski diagram

The processes that occur between the absorption and emission of light are usually illustrated by the Jablonski diagram. A typical Jablonski diagram (basically an energy diagram) is shown in the figure. The singlet ground, first and second excited electronic energy states are depicted by *S*0, *S*1 and *S*2, respectively. A molecular electronic energy state in which all electron spins are paired is called a *singlet state*. Each of the electronic energy states (ground or excited) has a number of vibrational energy levels, depicted by 0, 1, 2, etc. The lowest vibrational energy level for each electronic energy state is designated as 0, and the levels above it are successively 1, 2, etc. Each of these vibrational energy levels can be subdivided even further into rotational energy levels.

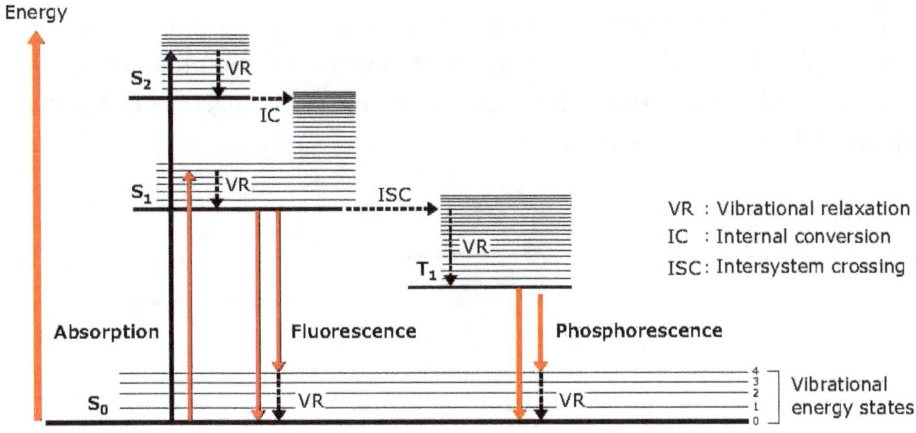

Fig. 5: Jablonski diagram.

The lowest thick horizontal line represents the ground electronic energy states of the molecule, which is normally a singlet state and is labeled *S*0. The upper thick lines are energy levels for the ground vibrational states of three excited electronic states. The two lines on the left represent the first (*S*1) and second (*S*2) electronic singlet states. The one on the right (*T*1) represents the energy of the first electronic triplet state. Numerous vibrational energy levels are associated with each of the four electronic states as represented by the thin horizontal lines. An excited molecule in a singlet or triplet excited energy state can undergo various modes of nonradiative as well as radiative decay. The light emitted by the radiative, singlet excited-to-ground state transition is called fluorescence (emission between states of same spin). The light emitted by the radiative, triplet-to-ground state transition is called phosphorescence (emission between states of different spin).

The absorbance of light of a particular wavelength by the molecule of interest causes transition of an electron from a lower electronic energy state to a higher electronic energy state. Only certain wavelengths of light are absorbed, that is, wavelengths that have energies that correspond to the energy difference between two different energy states of the particular molecule. The promotion of electrons to an excited energy state is called *excitation*. It is possible to excite the molecule to higher vibrational levels within the excited electronic states, so there are many possible absorption transitions.

Once an electron is in excited state, there are several ways that energy may be dissipated. The first is through a nonradiative process called vibrational relaxation. This is indicated on the Jablonski diagram as a dotted arrow between vibrational energy levels. This relaxation occurs between vibrational energy levels of one electronic energy state. Electrons will not change the electronic energy states.

However, if vibrational energy levels strongly overlap electronic energy state, a possibility exists that the excited electron can move from a vibration energy level of an excited electronic energy state to another vibration energy level in a lower energy state. This process is called internal conversion, a non-radiative process. It is mechanistically identical to vibrational relaxation but occurs between two vibrational energy levels in different electronic energy states. Internal conversion occurs because of the overlap of vibrational and electronic energy states.

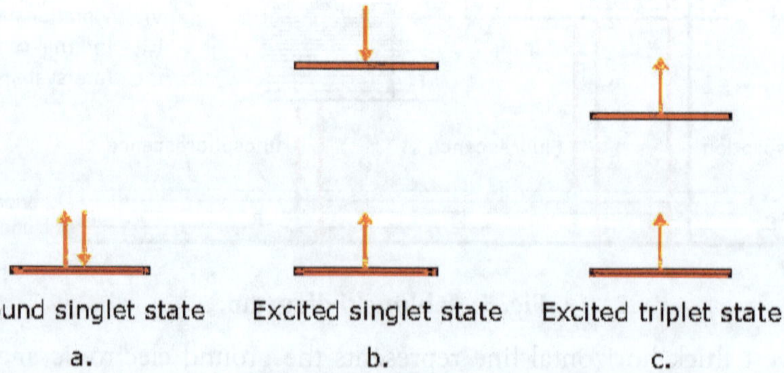

Fig. 6: Electronic spin states of molecules. In a, the ground electronic state is shown. In the ground state, the spins are always paired, and the state is said to be a singlet state. In b and c, excited electronic states are shown. If the spins remain paired in the excited state, the molecule is in an excited singlet state b. If the spins become unpaired, the molecule is in an excited triplet state c.

An excited state may return to the ground state by emitting light. The emission of light may occur during the transition of electrons from a singlet first excited energy state (*S*1) to a singlet ground state (*S*0). This radiative process is called fluorescence (emission involving states of the same spin multiplicity). The energy of fluorescent light is always less than that of the exciting light. This difference is because in the excited state, some energy is always lost by non-radiative processes (such as transitions between vibrational states). Therefore, the energy of the emitted light is always less than that of the absorbed light, and hence, greater wavelengths than the absorbed light. The difference between the absorption (excitation) and emission wavelengths is called the Stokes shift.

Molecules in the singlet first excited energy state (*S*1) can also undergo a spin conversion to the first triplet state, *T*1. When one electron of a pair of electrons of a molecule is excited to a higher energy level, a singlet or a triplet state is formed. In the excited singlet state, the spin of the promoted electron is still paired with the ground state electron. In the triplet state, however, the spins of the two electrons have become unpaired and are thus parallel. Conversion of *S*1 to *T*1 or vice versa is called intersystem crossing, a nonradiative process. It is a spin-dependent internal conversion process where the electron changes spin multiplicity. Intersystem crossing leads to several interesting routes back to the singlet ground energy state (*S*0). One direct transition is phosphorescence (emission involving states of different spin multiplicity), where a radiative transition from *T*1 to *S*0 occurs. This is also a very slow, forbidden transition.

Another possibility is delayed fluorescence, the transition back to the S1, leading to the radiative transition to the S0. Emission involving two states of the same spin is called fluorescence, while emission between states of different spin is phosphorescence. Emission from a triplet excited electronic state to a singlet ground state is thus phosphorescence.

Stokes shift

When a molecule absorbs a photon, it gains energy and enters an excited state. One way for the molecule to relax is to emit a photon, thus losing its energy. When the emitted photon has less energy than the absorbed photon, this energy difference is the Stokes shift. It is the difference (in wavelength units) between positions of the band maxima of the absorption and emission spectra of the same electronic transition.

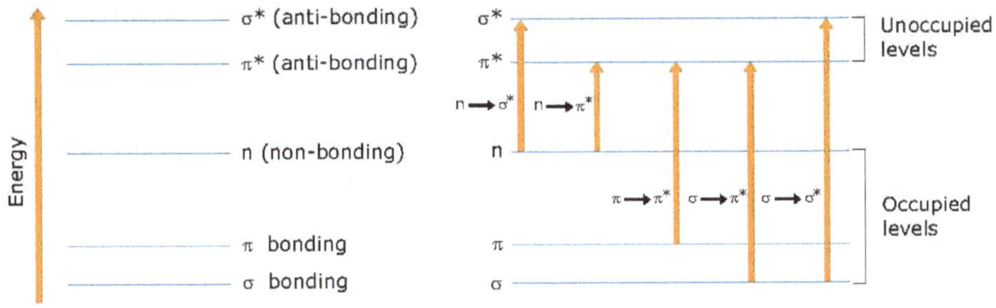

Fig. 7: The possible electron jumps that light might cause are shown here

In each possible case, an electron is excited from a full orbital into an empty anti-bonding orbital. Each jump takes energy from the light, and a big jump obviously needs more energy than a small one. Each wavelength of light has a particular energy associated with it. If that particular amount of energy is just right for making one of these energy jumps, then that wavelength will be absorbed – its energy will be used in promoting an electron. The energy required to bring about transitions from the highest occupied energy level (HOMO) in the ground state to the lowest unoccupied energy level (LUMO) is less than the energy required to bring about a transition from a lower occupied energy level.

When electromagnetic radiation (light) passes through a compound, energy from the radiation is used to promote an electron from a bonding or non-bonding orbital into one of the empty anti-bonding orbitals. The possible electron jumps that electromagnetic radiation might cause are shown in figure 7. In each possible case, an electron is excited from a full orbital into an empty anti-bonding orbital. The energy gaps between these levels determine the wavelength of the electromagnetic radiation absorbed, and these gaps will be different in different compounds. The larger the gap between the energy levels, the greater the energy required to promote the electron to the higher energy level; resulting in light of higher frequency, and therefore shorter wavelength, being absorbed. Each wavelength of electromagnetic radiation has a particular energy associated with it. If that particular amount of energy is just right for making one of these energy jumps, then that wavelength will be absorbed – its energy will have been used in promoting an electron. An absorption of electromagnetic radiation in UV and visible region (200 to 700 nm) cause only a limited number of the possible electron jumps. These jumps are from π bonding orbitals to π anti-bonding orbitals (π to π*) and from non-bonding orbitals, n, to π anti-bonding orbitals (n to π*). This means that in order to absorb electromagnetic radiation in the UV and visible region, the molecule must contain either π bonds or atoms with non-bonding orbitals. Both *n* to π* and π to π* transitions require the presence of an unsaturated functional group to provide the p-orbitals. Molecules containing such functional groups and capable of absorbing UV and visible radiation are called chromophores, Greek words meaning 'to bear color'.

Molecules that show increasing degrees of conjugation require less energy for excitation and as a result absorb radiation of longer wavelengths. Molecules that contain conjugated systems, i.e. alternating single and double bonds, will have their electrons *delocalized* due to overlap of the p-orbitals in the double bonds. As the amount of delocalization in the molecule increases, the energy gap between the π bonding orbitals and π anti-bonding orbitals gets smaller; and therefore, light of lower energy, and longer wavelength, is absorbed.

5.3.2 IR absorption Spectroscopy

The term infrared (IR) covers the range of the electromagnetic spectrum between 1 micrometer to 100 micrometers. It is commonly divided into three sub-regions – near-IR, mid-IR and far-IR. Each frequency (the number of wavelengths that pass a point per unit time) of infrared radiation has a specified amount of energy. If a particular frequency is being absorbed as it passes through the compound being investigated, it must mean that its energy is being transferred to the compound. As with other types of electromagnetic radiation absorption, molecules are excited to a higher energy state when they absorb IR radiation. A molecule absorbs only selected frequencies of IR radiation. The absorption of IR radiation corresponds to energy changes in the order of 8 to 40 kJ/mole. IR radiation does not have enough energy to induce electronic transitions as seen with UV and visible radiations. The energies of IR radiation correspond to the energies involved in bond vibrations.

Physical Basis of IR Spectroscopy

IR spectroscopy is one type of vibrational spectroscopy. At temperatures above absolute zero, the bonds within molecules all vibrate. There are two main types of bond vibrations – stretching and bending. A stretching vibration occurs along the line of the chemical bond whereas a bending vibration is any vibration that does not occur along the line of the chemical bond. Different bonds vibrate with characteristic frequencies.

Consider, for example, the C–H bonds in a typical organic compound. These bonds undergo various stretching and bending vibrations. Let us understand the situation by considering the C–H stretching bond vibration. This vibration takes place with a certain frequency v; that is, it occurs a certain number of times per second. Suppose that a C–H bond has a stretching frequency of 9 × 1013 times per second. A wave of electromagnetic radiation can transfer its energy to the vibrational wave motion of the C–H bond only if *there is an exact match between the frequency of the radiation and the frequency of the vibration*. Thus, if a C–H vibration has a frequency of 9 × 1013 times per second, then it will absorb energy from radiation with the same frequency. When radiation of this frequency interacts with a vibrating C–H bond, energy is absorbed and both the frequency and the intensity of the bond vibration increase. That is, after absorbing energy, the bond vibrates with a greater frequency and with a larger amplitude (a larger stretch and tighter compression). This absorption gives rise to the peak in the IR spectrum. Eventually, the bond returns to its normal vibration frequency, energy is released in the form of heat.

Thus, when the natural frequency of bond vibration in a molecule is equal to the frequency of the IR radiation directed on the molecule, the molecule absorbs the radiation. Different bonds have different vibrational frequencies and therefore different IR absorption frequencies. When IR radiation is absorbed, both the frequency and the intensity of the bond vibration increase. That is, after absorbing energy, the bond vibrates with a greater frequency and with a larger amplitude.

Not all bonds of a molecule are capable of absorbing IR radiation, even if the frequency of the IR radiation exactly matches that of the bond vibrational frequency. To absorb IR radiation, the molecule must have a dipole moment and that must change during vibration. A polar chemical bond has a bond dipole. If a polar bond vibrates with a particular frequency, its bond dipole vibrates with the same frequency. The magnitude of the bond dipole is proportional not only to the amount of charge on each bonded atom but also to the distance between the atoms, that is, the *bond length*. Therefore, as the bond stretches, the bond dipole increases, and, as the bond compresses, the bond dipole decreases. When the frequencies of the IR radiation and the vibrating bond dipole match, the bond dipole absorbs energy from the IR radiation.

Thus, a bond must have a dipole moment that is changing at the same frequency as the incoming radiation for energy to be transferred. Molecular vibrations that occur but do not give rise to IR absorptions are said to be infrared-inactive. In contrast, any vibration that gives rise to an IR absorption is said to be infrared-active. Diatomic molecules such as H2, N2 and O2 (having two identical atoms) have zero dipole moment and vibration of the bonds also do not produce it. These molecules are said to be infrared-inactive. For diatomic molecules, having two different atoms, usually have a permanent dipole moment due to a difference in the electronegativity of the bonded atoms. A greater electronegativity difference between the two atoms usually leads to a larger dipole moment for that bond. However, all molecules having a permanent dipole moment are also not infrared-active. In order to be infrared-active, the bond vibration (stretching or bending) must cause a change in the dipole moment of the molecule.

It is also possible for a molecule that has a zero dipole moment but a specific molecular vibration creates a temporary dipole moment in the molecule. Consider the CO2 molecule as an example. It is a linear symmetric molecule in which the three atoms are arranged linearly with a partial positive charge on carbon and partial negative charge on oxygens. Its equilibrium dipole moment is zero. In the case of symmetrical stretching, both bonds are compressed or stretched to the same extent. As a result dipole moment remains zero. Thus, symmetric stretching vibrations do not absorb IR radiation and said to be *infrared-inactive*. But in asymmetrical stretching, one of the bonds is compressed while the other is stretched. Hence, there is change in bond length and dipole moment. So it is *infrared-active*. Bending vibration is also allowed to this molecule. This also makes molecule infrared-active.

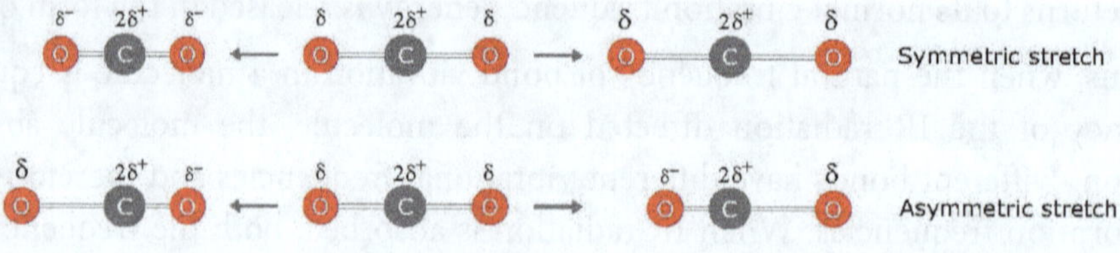

Fig. 8: Symmetric and asymmetric stretching vibration of the carbon dioxide molecule. In the symmetrical stretch, the two C=O bonds stretched (or compressed) at the same time so that the molecule maintains its symmetry. In the asymmetrical stretch, one C=O bond shortens when the other lengthens.

In case of the symmetrical stretch, both bonds are increased; but, because they are exactly equal and oppose each other, the dipole moment remains zero. Hence, the symmetrical stretching vibration is infrared- inactive. In an asymmetrical stretch, one C=O bond is reduced in length while the other is increased.

Because the 'long' C=O bond has a greater bond dipole than the 'short' C=O bond, the two bond dipoles no longer cancel. Thus, the asymmetrical stretch imparts a temporary dipole moment to the CO2 molecule.

Infrared spectra

An infrared spectrum, like any absorption spectrum, is a record of the IR radiation absorbed by a compound as a function of wavelength. The instrument that determines the IR absorption spectrum for a compound is called an infrared spectrometer. An infrared spectrum provides information about what functional groups are present in a compound.

Two aspects of IR spectra are particularly important. First is the *position* of the peak—the wavenumber or wavelength at which it occurs. Second is the *intensity* of the peak—how strong it is.

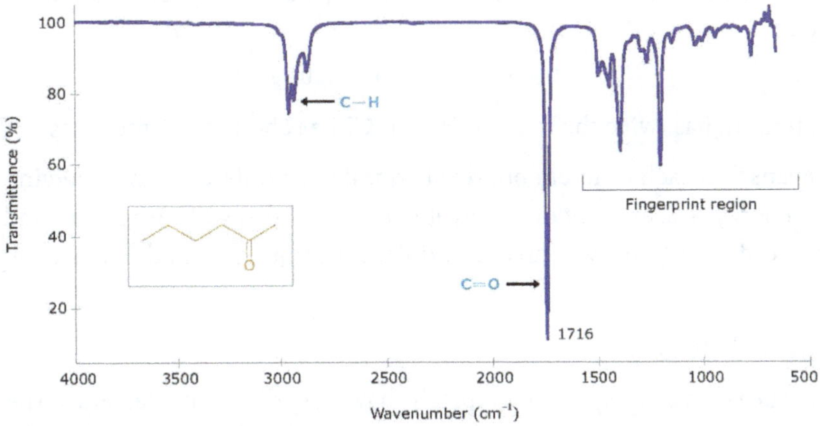

Fig. 9: Infrared spectrum of 2-hexanone. The light transmitted through a sample of 2-hexanone is plotted as a function of wavelength or wavenumber. The wavenumber is simply the inverse of the wavelength. Absorptions are indicated by the inverted peaks. Each peak in the IR spectrum of a molecule corresponds to the absorption of energy by the vibration of a particular bond or group of bonds. A larger value of absorbance is associated with a stronger peak and a smaller value of absorbance with a weaker peak.

IR Absorption Position

The energy involved in bond vibration (i.e. position of the peak in an infrared spectrum) depends on: Strength of the bond: A simple measure of bond strength is the energy required to break the bond, which is the *bond dissociation energy. The higher the bond dissociation energy, the stronger the bond.* Thus, *the IR absorptions of stronger bonds (e.g. bonds with greater bond dissociation energies) occur at higher wavenumber.* Triple bonds have higher stretching frequencies than corresponding double bonds, which in turn have higher frequencies than single bonds.

Masses of the atoms involved in the bond: *The vibration frequency for a bond between two atoms of different mass depends more on the mass of the lighter object than on the mass of the heavier one.* Bonds to lighter atoms (such as hydrogen atom) have higher stretching frequencies than those to heavier atoms.

Type of vibration being observed: The absorption frequency also depends on the type of vibration. The two general types of vibrations in molecules are stretching vibrations and bending vibrations.

In general, bending vibrations occur at lower frequencies (higher wavelengths) than stretching vibrations of the same groups. It is easier to bend a bond than to stretch or compress it.

Values chiefly affected by mass of atoms (lighter atom, higher frequency)

C-H	C-D	C-O	C-Cl
3000 cm^{-1}	2200 cm^{-1}	1100 cm^{-1}	700 cm^{-1}

Values chiefly affected by mass of atoms (stronger atom, higher frequency)

C≡O	C=O	C–O
2143 cm^{-1}	1715 cm^{-1}	1100 cm^{-1}

D, deuterium, has twice the mass of H, and CI has about twice the mass of O.

This means that each different bond will vibrate in a different way, involving different amounts of energy. The amount of energy it needs to do this will vary from bond to bond, and so each different bond will absorb a different frequency (and hence energy) of IR radiation.

IR absorption intensity

The different peaks in an IR spectrum typically have very different intensities. The intensity of an IR absorption depends on: Concentration of molecules in the sample: The greater the number of molecules per unit volume, the greater is the intensity of IR absorption.

Magnitude of dipole moment: The intensity of IR absorption varies with the change of dipole moment. *The greater the change in the dipole moment, the stronger will be the absorption.*

If the bond is perfectly symmetrical, there is no change in dipole moment and there is no IR absorption. As we know, the magnitude of the bond dipole is proportional not only to the amount of charge on each bonded atom but also to the distance between the atoms—that is, the bond length.

Infrared absorption and chemical structure

The most common application of IR spectroscopy is perhaps to identify the functional groups. This is possible because different functional groups vibrate at different frequencies allowing their identification. *In all compounds, a given type of functional group absorbs IR radiation in the same general region of the IR spectrum.* The frequency of vibration, however, depends on additional factors such as delocalization of electrons, H-bonding, and substitutions at the nearby groups.

An IR spectrum usually extends from radiation around 4000 cm–1 to 650 cm–1 and can be split into the functional group region and the fingerprint region. Many functional groups absorb infrared radiation at about the same frequency, regardless of the structure of the rest of the molecule. The absorption bands in the 4000 – 1500 cm–1 region allow identification of functional groups; this region therefore is also termed the *functional group region* of the IR spectrum. For example, C–H stretching vibrations usually appear between 3200 and 2800 cm–1; and carbonyl (C=O) stretching vibrations usually appear between 1800 and 1600 cm–1. This makes these bands diagnostic markers for the presence of a functional group in a sample. These types of infrared bands are called *group frequencies* because they tell us about the presence or absence of specific functional groups in a sample.

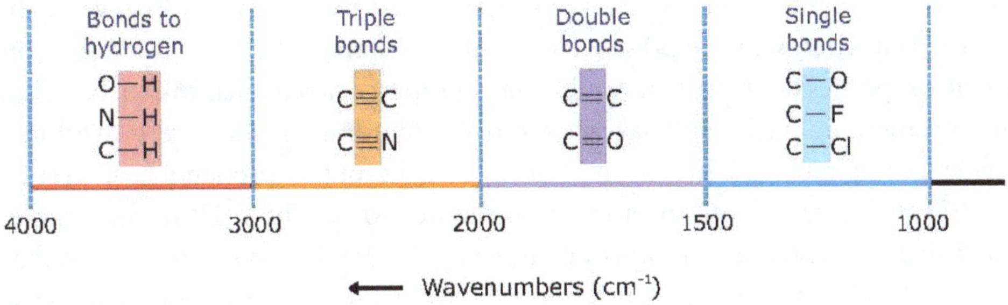

Fig. 10: Infrared spectrum. The approximate regions where various common types of bonds absorb for stretching vibrations only. The first region, from 4000 to 2500 cm−1 is the region for C-H, N-H, and O-H bond stretching. The triple bond region from about 2500 to 2000 cm^{-1}. Double bonds appear about 2000–1500 cm−1 and single bonds come below 1500 cm^{-1}.

The region of the infrared spectrum from 1200 to 700 cm−1 is called the fingerprint region. This region usually contains a very complicated series of absorptions. Many different vibrations, including C—O, C—C and C—N single bond stretches, C—H bending vibrations, and some bands due to benzene rings are found in this region. The fingerprint region is often the most complex and confusing region to interpret. This region is different for each molecule just like a fingerprint is different for each person. Two different molecules may have similar functional group regions because they have similar functional groups, but they will always have a different fingerprint region. Since every type of bond has a different natural frequency of vibration, and since two of the same types of bonds in two different compounds are in two slightly different environments, no two molecules of different structure have exactly the same infrared absorption pattern or infrared spectrum. By comparing the infrared spectra of two substances, one can establish whether they are identical or non-identical. A second use of the infrared spectrum is to determine structural information about a molecule. The absorptions of each type of bond (N—H, C—H, O—H, C—O, C—C and other) are found only in certain small portions of the infrared region. Thus, a small range of absorption can be defined for each type of bond.

In infrared spectra:

Position of band depends on	Bond strength – Strong bonds give high frequency Mass of atoms – Light atoms give high frequency.
Strength of band depends on Change in dipole moment	Strength of band depends on Change in dipole moment A large change in dipole moment gives strong absorption.
Width of band depends on Hydrogen bonding	Width of band depends on Hydrogen bonding Strong H-bond gives wide peak.

5.3.3 Nuclear Magnetic Resonance

Nuclear magnetic resonance (NMR) is a spectroscopic technique that involves a change in nuclear spin energy in the presence of an external magnetic field. It is based on the magnetic properties of nuclei that result from a property called *nuclear spin*. NMR allows us to detect atomic nuclei and say what sort of environment they are in, within their molecule. For example, the hydrogen of the hydroxyl group in propanol is different from the hydrogens of its carbon skeleton. With the help of a proton NMR, one can easily distinguish between these two sorts of hydrogens. '1H NMR' and skeleton. With the help of a proton NMR, one can easily distinguish between these two sorts of hydrogens. '1H NMR' and 'proton NMR' are interchangeable terms. The hydrogen atom is denoted by 1H. Likewise, **carbon NMR** (13C NMR) can easily distinguish between the three different carbon atoms in propanol. 'proton

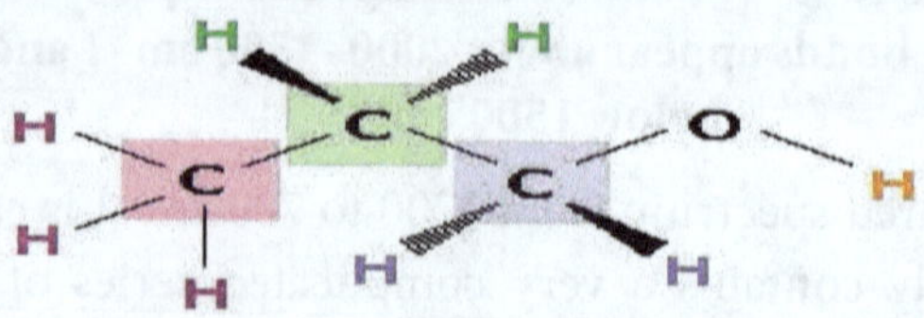

Figure 11 1H NMR distinguishes the coloured hydrogens and 13C NMR distinguishes the boxed carbons.

NMR spectroscopy is used to detect nuclei, but only those nuclei that have a magnetic property as a result of nuclear spin. Spin comes in multiples of 1/2 and can be + or –. Not all atomic nuclei have *nuclear spin*. The rules for determining the net spin of a nucleus are as follows;

- If the number of neutrons and the number of protons are both *even*, then individual spins are paired and the overall spin becomes zero (i.e. nuclei have no spin).
- If the number of neutrons and the number of protons are both *odd*, then the nucleus has an integer spin (i.e. 1, 2, 3).
- If the number of neutrons plus the number of protons is *odd*, then the nucleus has a halfinteger spin (i.e. 1/2, 3/2, 5/2).

Table 3.2 Number of protons, number of neutrons and nuclear spin quantum number of some elements

Number of protons	Number of neutrons	Spin quantum number (I)	Element
Even	Even	0	^{12}C, ^{16}O
Odd	Odd	Integer (1,2,...)	^{14}N
Even	Odd	Half-integer (1/2, 3/2,...)	^{13}C
Odd	Even	Half-integer (1/2, 3/2,...)	^{19}F, ^{15}N

For each nucleus with spin, the number of allowed spin states – it may adopt – is determined by its **nuclear spin quantum number** (l). A nucleus of spin quantum number l has $2l + 1$ allowed spin states. For example, 1H has the nuclear spin quantum number $l = 1/2$ and has two allowed spin states [2(1/2) + 1 = 2] for its nucleus, −1/2 and +1/2. For the chlorine nucleus, $l = 3/2$ and there are four allowed spin states [2(3/2) + 1 = 4] for its nucleus, −3/2, −1/2, +1/2 and +3/2.

Table 3.3 Spin quantum numbers and number of spin states of some common element

Element	$^{1}_{1}H$	$^{2}_{1}H$	$^{12}_{6}C$	$^{13}_{6}C$	$^{14}_{7}N$	$^{16}_{8}O$	$^{17}_{8}O$	$^{19}_{9}F$
Spin quantum number	½	1	0	½	1	0	5/2	½
Number of spin states	2	3	0	2	3	0	6	2

Physical basis of NMR

NMR spectroscopy is based on the magnetic properties of nuclei that result from a property called *nuclear spin.* All nucleus with spin act like a tiny magnet. We can compare the behaviour of a nucleus acting like a tiny magnet with magnetic compass placed in an external magnetic field. Imagine for a moment that we were able to 'switch off' the Earth's magnetic field. Compass needle (made of a magnetic material) will point randomly in any direction. However, as soon as we switched on the Earth's magnetic field back, the needle would point North—their *lowest* energy state. If we wanted needle to point South, we would have to apply force (energy). But after removal of external force, the needle would return to its lowest energy state, pointing North. How hard it is to turn the compass needle depends on how strong the magnetic field is and also on how well the needle is magnetized. If the needle is not magnetized at all, it is free to rotate.

Never theless, there is an important difference between a compass needle and the atomic nucleus acting like a tiny magnet. A real compass needle can rotate through 360° and have a virtually infinite number of different energy levels, all higher in energy than the lowest energy state (pointing North). When atomic nuclei acting like tiny magnets are placed in an external magnetic field, they have different energy levels. Fortunately, the number of energy levels an atomic nucleus can adopt is very less. For example, a 1H or C13 nucleus in a magnetic field can have two energy levels.

Let's take an example of the hydrogen nuclei in a chemical sample. In the discussion, the word 'proton' is used for '1H nuclei'. In the absence of an external magnetic field, the nuclear magnetic poles are oriented randomly. When an external magnetic field is applied to 1H nuclei, they can either align themselves parallel to the field, which would be the lowest energy state, or they can align themselves antiparallel to the field, which is higher in energy. The magnetic poles of nuclei with a spin of +½ are oriented *parallel* to the applied field, and those nuclei with a spin of -½ are oriented *antiparallel* to the applied field.

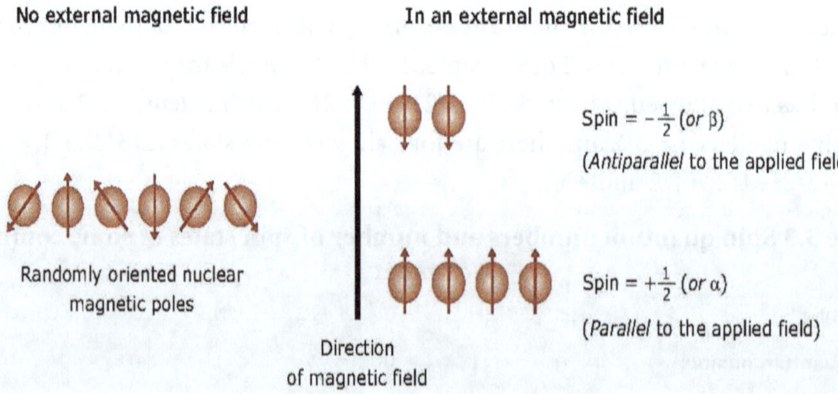

Fig. 12: In the absence of an external magnetic field, the nuclear magnetic poles are oriented randomly (with arrows indicating their magnetic, north-south, polarity).

In an external magnetic field or applied magnetic field, the spin state +½ is aligned parallel to the field, while the spin state −½ is aligned antiparallel to the applied field. More nuclei are oriented along with the applied field because this arrangement is lower in energy.

In the absence of an external magnetic field, the two spin states have the *same* energy. But when an external magnetic field is applied, the two spin states have *different* energies: the +½ spin state has lower energy than the -½ spin state. The energy difference between two spin states is a function of the strength of the applied magnetic field. The stronger the applied magnetic field, the higher the energy difference between the possible spin states. However, the energy difference between the two nuclear spin states is very small—so small that a very, very strong magnetic field is required to see any difference at all.

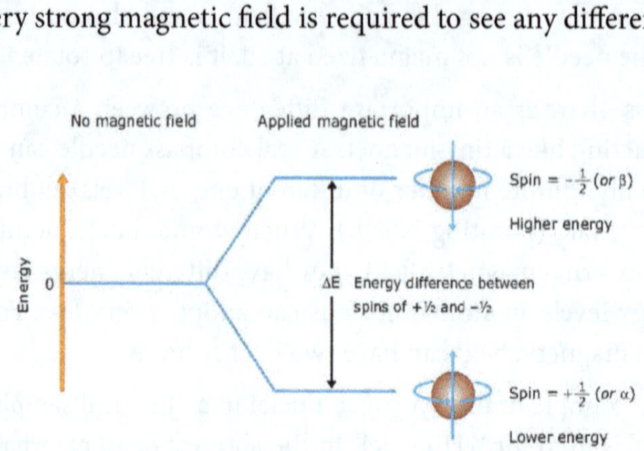

Fig. 13: Effect of increasing magnetic field at a 1H nucleus on the energy difference between its +½ and -½ spin states. The two spin states have identical energies when the external magnetic field is absent. The difference in energy between the two spin states depends on how strong the applied external magnetic field is, and also on the

properties of the nucleus itself. The energy difference between the two spin states grows with increasing field. The stronger the magnetic field, the higher the energy difference between the two states.

Thus, for a nucleus in an external magnetic field, the difference in energy between two spin states in an applied magnetic field depends on:

- How strong the magnetic field is, and
- The magnetic properties of the nucleus itself.

The energy difference between the two spin states is given by the fundamental equation of NMR:

$$\Delta E = \frac{h\gamma_H}{2\pi} B_p$$

where, *h* is Planck's constant,

B_p is the magnitude of the magnetic field at the H_1 nucleus (proton), in gauss; and γ_H is a fundamental constant of the proton, called the gyromagnetic ratio. The value of this constant is 26,753 radians gauss^{-1} s^{-1}.

This equation shows that when the external magnetic field is zero, there is no energy difference between the spin states and as the magnetic field is increased, the energy difference between the two spin states increases.

When the 1H nuclei in a chemical sample are subjected in a magnetic field; each 1H nucleus is in one of two spin states that differ in energy by an amount ΔE, and; a small excess of 1H nuclei have spin +½. Even at a very strong magnetic field, the difference in the populations of the two spin energy states is very small. If the sample is now subjected to electromagnetic radiation with energy *exactly equal* to ΔE, this energy is absorbed by some of the 1H nuclei in the +½ spin state. The absorbed energy causes these protons to 'flip' their spins and assume a more energetic state with a spin −½. Since the energy difference between the two states even in a very very strong external magnetic field is so small, the amount of energy needed to flip the nuclei can be provided by electromagnetic radiation of radio wave frequency. Radio waves flip the nucleus from the lower energy state to the higher state. This absorption phenomenon, called nuclear magnetic resonance, can be detected in a type of absorption spectrometer called an NMR spectrometer.

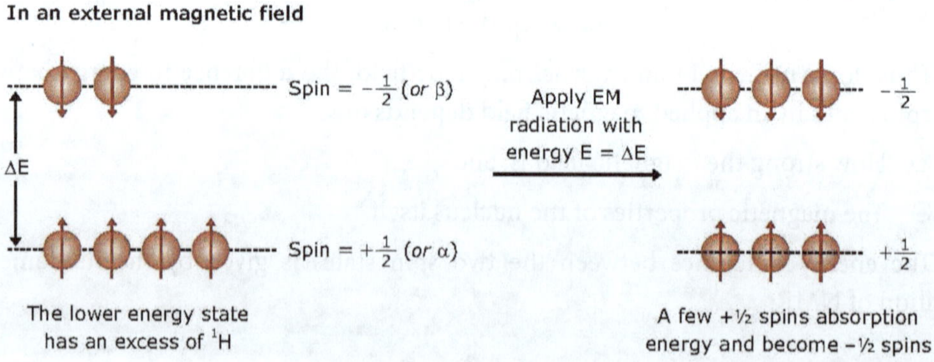

Fig. 14: A chemical sample is subjected in a magnetic field; each 1H nucleus is in one of two spin states that differ in energy by an amount ΔE, and; a small excess of 1H nuclei have spin +½. The lower energy level contains slightly more nuclei than the higher level. The amount of energy needed to flip the nucleus can be provided by electromagnetic radiation (EM) of radio wave frequency. Radio waves flip the nuclei from the lower energy state to the higher state.

Chemical shift

To understand the chemical shift, let us analyze the proton NMR (1H NMR) spectrum of dimethoxymethane. There are two absorption peaks in the protons NMR spectrum. Each peak represents a different kind of protons (hydrogen nuclei): each one absorbs energy at a different frequency. But why should protons present in the same molecule be different? We might expect all protons to resonate at one particular frequency. But they don't. In an applied magnetic field, not all protons in a molecule resonate at exactly the same frequency. This variability is due to the fact that the hydrogens in a molecule exist in slightly different electronic environments from one another.

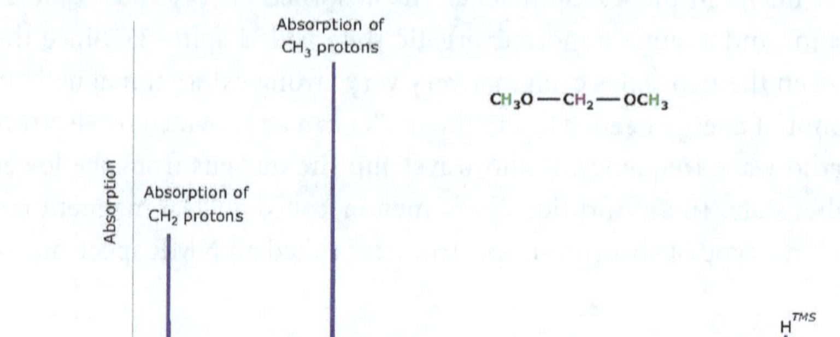

Fig. 15: The proton NMR (1H NMR) spectrum of dimethoxymethane. This spectrum is a plot of energy absorption on the y-axis versus relative frequency of the radiation

from which energy is absorbed on the x-axis. Peaks in an NMR spectrum are called *absorptions*, or *lines*. The position of absorption on the horizontal axis is called its *chemical shift*. We usually express the peak positions in ppm. The small peak at the far right is from the protons of tetramethylsilane (TMS), a reference standard.

Each nucleus is surrounded by electrons. The valence-shell electron densities vary from one hydrogen to another. In an applied magnetic field, the valence-shell electrons are caused to circulate. This circulation, called a local diamagnetic current, generates own magnetic field that opposes the applied magnetic field (B0). Circulation of electrons around a nucleus can be viewed as being similar to the flow of an electric current in an electric wire. In physics, we know that the flow of a current through a wire induces a magnetic field.

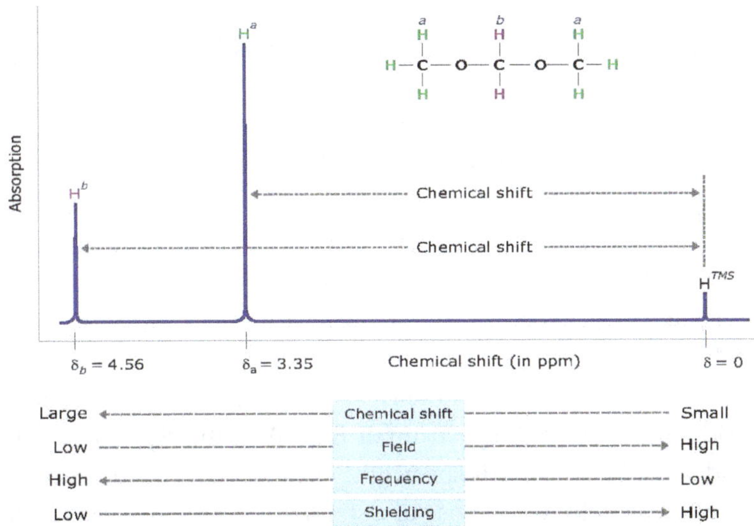

Fig. 16: The proton NMR spectrum of dimethoxymethane.

This spectrum is a plot of energy absorption on the y-axis versus relative frequency of the radiation from which energy is absorbed on the x-axis. Peaks in an NMR spectrum are called absorptions or lines. The position of an absorption peak on the horizontal axis is called its chemical shift. We express the peak positions in parts per million or ppm. The chemical shift of an absorption is written with a δ followed by the numerical value of the peak position. The absorption position of TMS (a reference standard) defines the $\delta = 0$ position on the x-axis of the spectrum. The chemical shift scale runs to the left from zero (where TMS resonates). The other two peaks—the ones at $\delta = 3.35$ and $\delta = 4.56$—are the NMR absorptions of the protons of dimethoxymethane. We can say that the spectrum contains peaks at 0, 3.35 and 4.56 ppm. Two absorptions ($\delta = 3.35$ and $\delta = 4.56$) indicate that there are two chemically distinguishable sets of protons in dimethoxymethane, the CH2 protons and the CH3 protons. The two peaks have different sizes because different numbers of protons contribute to each absorption. The resonance at $\delta = 3.35$ is larger because more protons (six) contribute to this resonance than to the resonance at $\delta = 4.56$ (two).

In summary, proton NMR provides four types of information:

i. The number of sets of chemically nonequivalent protons
ii. The chemical environments of each set of protons (chemical shift)
iii. The number of protons within each set
iv. The number of protons in adjacent sets

The reference sample — Tetramethylsilane

Unlike IR and UV/Vis spectroscopy, where the signals are fixed at particular frequencies or wavelengths, in NMR, the signal is dependent on the field strength. Since no two magnets have the same field, the frequency at which signals are obtained would vary correspondingly. Therefore, there was a need to characterize and specify the location of the signals. In order to avoid this situations, it has been decided to eliminate the problems by keeping a standard reference with respect to which a numerical value can be assigned. The compound we use as a reference sample in 1H and 13C NMR is usually tetramethylsilane, TMS. This is silane (SiH4) with each of the hydrogen atoms replaced by methyl groups to give Si(CH3)4. Because of molecular symmetry, all 12 protons of TMS absorb at the same frequency and all 4 carbons absorb at the same frequency. The frequency of absorption for a nucleus of interest relative to the frequency of absorption of a standard is called the *chemical shift* of the nucleus. The chemical shift of the 1H nuclei in the 1H NMR spectrum or 13C nuclei in the 13C NMR spectrum of TMS appear at 0 ppm. Typically, it increases from 0 on the right-hand side of the spectrum to 10 ppm on the left hand side of an 1H NMR spectrum or from 0 on the right hand side to 200 ppm on the left hand side of a 13C NMR spectrum. The reason frequencies of absorption are recorded on the δ-scale relative to those of a standard molecule. It makes the position of absorption independent of the spectrometer used to record the spectrum, in particular independent of the strength of the magnetic field of the spectrometer.

5.4 CIRCULAR DICHROISM

Light is a transverse electromagnetic wave. The electric and magnetic fields in an electromagnetic wave oscillate along directions perpendicular to the propagation direction of the wave. Light can be *unpolarized* or *polarized* depending on how the electric field is oriented. In unpolarized light, the electric field vector oscillates in more than one plane; all perpendicular to the direction of propagation. If the electric field vector oscillates in single plane perpendicular to the direction of propagation, it is called linearly or plane polarized light. The orientation of a linearly polarized light is defined by the direction of the electric field vector. Linear polarization is obtained by passing unpolarized light through a polarizer that absorbs all electric field vectors not lying along a particular plane.

If the electric field vector rotates around the propagation axis maintaining a constant magnitude, it is called circularly polarized light. Circularly polarized light is the antithesis of linearly polarized light. In linear polarized light, the direction of the vector stays constant and the magnitude oscillates whereas in circularly polarized light, the magnitude stays constant while the direction oscillates. Circularly polarized light is obtained by superimposing two plane polarized light of same wavelengths and amplitudes but having a phase difference of 90°. A phase difference of 90° means that when one wave is at its peak, then the other one is just crossing the zero line.

crossing the zero line.

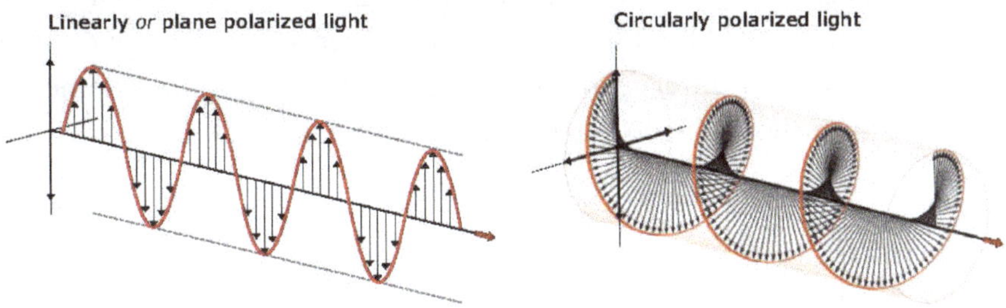

Fig. 17: Linearly polarized light results when the direction of the electric field vector is restricted to a plane perpendicular to the direction of propagation while its magnitude oscillates. In circularly polarized light, the magnitude of the oscillation is constant and the direction oscillates.

A circularly plane polarized light may be right or left-handed circularly polarized depending on the rotation direction. For left circularly polarized light with propagation towards the observer, the electric field vector rotates counter-clockwise. For right circularly polarized light, the electric field vector rotates clockwise. The superposition of left and right circularly polarized light beams of equal amplitudes can result in linearly polarized light. Thus, linear polarized light can be viewed as a superposition of opposite circular polarized light of equal amplitude and phase.

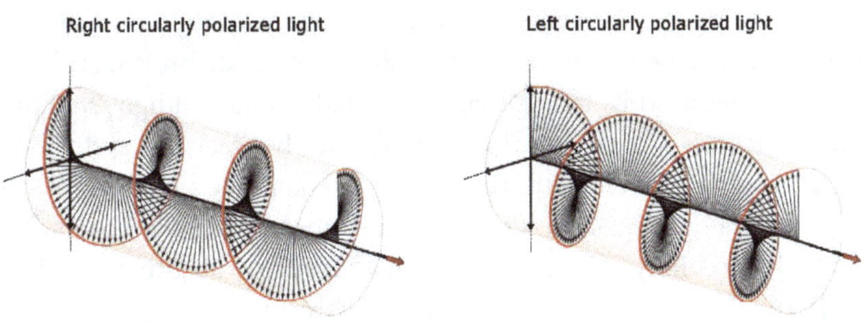

Fig. 18: Schematic representation of right circularly polarized and left circularly polarized light. In both cases, the length of the vector remains constant. In linearly polarized light, the electric vector stays in the same plane but its length changes.

Optical rotatory dispersion (ORD) and circular dichroism (CD) are two phenomena that result when *chiral molecules* (molecules that have no plane and center of symmetry) interact with polarized light. Most of biological molecules are chiral molecules and thus they are optically active. An optically active compound shows a phenomenon called optical rotation. It means that the plane of polarization of a linearly polarized light rotates as it passes through an optically active medium. The technique of ORD measures the ability of optically active compounds to rotate plane polarized light as a function of the wavelength (i.e. wavelength dependence of optical rotation).

Chiral molecules Most of chiral molecules contain one or more *chiral center*(s). Any tetrahedral carbon atom that has four different substituents is a *chiral center*. A molecule is not chiral (even if it has chiral centers) if it has a plane or center of symmetry. To be a chiral molecule there should be no plane and center of symmetry

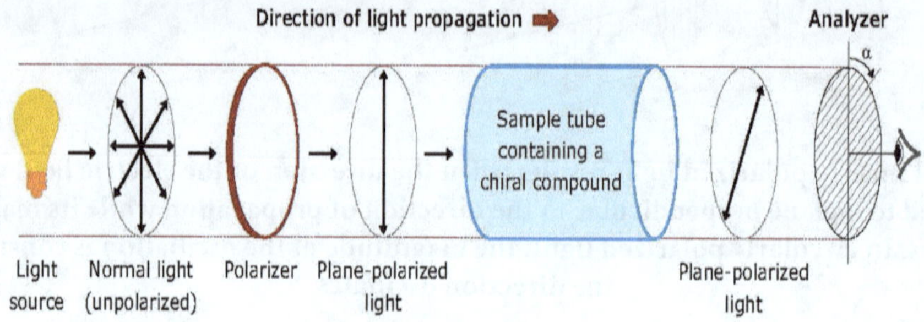

Fig. 19: When plane-polarized light is passed through a solution containing an optically active compound, there is net rotation of the plane polarized light. The plane-polarized light is rotated either clockwise (dextrorotatory) or counterclockwise (levorotatory) by an angle.

Circular Dichroism is an absorption spectroscopy. Dichroism is a word derived from Greek which means two-colors, because the sample under analysis has one color if illuminated with the right polarized light and a different color if illuminated with the left one. The color, in fact, depends on light absorption. Circular dichroism is observed when optically active matter absorbs left and right circularly polarized light slightly different. In fact, circular dichroism is the absorption difference between left and right circularly polarized light at a given wavelength. Absorption is quantitated by the molar extinction coefficient (ε). Optically active samples have distinct molar extinction coefficients for left (εL) and right (εR) circularly polarized light. The difference in absorbance of left and right circularly polarized light is a measure of circular dichroism.

6

MASS SPECTROMETRY

6.1 INTRODUCTION

Mass spectrometry is an analytical tool useful for measuring the mass-to-charge ratio (m/z) of one or more molecules present in a sample. These measurements can often be used to calculate the exact molecular weight of the sample components as well.

Mass spectrometry is an analytical technique that determines the mass of an ionic species by measuring the mass-to-charge (m/z) ratio. In mass spectrometry, an instrument called a mass spectrometer is used to determines the m/z ratio by performing four essential steps – ionization, acceleration, deflection or separation, and detection. In the process of ionization, a sample is first converted into gas-phase ions. Once the gas-phase ions are formed, the ions are accelerated so that they all have the same kinetic energy (acceleration). A magnetic field then deflects the ions according to their masses (deflection). Finally, the beam of ions passing throu gh the machine is detected electrically (detection). A mass spectrometer has three basic components: the ionization source (generates gas-phase ions), the mass analyzer (causes separation of ions) and the detector (detects the relative number of ions of each mass).

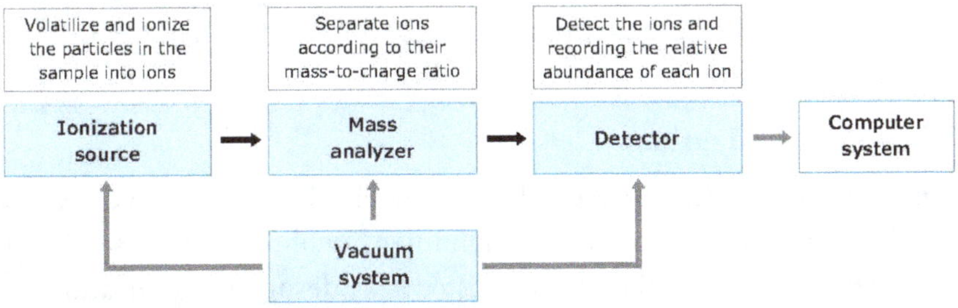

Fig.1 Basic Components of Mass Spectrometer

The operation of the *mass spectrometer* involves the following steps:

i. The particles in the sample (atoms or molecules) are converted in to gas-phase ions.
ii. The ions are *accelerated* so that they all have the same kinetic energy.
iii. Different ions are *deflected* or *separated* by the magnetic field. The amount of deflection depends on: the *mass* of the ions (lighter ions are deflected more than heavier ones) and the *charge* on the ions. These two factors are combined into the mass-to-charge ratio.
iv. Detection of the relative number of ions of each mass.

6.2 IONIZATION PROCESS

The particles in the sample (atoms or molecules) studied by mass spectrometer must be converted to gas phase-charged particles by the ionization process before they can be analyzed and detected. However, the term *ionization* is misleading because most ionization processes do not perform the ionization of sample particles per se. Instead, the term ionization relates to the transfer to the gas-phase of a sample while maintaining its charge and/or acquiring a charge from the sample environment, typically in the form of a proton. Several methods are used for converting the sample to ions. These methods include – electron ionization, chemical ionization, desorption ionization and electrospray ionization. All these ionization processes are fundamentally different techniques, but they achieve essentially the same end result — the generation of gas-phase ions via non-destructive vaporization and ionization.

6.2.1 Electron Ionization

In electron ionization, a beam of high-energy electrons strikes the molecules. The high-energy electrons cause the ejection of an electron from the molecule, creating a radical cation. For example, if CH4 is treated with high electrons, it loses an electron from one of the C–H bonds.

6.2.2 Chemical Ionization

In chemical ionization, the sample molecules are combined with an ionized reagent gas. When the sample molecules collide with the ionized reagent gas, some of the sample molecules are ionized by various mechanisms like proton transfer, electron transfer and adduct formation.

6.2.3 Desorption Ionization

In desorption ionization, the sample to be analyzed is dissolved in a matrix and placed in the path of high energy beam of ions or high intensity photons. In case of high intensity photons, it is termed as MALDI (matrix-assisted laser desorption/ionization). In MALDI, analytes are placed in a light-absorbing solid matrix. A matrix is used to protect the analytes from being destroyed by direct laser beam. Solid matrix strongly absorbs the laser radiation and acts as a receptacle for energy deposition. With a short pulse of laser light, the analytes are ionized and then desorbed from the matrix into the vacuum system. The exact mechanism of the origin of ions in MALDI is still not completely understood. Commonly used matrix materials are aromatic compounds that contain carboxylic acid functional groups. The aromatic ring of the matrix acts as a chromophore for the absorption of laser irradiation leading to the desorption of matrix and analytes into the gas phase.

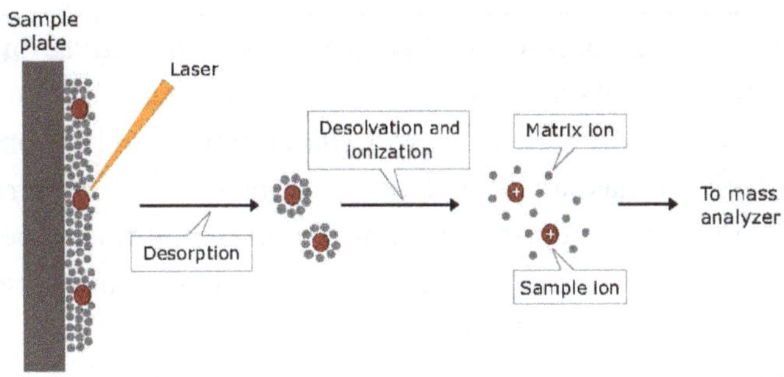

Fig. 2: Schematic representation of MALDI. It is a mass spectrometry ionization method for directly vaporizing and ionizing non-volatile samples. This method involves three steps. First, the heat-sensitive sample is mixed with the matrix and applied to the metal plate. Secondly, a pulsed laser is used to evaporate the matrix and the sample into the gas phase. Finally, ionization takes place in the gas phase. This is achieved by proton transfer, leading to predominantly [M + H]+ species being produced.

6.2.4 Electrospray Ionization

During standard electrospray ionization, the sample is dissolved in a polar, volatile solvent and pumped through a narrow, stainless steel capillary. A high voltage of 3 or 4 kV is applied to the tip of the capillary. As a consequence of this strong electric field, the sample emerging from the tip is dispersed into an aerosol of highly charged droplets. The charged droplets diminish in size by solvent evaporation, assisted by a flow of drying gas. Eventually charged sample ions, free from solvent, are released from the droplets.

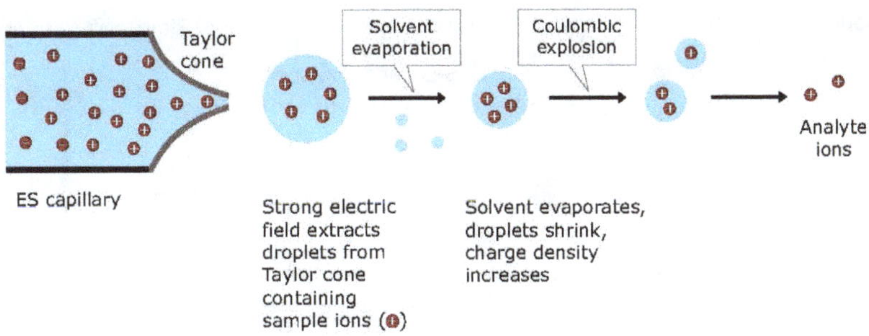

Fig. 3: Electrospray ionization:

The liquid effluent containing the analytes is electrostatically dispersed. This generates highly charged droplets, which are normally positively charged. Once the droplets are airborne, the solvent evaporates, which decreases the size and increases the charge density of the droplets. Desolvated ions are generated by the desorption of sample ions from the droplet surface due to high electrical fields and/or the formation of very small droplets due to repetitive droplet fission until each droplet contains, on average, only one sample ion.

6.3 MASS ANALYSIS

Once the sample has been ionized, the beam of ions is *accelerated* by an electric field and then passes into the mass analyzer, the region of the mass spectrometer where the ions are separated according to their mass-to-charge ratio. Just like ionization methods, there are several types of mass analyzer. Some examples include quadrupoles, ion traps, orbitraps, Fourier transform and time-of-flight. The most common type of mass analyzer is time-offlight (TOF) mass analyzer. The TOF mass analyzer measures ion flight time. It is based on the simple idea that velocities of two ions with the same kinetic energy will vary depending on the mass of an ion – the lighter ion will have higher velocity. Because the ions have different velocities, the ions reach the detector at different times. The smaller ions reach the detector first because of their greater velocity as compared to the larger ions. Hence, the analyzer is called TOF because the mass is determined from the ion's time-of-flight.

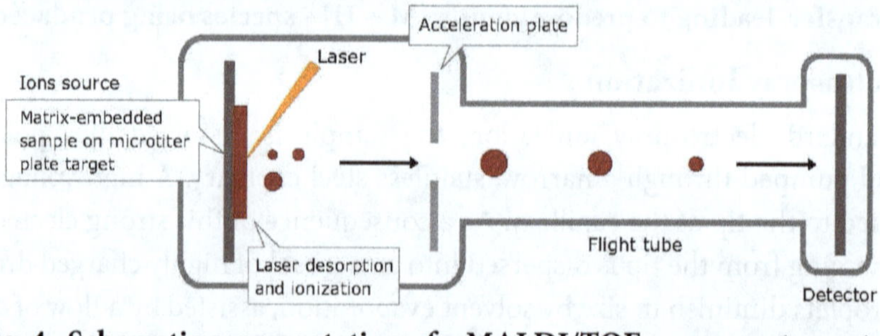

Fig. 4: Schematic representation of a MALDI/TOF mass spectrometer.

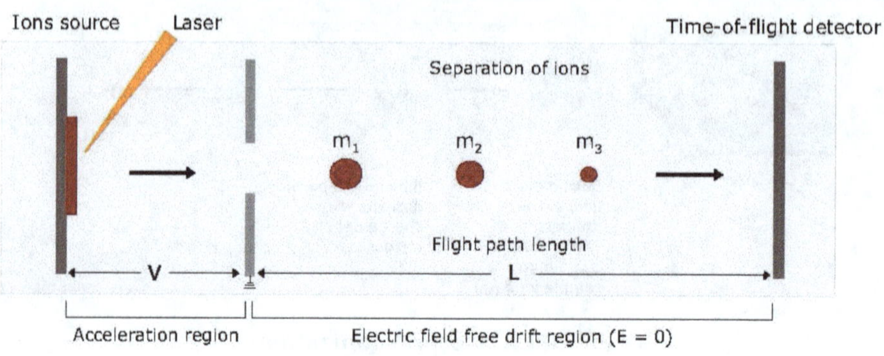

Fig. 5: In MALDI-TOF mass spectrometry, first desorption and ionization of the sample occurs. The ions formed are then accelerated by an electric field, so that they all have the

same kinetic energy. Accelerated ions pass through electric field free drift region. Ions with smaller m/z value move faster through the drift region. Consequently, the time of ions flight differs according to the mass-to-charge ratio of the ions. TOF mass analyzer measures the time these ions take to reach a detector.

Quantification of m/z

Mass-to-charge ratios are determined by measuring the time ions take to move through a field-free region between the source and the detector. Based on classical physics, ions with lower m/z will travel the fastest and arrive at the detector first, while ions with larger m/z will travel the slowest and arrive at the detector last. Once the ion source generates gas-phase ions, these ions are accelerated so that they all have the same kinetic energy. Thus, the potential energy given to ions in the accelerated regions is converted to kinetic energy for all ions.

Potential energy of a charged particle in the electric field = zeV

where, z is the number of charges, e is the amount of charge on an electron and V is the voltage When an ion is accelerated into a flight tube (or TOF tube) by the voltage V, its potential energy is converted to kinetic energy.

$$\text{Kinetic energy} = \frac{1}{2}\,mv^2$$

where, v = velocity of sample ion and m = mass of sample ion

The potential energy is converted to kinetic energy.

$$\text{Hence, } zeV = \frac{1}{2}\,mv^2$$

$$\text{The velocity of the sample ion, } v = \frac{\text{Flight path length (L)}}{\text{Time required to reach the detector (t)}}$$

$$zeV = \frac{1}{2}\,mv^2 = \frac{1}{2}\,m\,(L/t)^2$$

$$\frac{m}{z} = 2eV\left[\frac{t}{L}\right]^2$$

Because V and L are known, the *m*/*z* of an ion can be determined from the time the ion takes to reach the detector.

6.4 APPLICATION OF MASS SPECTROMETRY

Mass spectrometry is a powerful analytical technique used to determine the molecular mass of a compound. As an essential analytical tool, it has both qualitative and quantitative uses. It is used for three purposes: to determine the mass of an unknown compound, to determine the structure (or a partial structure) of an unknown compound and to confirm the Numerous approaches are used for the characterization of glycoprotein through mass spectrometry.

However, two approaches are most common

- Removal of carbohydrates from glycoprotein by chemical or enzymatic treatment and analysis of the deglycosylated protein and the released carbohydrates separately. The differences in protein–sugar linkage chemistry of *N*- and *O*-glycans require different approaches for the release of covalently-linked carbohydrates.
- Proteolytic digestion of the glycoprotein and analysis of the resulting glycopeptides.

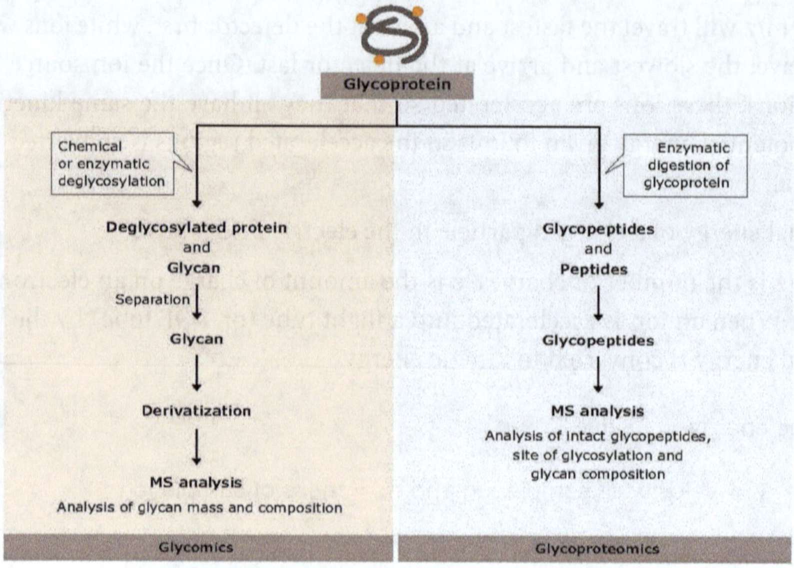

Fig.6: Glycomics and Giycoproteomics

7

BIOPHYSICS, CENTRIFUGATION AND MICROSCOPY

7.1 CENTRIFUGATION

Centrifugation is a technique used to separate or concentrate materials suspended in a liquid medium by applying centrifugal force. The *separation* of particles is based on their differences in sedimentation rate under the centrifugal field. The sedimentation rate of particles depends on a number of different factors, including size, density, and shape. A device that generates centrifugal force by spinning the fluid at high speed and separate particles is called a centrifuge.

In addition to the separation of molecules, it is also used to measure the physical properties (such as molecular weight, density, and shape). If centrifugation is used to separate one type of material from others, it is termed preparative centrifugation; whereas if it is used for measurement of physical properties of materials, then termed analytical centrifugation. Principle of centrifugation Particles suspended in a solution are pulled downward by Earth's gravitational force. Gravitational force depends only on the mass of the particles and not on charge, shape, and chemical composition. Because the Earth's gravitational field is weak, a solution containing particles of very small masses usually remains suspended due to random thermal motion. Hence, forces much larger than Earth's gravitational force are required to cause appreciable sedimentation of such small masses. Such forces can be obtained by subjecting particles to centrifugation.

By applying centrifugal forces, the sedimentation of these particles can be enhanced. A centrifuge does the same thing. It increases the sedimentation by generating centrifugal forces as great as 10,00,000 times the force of gravity. Let us understand this principle by considering a solution being spun in a centrifuge tube. The centrifugal force acting on a solute particle of mass m, Centrifugal force = mω2r where, ω is the angular velocity in radians per second, r is the distance from the center of rotation to the particle, and ω2r is the centrifugal acceleration.

A particle will move through a liquid medium when subjected to a centrifugal force. Hence, we must also consider the particle's buoyancy due to the displacement of the solvent molecules by the particle. This buoyancy reduces the force on the particle by ω2r times the mass of the displaced solvent.

$$\text{Buoyant force} = m_0\omega_2 r$$

where, m_0 is the mass of fluid displaced by the particle

$$m_0 = m v \rho$$

Where, ρ is the density of the solution (g/mL) and v is the *partial specific volume* of the particle. As measuring the volume of a very small particle is difficult, for convenience we use a term called partial specific volume. It is the volume in mL that each gram of the particle occupies in solution. Thus, the net force acting on the particle downward is given by, Net force = Centrifugal force – Buoyant force

$$= m\omega_2 r - m0\omega_2 r$$

$$= m\omega_2 r - m v \rho \omega_2 r$$

When particles move downward through the solution, the motion is also opposed by the frictional force. The frictional force is equal to the product of the *frictional coefficient*, *f* (Nm–1s) and the sedimentation velocity, v. It acts in the opposite direction to the net force.

$$F\text{friction} = f \,.\, v$$

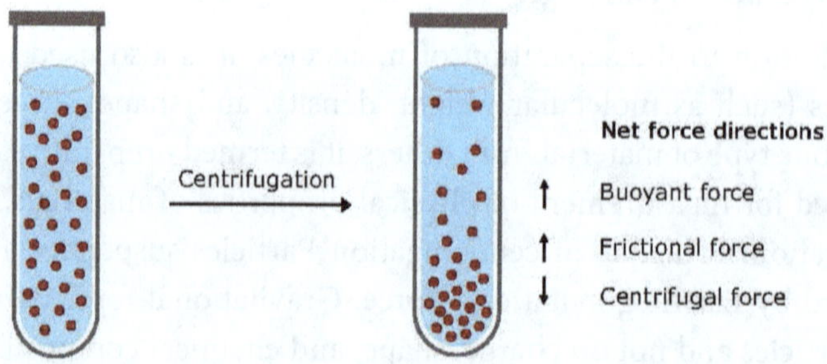

Fig. 1: Direction of different forces and distribution of solute particles during centrifugation. At steady state, then, the *frictional force* is equal to the net force and the molecule moves with velocity v downward:

$$f v = m\,\omega^2 r - m\bar{v}\rho\,\omega^2 r$$

$$f v = m\,\omega^2 r\,(1 - \bar{v}\rho)$$

Since, m (mass of a single particle in grams) = M/N.

Where, M is the molar weight of the solute in g/mol and N is the Abogadro's number.

$$f v = \frac{M}{N} \omega^2 r (1 - \bar{v}\rho)$$

$$v = \frac{M \omega^2 r (1 - \bar{v}\rho)}{N f}$$

$$s = \frac{v}{\omega^2 r} = \frac{M}{N} \frac{(1 - \bar{v}\rho)}{N f}$$

The sedimentation coefficient is the ratio of a velocity to the centrifugal acceleration. The sedimentation coefficient (s) has units of second. A sedimentation coefficient of 1 × 10–13 second is defined as one Svedberg, S, (1S = 10–13 second). This unit is named for The Svedberg, a pioneer in the field of centrifugation.

RCF describes and compares the strength of the fields generated by different-sized rotors and different operating speeds. A centrifuge with a higher radius of rotation will spin samples at a higher RCF. For example, when revolving at 2000 rpm, a larger centrifuge with a higher radius of rotation will spin samples at a higher RCF than a smaller centrifuge with a shorter radius of rotation.

Fixed-angle or swing-bucket rotors

Two types of rotors are commonly used in the laboratory – *fixed-angle* and *swing-bucket* rotors. However, these rotors are fundamentally different in their geometry and consequently in their sedimentation properties. In the following diagram, major differences between these rotors have been shown.

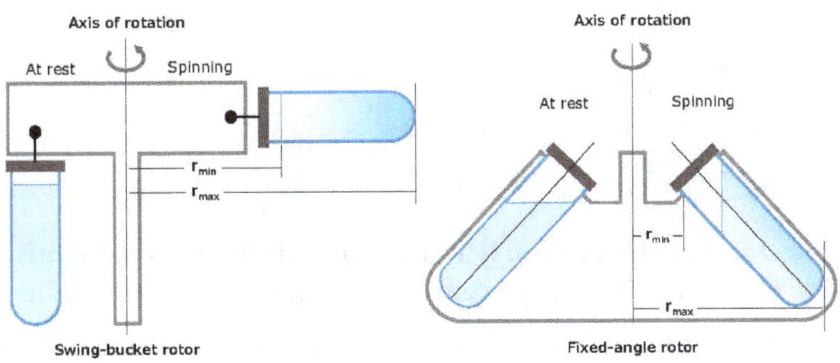

Fig. 2: Schematic sketches of the two types of rotors: swinging bucket and fixed angle rotors.

Differential centrifugation Differential centrifugation is the simplest form of centrifugation. Particles of different densities or sizes present in a suspension will sediment at different rates, with the largest and most dense particles sedimenting fast followed by less dense and smaller particles. These sedimentation rates can be increased by using centrifugal force. This centrifugation process is used mainly for the separation of sub-cellular components.

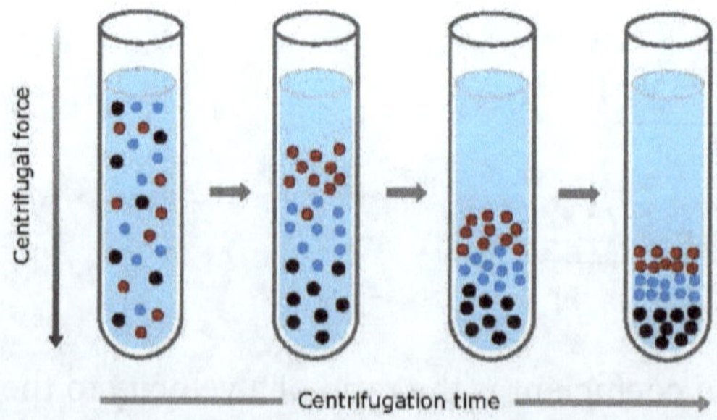

Fig. 3: Differential centrifugation separates particles based on differences in sedimentation rate, which reflect differences in size or density. Particles which are large or dense sediment rapidly, those that are intermediate in size or density sediment less rapidly, and the smallest or least dense particles sediment very slowly. Eventually, all of the particles reach the bottom of the tube.

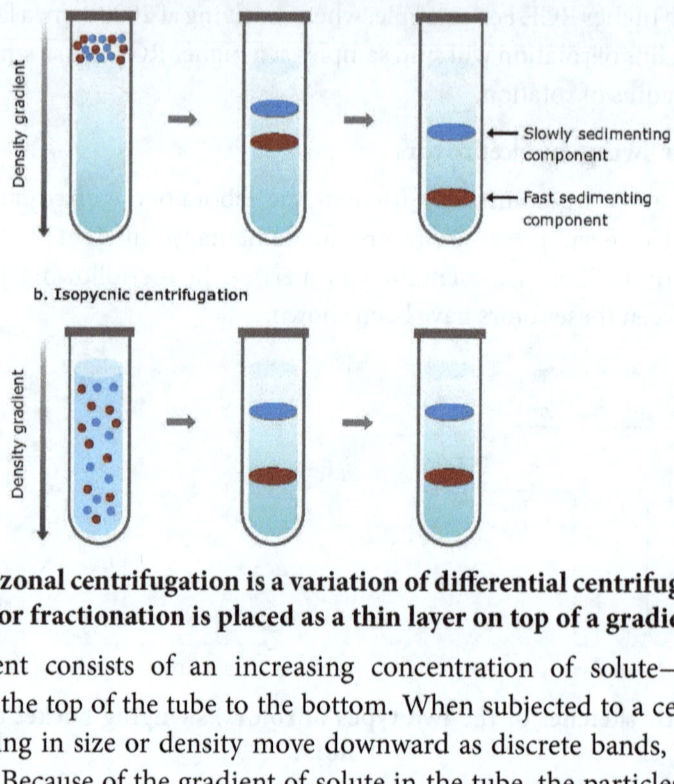

Fig. 4:(a) Rate-zonal centrifugation is a variation of differential centrifugation in which the sample for fractionation is placed as a thin layer on top of a gradient of solute.

The gradient consists of an increasing concentration of solute—and therefore density—from the top of the tube to the bottom. When subjected to a centrifugal force, particles differing in size or density move downward as discrete bands, that migrate at different rates. Because of the gradient of solute in the tube, the particles at the leading edge of each zone continually encounter a slightly denser solution and are, therefore, slightly impeded. As a result, each zone remains very compact, maximizing the resolution of different particles. (b) Isopycnic centrifugation also includes a gradient of solute that increases in concentration and density, but in this case the solute is concentrated so that the density gradient spans the range of densities of the particles about to be separated. During centrifugation, the particles move into the gradient until each reaches its equilibrium (or buoyant) density—the point in the gradient at which the density of the particle is exactly equal to the density of the gradient.

When macromolecules present in the CsCl solution are centrifuged, they form bands at distinct points in the gradient and exactly where a particular macromolecule forms band will depend on its buoyant density. DNA has a buoyant density of about 1.70 g/cm3, and therefore it will migrate to the point in the gradient where the CsCl density is also 1.70 g/cm3. In contrast, protein molecules have much lower buoyant densities and so float at the top of the tube, whereas RNA forms a pellet at the bottom of the tube due to high buoyant density. Isopycnic centrifugation can therefore separate DNA, RNA and protein.

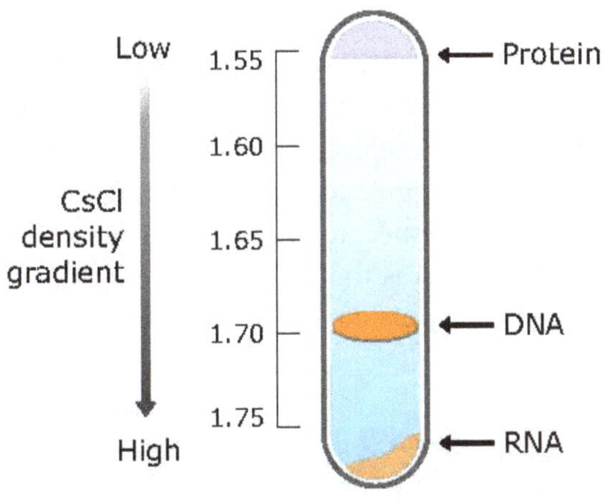

Fig. 5: CsCl equilibrium density gradient centrifugation. Separation of protein, DNA and RNA in a density gradient.

7.2 MICROSCOPY

Microscopy is a technique for making very small things visible to the unaided eye. An instrument used to make the small things visible to the naked (unaided) eye is called a microscope. There are two fundamentally different types of microscopes based on the source of illumination: the light microscope and the electron microscope.

7.2. 1 Light Microscope

Light or *optical microscope* uses visible light as a source of illumination. Because the light travels through the specimen, this instrument can also be called a *transmission light microscope*. The simplest form of light microscope consists of a single lens, a magnifying glass. Microscope made up of more than one glass lens in combination is termed compound microscope. Compound microscope includes the *condenser lens*, the *objective lens* and the *eyepiece lens.* Condenser lens focuses the light from the light source at the specimen. The one facing the object is called the objective and the one close to the eye is called the eyepiece. The objective has a smaller aperture and smaller focal length than those of the eyepiece (also referred to as the ocular). A compound microscope with a single eyepiece is said to be monocular and one with two eyepieces is said to be binocular. The word 'compound' refers to the fact that two lenses, the objective lens and the eyepiece, work together to produce the final magnified image.

The objective lens is responsible for producing the magnified image. It is available in different varieties (4x, 10x, 20x, 40x, 60x, 100x). The power of a lens is described with a number followed by the letter 'x'. For example, if through a microscope one can see something 25 times larger than actual size, its magnification power is 25x. The eyepiece works in combination with the objective lens to further magnify the image. Eyepieces usually magnify by 10x, since an eyepiece of higher magnification merely enlarges the image, with no improvement in resolution.

In *compound microscope,* the objective lens forms a real and inverted magnified image of the object (called the *real intermediate image*) in the focal plane of eyepiece. This image works as an object for the eyepiece. It magnifies the intermediate image. The image produced by the eyepiece is a magnified virtual image. It is erect with respect to the first image and hence, inverted with respect to the object. The eye views the virtual image created by the eyepiece, which serves as the object for the lens in the eye. The virtual image formed by the eyepiece is well outside the focal length of the eye, so the eye forms a real image on the retina.

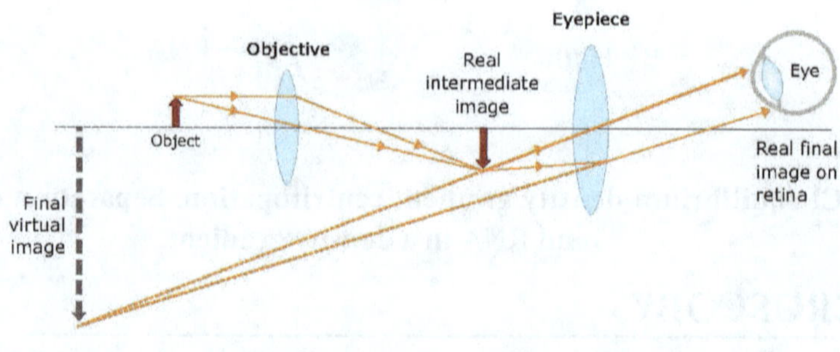

Fig. 6: Compound microscope and the ray diagram for image formation. It consists of two converging lenses arranged coaxially. The one facing the object is called the objective and the other close to eye is called the eyepiece.

7.2.2 Magnification

The magnification of a compound microscope is the product of the magnification of the objective and the eyepiece. The magnification produced by the *objective lens* is called the linear magnification of the compound microscope, because it is measured in linear dimensions. It is the ratio of the image size to the object size.

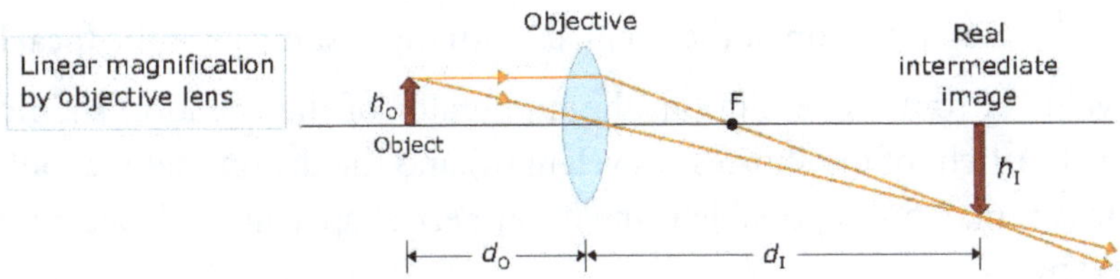

$$\text{Linear magnification} = \frac{\text{Image size}}{\text{Object size}} = \frac{h_I}{h_O} = \frac{d_I}{d_O}$$

The magnification produced by the *eyepiece* is called the angular magnification. This is also known as the magnifying power. It is the ratio of the angle subtended by the image on the eye when the microscope is used and the angle subtended by the object without microscope (i.e. naked eye).

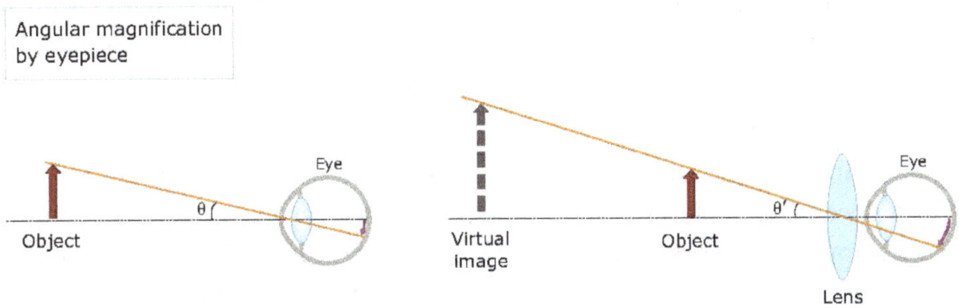

$$\text{Angular magnification} = \frac{\text{Angle subtended by image } (\theta')}{\text{Angle subtended by object } (\theta)}$$

The *maximum magnification* of the compound light microscopes is usually 1500x and has a limit of resolution of about 0.2 μm.

Th magnification produced by the *eyepiece* is called the angular magnification. This is also known as the magnifying power. It is the ratio of the angle subtended by the image on the eye when the microscope is used and the angle subtended by the object without microscope (i.e. naked eye).

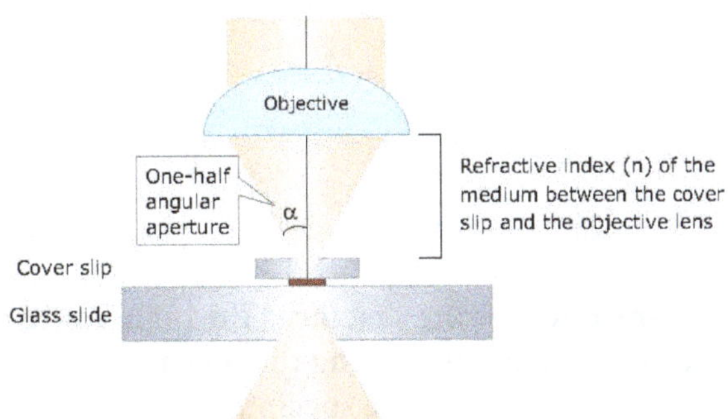

The overall magnification is the product of the linear magnification of the objective lens and the angular magnification of the eyepiece with the first image at the focal length.

For small value of limit of resolution, the numerator of the equation should be as small as possible (i.e. light of the shortest wavelength) and the denominator should be as large as possible (i.e. maximum possible value of numerical aperture). Hence, resolution can be improved by:

i. Increasing the *one-half angular aperture* (α)
ii. Increasing the *refractive index* (n) of medium between the objective lens and the specimen
iii. Shortening the *wavelength* (λ) of the illuminating light

7.2.3 Angular Aperture

The angular aperture is the angle of the cone of light collected by the objective lens. It varies with the size and shape of the illumination cone entering the objective lens. As the light cones grow larger, the angular aperture increases. The light cone can be increased either by shortening the distance between the objective lens and the specimen or by increasing the diameter of the lens.

Theoretically, the highest angular aperture possible with a standard microscope objective lens would be 180°, resulting in a value of 90° for the one-half angular aperture. Thus, maximum value of sin α will be 1. The working value of one-half angular aperture for the best objective lenses is about 70°. Hence, the maximum value for sin α is about 0.94. Since the refractive index of air is about 1.0, no lens working in the air can have a numerical aperture greater than 1.0.

7.2.4 Refractive index

The refractive index of air is about 1.0, so for a lens designed for use in air, the maximum numerical aperture is about 1.0. To increase the numerical aperture some microscope lenses are designed to be used with a layer of immersion oil between the objective lens and the specimen. Immersion oil has a higher refractive index (about 1.5).

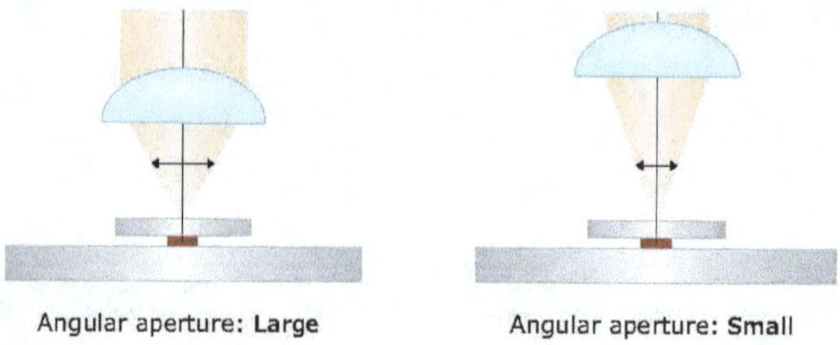

Fig. : Angular aperture and working distance. The higher value of the one-half angular aperture (α) can be obtained by shortening the distance between the objective lens and the specimen.

7.3 TYPES OF LIGHT MICROSCOPE

7.3.1 Bright-Field Microscopy

It is the original and most commonly used form of microscopy. When a specimen is illuminated with light, the transmitted light may undergo change in amplitude, phase and polarization. In bright-field microscopy, illuminating light is transmitted through the specimen and the contrast is generated due to differential absorption of light. In this microscopy, stained specimens or specimens that have natural pigmentation, give color contrast due to differential absorption of illuminating light. Absorption of light may also occur if specimens are thick enough to absorb a significant amount of light despite being colorless (*brightness contrast*). Absorption of light causes decrease in the amplitude of the transmitted light, resulting in amplitude contrast.

7.3.2 Dark-Field Microscopy

A dark-field microscope is a type of microscope in which objects are illuminated at a very low angle from the side so that the background appears dark and the objects show up against this dark background. It is a technique for improving the contrast of unstained, transparent specimens. Central circular disk stop that prevents direct condenser rays from entering the objective lens. To view a specimen in a dark field, a carefully aligned light source is used to minimize the quantity of directly transmitted light and collecting only the light scattered by the specimen. To achieve this an opaque disc is placed underneath the condenser lens, that blocks rays coming from condenser to enter the objective lens directly. Only those rays that have been scattered by the specimen enter the objective lens to generate the final image.

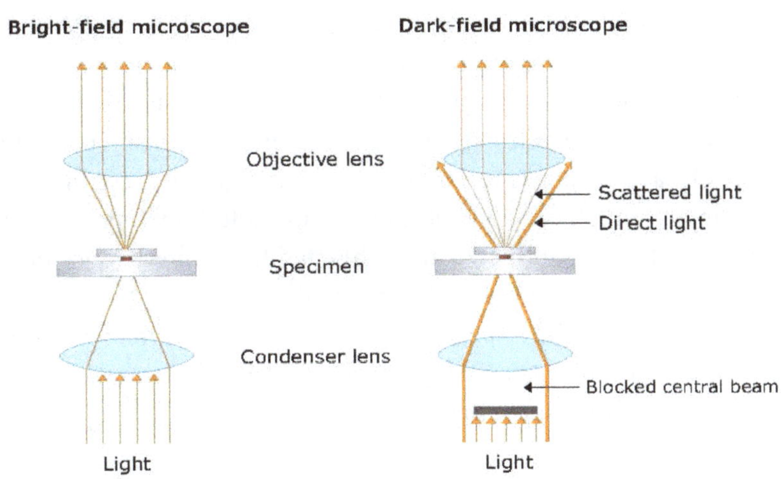

Fig. 1: In bright-field microscope, the specimen is evenly illuminated by light directly coming from condenser.

Light is transmitted through the specimen and the contrast is generated by the differential absorption of light. It forms a dark image against a brighter background. A dark-field microscope uses a dark-field condenser that contains an opaque disk, placed between light source and condenser. Due to presence of opaque disk, light can only pass through the outer edge of the condenser at a wide-angle. Thus, direct light is prevented from passing through the specimen. The specimen is illuminated by oblique light rays. Light passing through the specimen is scattered and only these scattered light reaches the objective lens for visualization. Direct light rays do not enter the objective thus the specimen is brightly illuminated on a dark background. Because there is no direct background light, the specimen appears light against a black background-the dark field.

7.3.4 Phase-Contrast Microscopy

Phase-contrast microscopy is used to study transparent and colourless specimens. This microscopic technique does not require cells to be killed, fixed or stained. When light passes through a living cell, the phase of the light wave changes due to variations in thickness and density within the cell. Light passing through a relatively thick or dense part of the cell, such as the nucleus, is retarded; its phase, consequently, is shifted relative to light that has passed through an adjacent thinner region of the cytoplasm.

Let us understand this by considering two beams of light from same source, one passing through an optically transparent material denser than air and other simply through air. When one beam of light passes through optically transparent material, its speed decreases and magnitude of change depends on both the density (as determined by the refractive index) and thickness of the material. The speed of second beam of light passing through air remain unchanged. Hence, these two beams, which was in phase are now out of phase; this difference is referred to as the phase difference. When transparent unstained specimens are used for phase contrast imaging, they do not absorb light and are called phase objects. However, when light passes through the specimens, diffraction or scattering of light occurs. The diffracted and the non-diffracted beam of light have phase difference. Non-diffracted light is referred to as direct light as it continues unchanged through the sample. In phase-contrast microscopy additional phase differences is introduced between the non-diffracted and diffracted light. Interference of the two sets of beams leads to overall differences in the amplitude, which can be detected by the eye as differences in brightness. The specimen shows different degrees of brightness and contrast. Thus, phase contrast microscopy transforms differences in the relative phase of light waves to amplitude differences in the image.

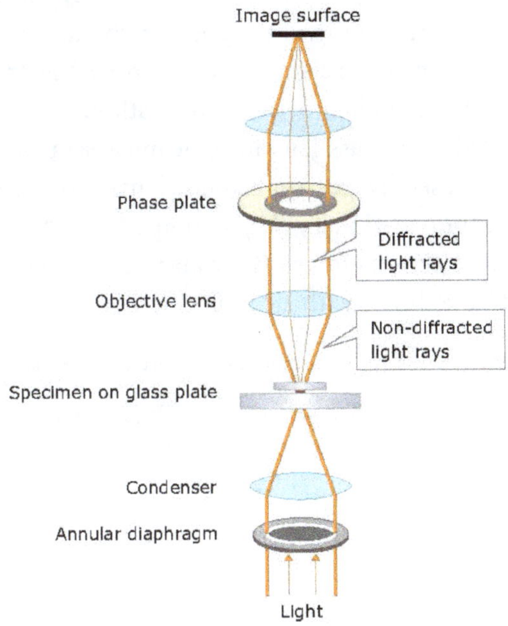

Fig. 8: Phase contrast microscopy (ray diagram). When light passes through a transparent and colourless specimen, the phase of the light wave changes due to variations in thickness and density within the specimen.

The thick lines represent the non-diffracted beams, while the diffracted beams are shown by thin lines. The diffracted light represents only a small part of the total light. This diffracted light beam arrives at the detector out of phase with the non-diffracted light. Phase plate introduces additional phase shift between the non-diffracted and diffracted light. These phase shifts are converted into changes in amplitude and can be observed as differences in image contrasts.

7.4 ELECTRON MICROSCOPE

The fundamental principles of electron microscopy are similar to those of light microscopy except one major difference of using electromagnetic lenses, rather than optical lenses to focus a high-velocity electron beam instead of visible light. The relationship between the limit of resolution and the wavelength of the illuminating radiation hold true for both - a beam of light or a beam of electrons. Due to the short wavelength of electrons, the resolving power of the electron microscope is very high.

Electron microscopes are of two basic types: *transmission electron microscope* and *scanning electron microscope*. The most commonly used type of electron microscope is called the transmission electron microscope (TEM), because it forms an image from electrons that are transmitted through the specimen being examined. Scanning electron microscope (SEM) is fundamentally different from TEM, because it produces image from electrons deflected from a specimen's outer surface (rather than electrons transmitted through the specimen).

Electron microscopes consist largely of a tall, hollow cylindrical column through which the electron beam passes. The top of the column contains the cathode, a tungsten wire filament that is heated to provide a source of electrons. Electrons are accelerated as a fine beam by the high voltage applied between the cathode and anode. Just as a beam of light rays can be focused by a glass lens in light microscopes, a beam of negatively charged electrons can be focused by electromagnetic lenses. The strength of the magnets is controlled by the current provided them. Air is pumped out of the column, producing a vacuum through which the electrons travel. If the air were not removed, electrons would be prematurely scattered by collision with air molecules.

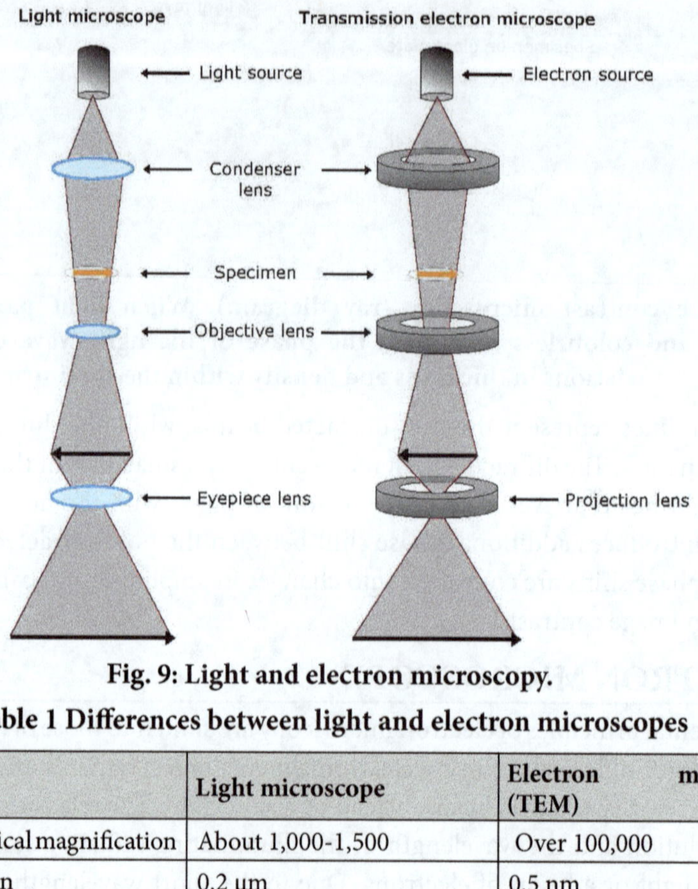

Fig. 9: Light and electron microscopy.

Table 1 Differences between light and electron microscopes

Feature	Light microscope	Electron microscope (TEM)
Highest practical magnification	About 1,000-1,500	Over 100,000
Best resolution	0.2 μm	0.5 nm
Radiation source	Visible light	Electron beam
Medium of travel	Air	High vacuum
Type of lens	Glass	Electromagnet
Source of contrast	Differential light absorption	Scattering of electrons
Specimen mount	Glass slide	Metal grid (usually copper)

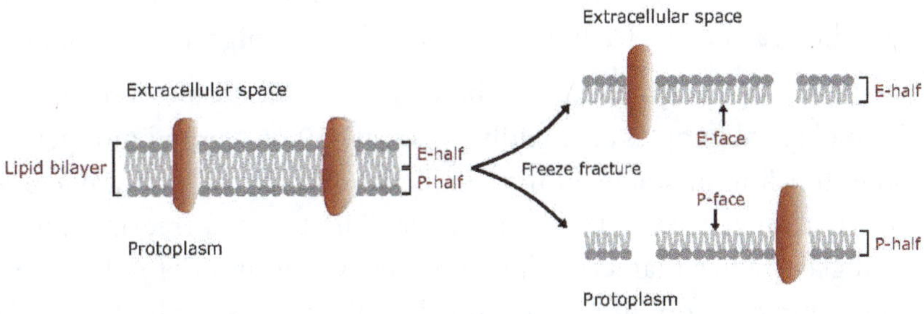

Fig. 10:Nomenclature for describing the aspects of the plasma membrane revealed by freeze fracture and etching. The membrane comprises a lipid bilayer with intercalated proteins.

The half-membrane leaflet adjacent to the extracellular space is termed the E-half and that adjacent to the protoplasm is termed the P-half. The interior views revealed when the membrane is split by freeze fracture are termed faces to distinguish them from the true surfaces. The P-face is the fracture face of the P-half of the membrane and will always have protoplasm beneath it. The E-face is the fracture face of the E-half of the membrane and has extracellular space beneath it.

7.5 CRYO-ELECTRON MICROSCOPY: RESOLUTION REVOLUTION

Cryo-electron microscopy (Cryo-EM) is a high-resolution structure determination technique of biomolecules in solution. It is an electron microscope imaging of frozen biological samples kept at cryogenic temperatures.

As we know, transmission electron microscopes (TEMs) use a beam of electrons to examine the structures of biomolecules such as proteins at the atomic scale. As the beam is transmitted through a biological sample, it interacts with the samples, which projects an image of the sample onto the detector. Image formation depends on the differential scattering of electrons by different parts of the sample. Due to the short wavelength of the electron beam, the resolving power of the electron microscope is very high. But biological samples are not compatible with the high-vacuum conditions and high-energy electron beams used in TEMs. There are two reasons for this. *First*, the high-energy electron beam burns and destroys the biological sample. *Second*, the water that surrounds the biological sample evaporates and the vapour from the sample destroys the vacuum required to run the microscope. To overcome this, the biological sample could be dried out before being examined, but this can alter their structure. Cryo-EM uses frozen samples and low-energy electron beams to overcome these problems. But low-energy electron beam would result in fuzzier images. This problem has been overcome by the use of a powerful image-analysis algorithm for merging multiple fuzzy 2D images from an electron microscope to create a 3D image of a molecule. This involves taking thousands of images of randomly oriented molecules and sorting them into groups of similar images. Each group is then processed to create a set of much sharper images. The spatial relationships between the groups are then calculated, leading to the assembly of a high-resolution 3D image.

Since liquid water evaporates in the case of TEM, thus to obtain an image using cryo-EM, biological samples are rapidly frozen at –190°C. Liquid nitrogen-cooled ethane freezes the thin film of water on the sample so quickly that the water molecules don't have time to arrange into a crystalline lattice. Due to rapid freezing, the water forms a disordered glass rather than crystalline ice. This vitrification of water is crucial because it does not cause electron diffraction and also allows the samples to retain their natural shape and organization. If the sample is frozen too slowly, then ice crystals form and the ordered ice crystals would strongly diffract the electron beam. Once the cryo-EM sample has been frozen, a focused beam of electrons reveals the shape of these very small, nanometer-sized 'vitrified' biological samples. Each image contains all the information required to determine the 3D structure. Specialized computer analysis then combines hundreds of thousands of individual, 2D snapshots from different angles into a composite that can be viewed in 3D.

Steps involved in structure determination by single particle cryo-EM Cryo-EM Vs X-ray crystallography X-ray crystallography gives very high-resolution structures of biomolecules. X-ray crystallography uses X-ray to determine the position and arrangement of atoms in a crystal. It is based on X-ray diffraction. When a beam of X-ray strikes a crystal, the beam may be diffracted. From the angles and intensities of these diffracted beams, a three-dimensional picture of the density of electrons within the crystal can be derived. But to get an x-ray diffraction structure, the biomolecule such as protein should be crystallised. Many proteins won't crystallise at all. And in some cases, the process of crystallisation alters the structure, so it's not representative of what the molecule looks like in real life. Cryo-EM doesn't require crystals, and it also enables scientists to see how biomolecules move and interact as they perform their functions.

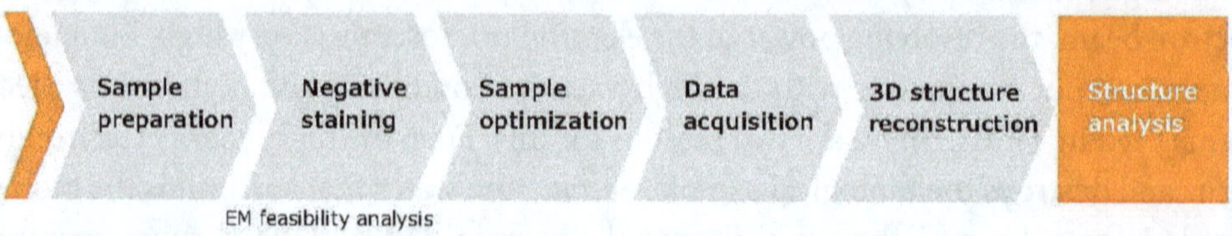

Cryo-EM Vs NMR

NMR can also give very detailed structures, and investigate conformational changes or binding of small biological molecules. But it's limited to relatively small molecular mass (<30 kDa) and usually soluble intracellular molecules, rather than those membrane bound. If someone want to study high molecular mass proteins, membrane-bound receptors, or complexes of several biomolecules together, cryo-EM is where it's at.

7.6 SCANNING PROBE MICROSCOPES

Scanning probe microscopes (SPMs) are a family of microscopes used to map topographical features of the sample surface at atomic level (resolution in nanometer scale). In all SPMs, a physical probe is used to scan the surface of a sample. During the scanning process, interactions of the probe tip with the surface of a sample are measured and data are used to generate an image of the surface. Various types of interaction between the probe tip and the surface are exploited in different types of SPMs. Hence, there are several types of SPMs. Scanning tunneling microscope (STM) and Atomic force microscope (AFM) are two major types of SPMs. AFMs measure the interactive forces between the probe tip and the sample whereas STMs measure the tunneling current flowing between the probe tip and the conducting sample. Application of the STM for surface imaging requires conductive sample. Hence, most biological samples due to poor electron conductivity are not suitable for STM imaging. Unlike STMs, the AFM does not need a conducting sample.

7.6.1 Scanning Tunneling Microscope

Historically, the first microscope in the family of SPMs is the scanning tunneling microscope (STM). It was invented by Binning and Rohrer in 1981. STM is based on the concept of electrons tunneling through a narrow potential barrier between a metal tip and a conducting sample in an external electric field. If the probe tip and sample are connected to a voltage source and the tip is close enough to the sample surface and the sample is electrically conductive, electrons will begin to leak or tunnel across the gap between the probe and the sample. This current can be measured, and the magnitude depends on the distance between the tip and the surface. The high spatial resolution of the STM is due to the exponential dependence of the tunneling current on the tip-sample distance.

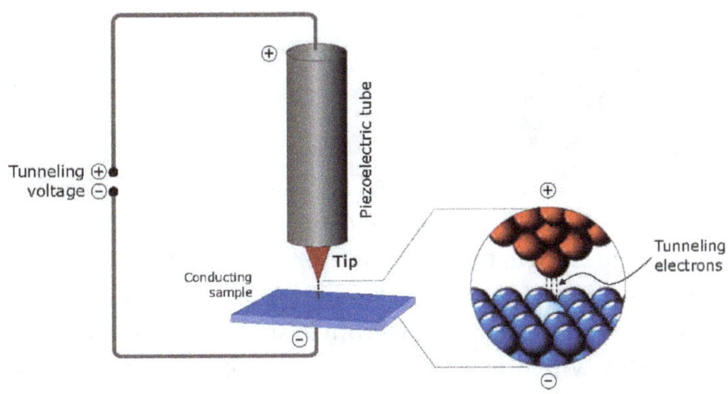

Fig. 11: The metal probe tip and the sample are both connected to a voltage supply. The sample is scanned by a very fine metallic probe tip mounted on a piezoelectric tube. A tunnelling current starts to flow when the tip is close to the sample. This feeble tunneling current is amplified and measured.

7.6.2 Atomic-Force Microscope

Atomic-force microscope (AFM) is based on the principle of the measurement of the interactive forces between a probe tip and a sample surface. The interactive forces (like capillary forces, electrostatic forces, van der Walls forces, etc.) have been classified as short- or long-range forces, attractive or repulsive forces. As the tip scans the surface, the interactions between the tip and the sample experience attractive or repulsive forces. This causes the tip to move up and down, which in turn causes the entire cantilever to deflect. The attractive force between the surface and the tip causes the cantilever to deflect towards the surface whereas repulsive force causes the cantilever to deflect away from the surface. By measuring the deflection of the cantilever, the topographical features of the surface can be mapped out. The methods used in AFM to acquire images can be categorized in three groups: *contact mode, tapping* or *intermittent mode* and *non-contact mode*. In contact mode, the probe tip directly touches the sample surface during scanning and the force acting between the atoms of probe tip and sample is counterbalanced by the elastic force. The probe tip predominately experiences repulsive forces. The contact mode may be carried out either at *constant force* (fixing the tip–sample interaction force through a feedback loop) or at *constant distance* between probe tip and sample (fixing the height of the scanner). A drawback of contact modes is the direct mechanical interaction of the probe tip with the sample. The tapping mode is similar to contact mode. However, in this mode intermittent contact occurs between the probe tip and the sample surface during scanning. The probe tip experiences both attractive and repulsive forces.

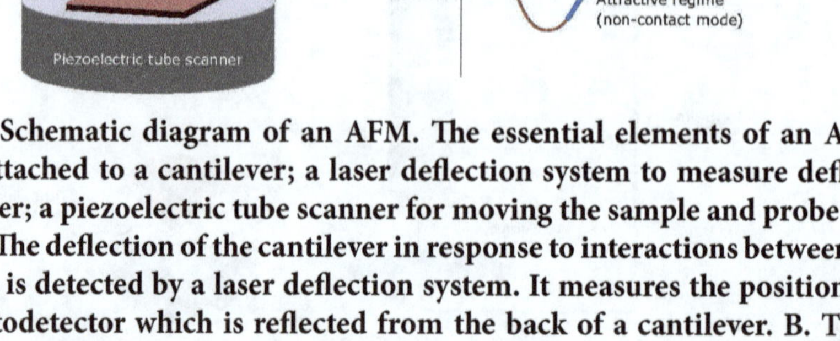

Fig. 12: A. Schematic diagram of an AFM. The essential elements of an AFM are: a probe tip attached to a cantilever; a laser deflection system to measure deflections of the cantilever; a piezoelectric tube scanner for moving the sample and probe relative to each other. The deflection of the cantilever in response to interactions between probe tip and surface is detected by a laser deflection system. It measures the position of a laser on the photodetector which is reflected from the back of a cantilever. B. Tip–sample interaction force curve. Graph shows different force regimes which correspond to inter-atomic forces. As seen in this figure, the atoms are separated by a large distance on the right side of the curve.

The cantilever is oscillated at its resonant frequency. *Resonant frequency* is a natural frequency of vibration determined by the physical parameters of the vibrating object. By maintaining a constant oscillation amplitude, a constant tip-sample interaction is maintained and an image of the surface is obtained. In non-contact mode, probe tip does not contact the sample surface, but oscillates above the surface during scanning. In this mode, the probe tip predominately experiences attractive forces.

As the atoms are brought close, they attract each other. This attraction increases until they come so close together that their electrons begin to repel each other electrostatically. This repulsive electrostatic force exceeds the attractive force as the inter-atomic distance decreases.

8

CONCEPTS OF ENERGY AND METABOLISM

8.1 INTRODUCTION: AN OVERVIEW

All living organisms need energy to grow and reproduce, maintain their structures, and respond to their environments. Metabolism is the set of life-sustaining chemical processes that enables organisms transform the chemical energy stored in molecules into energy that can be used for cellular processes. Animals consume food to replenish energy; their metabolism breaks down the carbohydrates, lipids, proteins, and nucleic acids to provide chemical energy for these processes. Plants convert light energy from the sun into chemical energy stored in molecules during the process of photosynthesis.

Scientists use the term bioenergetics to describe the concept of energy flow (Figure 1) through living systems, such as cells. Cellular processes such as the building and breaking down of complex molecules occur through stepwise chemical reactions. Some of these chemical reactions are spontaneous and release energy, whereas others require energy to proceed. Just as living things must continually consume food to replenish their energy supplies, cells must continually produce more energy to replenish that used by the many energy-requiring chemical reactions that constantly take place. Together, all of the chemical reactions that take place inside cells, including those that consume or generate energy, are referred to as the cell's metabolism.

Cells perform the functions of life through various chemical reactions. A cell's metabolism refers to the combination of chemical reactions that take place within it. Catabolic reactions break down complex chemicals into simpler ones and are associated with energy release. Anabolic processes build complex molecules out of simpler ones and require energy. In studying energy, the term system refers to the matter and environment involved in energy transfers. Entropy is a measure of the disorder of a system. The physical laws that describe the transfer of energy are the laws of thermodynamics. The first law states that the total amount of energy in the universe is constant. The second law of thermodynamics states that every energy transfer involves some loss of energy in an unusable form, such as heat energy. Energy comes in different forms: kinetic, potential, and free. The change in free energy of a reaction can be negative (releases energy, exergonic) or positive (consumes energy, endergonic). All reactions require an initial input of energy to proceed, called the activation energy. Enzymes are chemical catalysts that speed up chemical reactions by lowering their activation energy. Enzymes have an active site with a unique chemical environment that fits particular chemical reactants for that enzyme, called substrates. Enzymes and substrates are thought to bind according to an induced-fit model. Enzyme action is regulated to conserve resources and respond optimally to the environment.

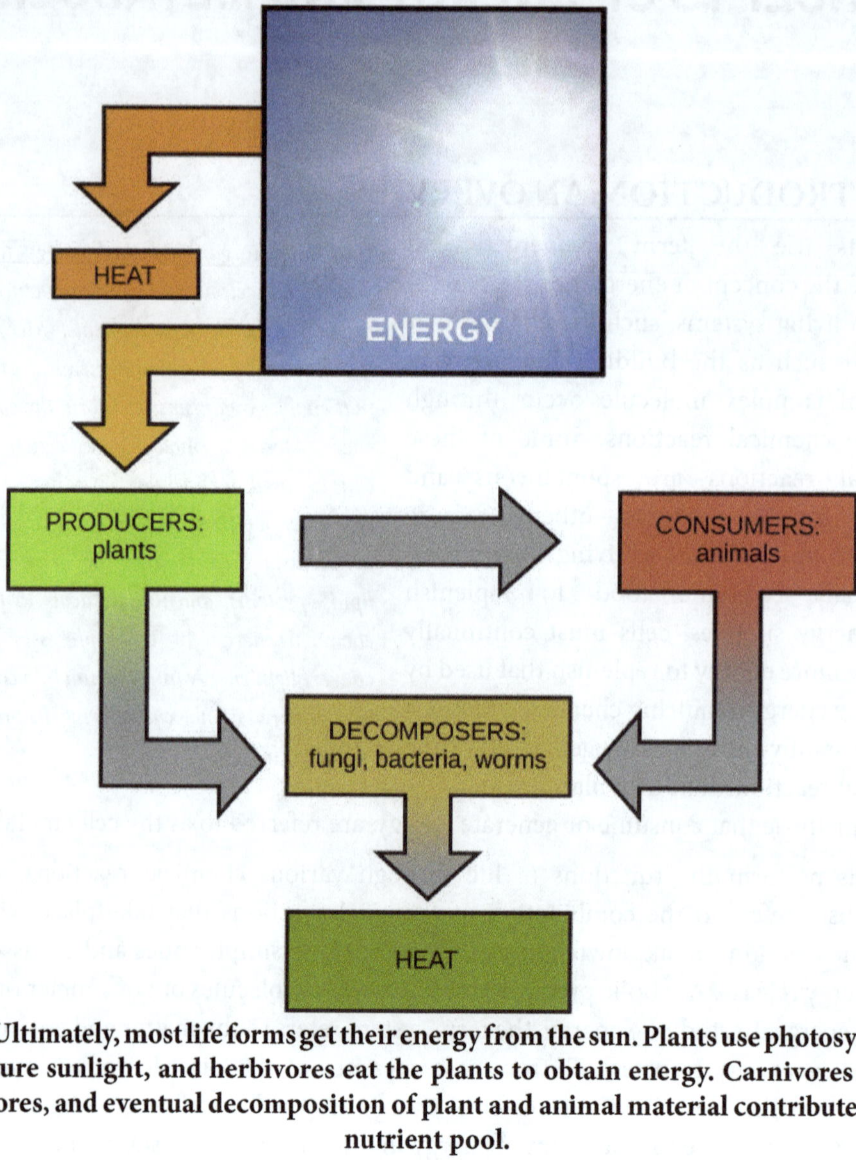

Fig. 1: Ultimately, most life forms get their energy from the sun. Plants use photosynthesis to capture sunlight, and herbivores eat the plants to obtain energy. Carnivores eat the herbivores, and eventual decomposition of plant and animal material contributes to the nutrient pool.

8.2 METABOLIC PATHWAYS

Consider the metabolism of sugar. This is a classic example of one of the many cellular processes that use and produce energy. Living things consume sugars as a major energy source, because sugar molecules have a great deal of energy stored within their bonds. For the most part, photosynthesizing organisms like plants produce these sugars. During photosynthesis, plants use energy (originally from sunlight) to convert carbon dioxide gas (CO_2) into sugar molecules (like glucose: $C_6H_{12}O_6$). They consume carbon dioxide and produce oxygen as a waste product. This reaction is summarized as:

$$6CO_2 + 6H_2O + \text{energy} \text{ ——-> } C_6H_{12}O_6 + 6O_2$$

Because this process involves synthesizing an energy-storing molecule, it requires energy input to proceed. During the light reactions of photosynthesis, **energy is provided by a molecule called adenosine triphosphate (ATP)**, which is the primary energy currency of all cells. Just as the dollar is used as currency to buy goods, cells use molecules of ATP as energy currency to perform immediate work. In contrast, energy-storage molecules such as glucose are consumed only to be broken down to use their energy. The reaction that harvests the energy of a sugar molecule in cells requiring oxygen to survive can be summarized by the reverse reaction to photosynthesis. In this reaction, oxygen is consumed and carbon dioxide is released as a waste product. The reaction is summarized as:

$$C_6H_{12}O_6 + 6O_2 \text{ ——> } 6CO_2 + 6H_2O + \text{energy}$$

Both of these reactions involve many steps.

The processes of making and breaking down sugar molecules illustrate two examples of metabolic pathways. A metabolic pathway is a series of chemical reactions that takes a starting molecule and modifies it, step-by-step, through a series of metabolic intermediates, eventually yielding a final product. In the example of sugar metabolism, the first metabolic pathway synthesized sugar from smaller molecules, and the other pathway broke sugar down into smaller molecules. These two opposite processes—the first requiring energy and the second producing energy—are referred to as **anabolic pathways (building polymers) and catabolic pathways (breaking down polymers into their monomers)**, respectively. Consequently, metabolism is composed of synthesis (anabolism) and degradation (catabolism) (Figure 2).

It is important to know that the chemical reactions of metabolic pathways do not take place on their own. Each reaction step is facilitated, or catalyzed, by a protein called an enzyme. **Enzymes are important for catalyzing all types of biological reactions**—those that require energy as well as those that release energy.

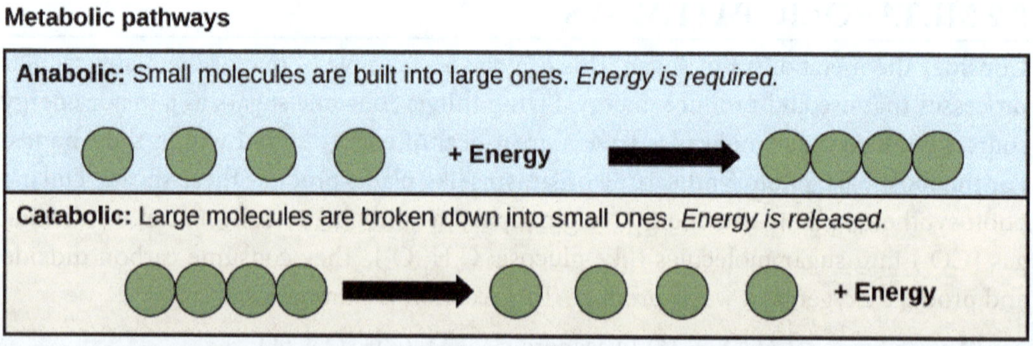

Fig. 2: Catabolic pathways are those that generate energy by breaking down larger molecules. Anabolic pathways are those that require energy to synthesize larger molecules. Both types of pathways are required for maintaining the cell's energy balance.

8.3 ENERGY

Thermodynamics refers to the study of energy and energy transfer involving physical matter. The matter relevant to a particular case of energy transfer is called a system, and everything outside of that matter is called the surroundings. For instance, when heating a pot of water on the stove, the system includes the stove, the pot, and the water. Energy is transferred within the system (between the stove, pot, and water). There are two types of systems: open and closed. In an open system, energy can be exchanged with its surroundings. The stovetop system is open because heat can be lost to the air. A closed system cannot exchange energy with its surroundings.

Biological organisms are open systems. Energy is exchanged between them and their surroundings as they use energy from the sun to perform photosynthesis or consume energy-storing molecules and release energy to the environment by doing work and releasing heat. Like all things in the physical world, energy is subject to physical laws. The laws of thermodynamics govern the transfer of energy in and among all systems in the universe.

In general, energy is defined as the ability to do work, or to create some kind of change. Energy exists in different forms. For example, electrical energy, light energy, and heat energy are all different types of energy. To appreciate the way energy flows into and out of biological systems, it is important to understand two of the physical laws that govern energy.

8.4 THERMODYNAMICS

The first law of thermodynamics states that the total amount of energy in the universe is constant and conserved. In other words, there has always been, and always will be, exactly the same amount of energy in the universe. **Energy exists in many different forms**. According to the first law of thermodynamics, energy may be transferred from place to place or transformed into different forms, **but it cannot be created or destroyed**.

The transfers and transformations of energy take place around us all the time. Light bulbs transform electrical energy into light and heat energy. Gas stoves transform chemical energy from natural gas into heat energy. Plants perform one of the most biologically useful energy transformations on earth: that of converting the energy of sunlight to chemical energy stored within organic molecules. Some examples of energy transformations are shown in Figure 3.

The challenge for all living organisms is to obtain energy from their surroundings in forms that they can transfer or transform into usable energy to do work. Living cells have evolved to meet this challenge. Chemical energy stored within organic molecules such as sugars and fats is transferred and transformed through a series of cellular chemical reactions into energy within molecules of ATP. Energy in ATP molecules is easily accessible to do work. Examples of the types of work that cells need to do include building complex molecules, transporting materials, powering the motion of cilia or flagella, and contracting muscle fibers to create movement.

Fig. 3: Shown are some examples of energy transferred and transformed from one system to another and from one form to another. The food we consume provides our cells with the energy required to carry out bodily functions, just as light energy provides plants with the means to create the chemical energy they need. (credit "ice cream": modification of work by D. Sharon Pruitt; credit "kids": modification of work by Max from Providence; credit "leaf": modification of work by Cory Zanker)

A living cell's primary tasks of obtaining, transforming, and using energy to do work may seem simple. However, the **second law of thermodynamics** explains why these tasks are harder than they appear. **All energy transfers and transformations are never completely efficient.** In every energy transfer, some amount of energy is lost in a form that is unusable. In most cases, this form is heat energy. Thermodynamically, heat energy is defined as the energy transferred from one system to another that is not work. For example, when a light bulb is turned on, some of the energy being converted from electrical energy into light energy is lost as heat energy. Likewise, some energy is lost as heat energy during cellular metabolic reactions.

An important concept in physical systems is that of order and disorder. The more energy that is lost by a system to its surroundings, the less ordered and more random the system is. Scientists refer to the measure of **randomness or disorder within a system as entropy**. High entropy means high disorder and low energy. Molecules and chemical reactions have varying entropy as well. For example, entropy increases as molecules at a high concentration in one place diffuse and spread out. The second law of thermodynamics says that energy will always be lost as heat in energy transfers or transformations. Living things are highly ordered, requiring constant energy input to be maintained in a state of low entropy.

8.5 POTENTIAL AND KINETIC ENERGY

When an object is in motion, there is energy associated with that object. Think of a wrecking ball. Even a slow-moving wrecking ball can do a great deal of damage to other objects. Energy associated with objects in motion is called kinetic energy. A speeding bullet, a walking person, and the rapid movement of molecules in the air (which produces heat) all have kinetic energy.

Now what if that same motionless wrecking ball is lifted two stories above ground with a crane? If the suspended wrecking ball is unmoving, is there energy associated with it? The answer is yes. The energy that was required to lift the wrecking ball did not disappear, but is now stored in the wrecking ball by virtue of its position and the force of gravity acting on it. This type of energy is called potential energy. If the ball were to fall, the potential energy would be transformed into kinetic energy until all of the potential energy was exhausted when the ball rested on the ground. Wrecking balls also swing like a pendulum; through the swing, there is a constant change of potential energy (highest at the top of the swing) to kinetic energy (highest at the bottom of the swing). Other examples of potential energy include the energy of water held behind a dam or a person about to skydive out of an airplane.

Fig. 4: Still water has potential energy; moving water, such as in a waterfall or a rapidly flowing river, has kinetic energy. (credit "dam": modification of work by "Pascal"/Flickr; credit "waterfall": modification of work by Frank Gualtieri)

Potential energy is not only associated with the location of matter, but also with the structure of matter. Even a spring on the ground has potential energy if it is compressed; so does a rubber band that is pulled taut. On a molecular level, the bonds that hold the atoms of molecules together exist in a particular structure that has potential energy. Remember that anabolic cellular pathways require energy to synthesize complex molecules from simpler ones and catabolic pathways release energy when complex molecules are broken down. The fact that energy can be released by the breakdown of certain chemical bonds implies that those bonds have potential energy. In fact, there is potential energy stored within the bonds of all the food molecules we eat, which is eventually harnessed for use. This is because these bonds can release energy when broken. The type of potential energy that exists within chemical bonds, and is released when those bonds are broken, is called chemical energy. Chemical energy is responsible for providing living cells with energy from food. The release of energy occurs when the molecular bonds within food molecules are broken.

8.6 FREE AND ACTIVATION ENERGY

After learning that chemical reactions release energy when energy-storing bonds are broken, an important next question is the following: How is the energy associated with these chemical reactions quantified and expressed? How can the energy released from one reaction be compared to that of another reaction? A measurement of free energy is used to quantify these energy transfers. Recall that according to the second law of thermodynamics, all energy transfers involve the loss of some amount of energy in an unusable form such as heat. Free energy specifically refers to the energy associated with a chemical reaction that is available after the losses are accounted for. In other words, free energy is usable energy, or energy that is available to do work.

If energy is released during a chemical reaction, then the change in free energy, signified as ΔG (delta G) will be a negative number. A negative change in free energy also means that the products of the reaction have less free energy than the reactants, because they release some free energy during the reaction. Reactions that have a negative change in free energy and consequently release free energy are called exergonic reactions. Think: ***ex*ergonic** means energy is *ex*iting the system. These reactions are also referred to as spontaneous reactions, and their products have less stored energy than the reactants. An important distinction must be drawn between the term spontaneous and the idea of a chemical reaction occurring immediately. Contrary to the everyday use of the term, a spontaneous reaction is not one that suddenly or quickly occurs. The rusting of iron is an example of a spontaneous reaction that occurs slowly, little by little, over time.

If a chemical reaction absorbs energy rather than releases energy on balance, then the ΔG for that reaction will be a positive value. In this case, the products have more free energy than the reactants. Thus, the products of these reactions can be thought of as energy-storing molecules. These chemical reactions are called **endergonic** reactions and they are **non-spontaneous**. An endergonic reaction will not take place on its own without the addition of free energy.

Fig. 5: Shown are some examples of endergonic processes (ones that require energy) and exergonic processes (ones that release energy). (credit a: modification of work by Natalie Maynor; credit b: modification of work by USDA; credit c: modification of work by Cory Zanker; credit d: modification of work by Harry Malsch) Look at each of the processes shown and decide if it is endergonic or exergonic.

There is another important concept that must be considered regarding endergonic and exergonic reactions. Exergonic reactions require a small amount of energy input to get going, before they can proceed with their energy-releasing steps. These reactions have a net release of energy, but still require some energy input in the beginning. This small amount of energy input necessary for all chemical reactions to occur is called the activation energy.

8.7 ENZYMES

A substance that helps a chemical reaction to occur is called a catalyst, and the molecules that catalyze biochemical reactions are called enzymes. Most enzymes are **proteins** and perform the critical task of **lowering the activation energies** of chemical reactions inside the cell. Most of the reactions critical to a living cell happen too slowly at normal temperatures to be of any use to the cell. Without enzymes to **speed up these reactions**, life could not persist. Enzymes do this by binding to the reactant molecules and holding them in such a way as to make the chemical bond-breaking and -forming processes take place more easily. It is important to remember that enzymes do not change whether a reaction is exergonic (spontaneous) or endergonic. This is because they do not change the free energy of the reactants or products. They only reduce the activation energy required for the reaction to go forward (Figure6). In addition, an enzyme itself is unchanged by the reaction it catalyzes. Once one reaction has been catalyzed, the enzyme is able to participate in other reactions.

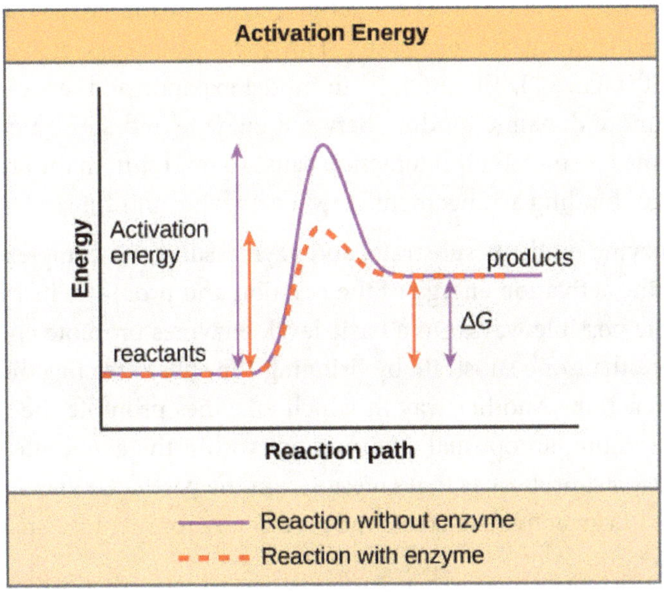

Fig. 6: Enzymes lower the activation energy of the reaction but do not change the free energy of the reaction.

The chemical reactants to which an enzyme binds are called the enzyme's substrates. There may be one or more substrates, depending on the particular chemical reaction.

In some reactions, a single reactant substrate is broken down into multiple products. In others, two substrates may come together to create one larger molecule. Two reactants might also enter a reaction and both become modified, but they leave the reaction as two products. The location within the enzyme where the substrate binds is called the enzyme's **active site**. The active site is where the "action" happens. Since enzymes are proteins, there is a unique combination of amino acid side chains within the active site. Each side chain is characterized by different properties. They can be large or small, weakly acidic or basic, hydrophilic or hydrophobic, positively or negatively charged, or neutral. The unique combination of side chains creates a very specific chemical environment within the active site. This specific environment is suited to bind to one specific chemical substrate (or substrates).

Active sites are subject to influences of the local environment. Increasing the environmental temperature generally increases reaction rates, enzyme-catalyzed or otherwise. However, temperatures outside of an optimal range reduce the rate at which an enzyme catalyzes a reaction. Hot temperatures will eventually cause enzymes to denature, an irreversible change in the three-dimensional shape and therefore the function of the enzyme. Enzymes are also suited to function best within a certain pH and salt concentration range, and, as with temperature, extreme pH, and salt concentrations can cause enzymes to denature.

For many years, scientists thought that enzyme-substrate binding took place in a simple "lock and key" fashion. This model asserted that the enzyme and substrate fit together perfectly in one instantaneous step. However, current research supports a model called induced fit (Figure 7). The induced-fit model expands on the lock-and-key model by describing a more dynamic binding between enzyme and substrate. As the enzyme and substrate come together, their interaction causes a mild shift in the enzyme's structure that forms an ideal binding arrangement between enzyme and substrate.

When an enzyme binds its substrate, an enzyme-substrate complex is formed. This complex lowers the activation energy of the reaction and promotes its rapid progression in one of multiple possible ways. On a basic level, enzymes promote chemical reactions that involve more than one substrate by bringing the substrates together in an optimal orientation for reaction. Another way in which enzymes promote the reaction of their substrates is by creating an optimal environment within the active site for the reaction to occur. The chemical properties that emerge from the particular arrangement of amino acid R groups within an active site create the perfect environment for an enzyme's specific substrates to react.

The enzyme-substrate complex can also lower activation energy by compromising the bond structure so that it is easier to break. Finally, enzymes can also lower activation energies by taking part in the chemical reaction itself. In these cases, it is important to remember that the enzyme will always return to its original state by the completion of the reaction. One of the hallmark properties of enzymes is that they remain ultimately unchanged by the reactions they catalyze. After an enzyme has catalyzed a reaction, it releases its product(s) and can catalyze a new reaction.

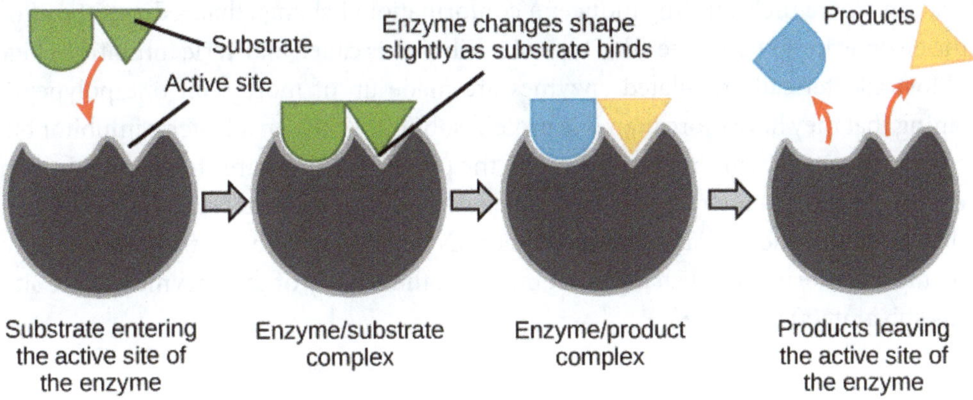

Fig. 7: The induced-fit model is an adjustment to the lock-and-key model and explains how enzymes and substrates undergo dynamic modifications during the transition state to increase the affinity of the substrate for the active site.

It would seem ideal to have a scenario in which all of an organism's enzymes existed in abundant supply and functioned optimally under all cellular conditions, in all cells, at all times. However, a variety of mechanisms ensures that this does not happen. Cellular needs and conditions constantly vary from cell to cell, and change within individual cells over time. The required enzymes of stomach cells differ from those of fat storage cells, skin cells, blood cells, and nerve cells. Furthermore, a digestive organ cell works much harder to process and break down nutrients during the time that closely follows a meal compared with many hours after a meal. As these cellular demands and conditions vary, so must the amounts and functionality of different enzymes.

Since the rates of biochemical reactions are controlled by activation energy, and enzymes lower and determine activation energies for chemical reactions, the relative amounts and functioning of the variety of enzymes within a cell ultimately determine which reactions will proceed and at what rates. This determination is tightly controlled in cells. In certain cellular environments, enzyme activity is partly controlled by environmental factors like pH, temperature, salt concentration, and, in some cases, cofactors or coenzymes.

Enzymes can also be regulated in ways that either promote or reduce enzyme activity. There are many kinds of molecules that inhibit or promote enzyme function, and various mechanisms by which they do so. In some cases of enzyme **inhibition**, an inhibitor molecule is similar enough to a substrate that it can bind to the active site and simply block the substrate from binding. When this happens, the enzyme is inhibited through **competitive inhibition**, because an inhibitor molecule competes with the substrate for binding to the active site.

On the other hand, in **noncompetitive inhibition**, an inhibitor molecule binds to the enzyme in a location other than the active site, called an **allosteric site**, but still manages to block substrate binding to the active site. Some inhibitor molecules bind to enzymes in a location where their binding induces a conformational change that reduces the affinity of the enzyme for its substrate. This type of inhibition is called allosteric inhibition (Figure 9). Most allosterically regulated enzymes are made up of more than one polypeptide, meaning that they have more than one protein subunit. When an allosteric inhibitor binds to a region on an enzyme, all active sites on the protein subunits are changed slightly such that they bind their substrates with less efficiency. There are allosteric activators as well as inhibitors. Allosteric activators bind to locations on an enzyme away from the active site, inducing a conformational change that increases the affinity of the enzyme's active site(s) for its substrate(s) (Figure 8).

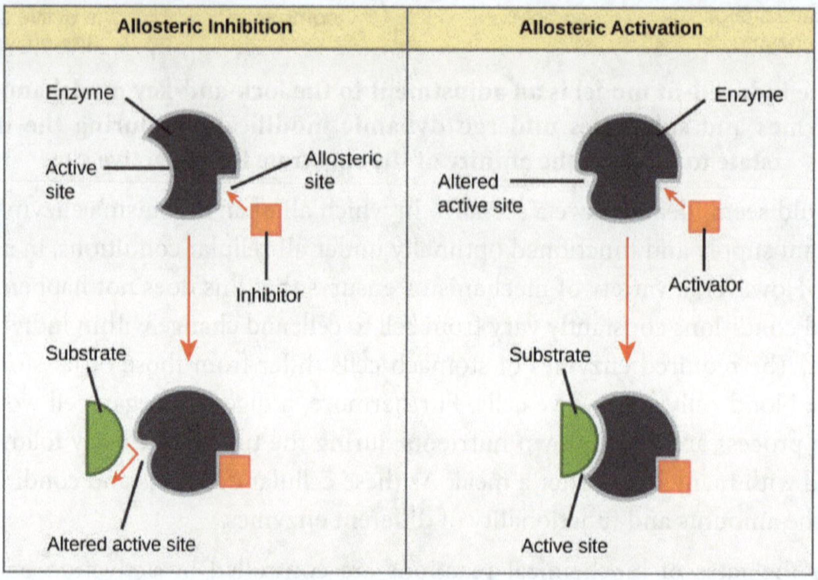

Figure 8 Allosteric inhibition works by indirectly inducing a conformational change to the active site such that the substrate no longer fits. In contrast, in allosteric activation, the activator molecule modifies the shape of the active site to allow a better fit of the substrate.

Through the Indigenous Lens

Plants cannot run or hide from their predators and have evolved many strategies to deter those who would eat them. Think of thorns, irritants and secondary metabolites: these are compounds that do not directly help the plant grow, but are made specifically to keep predators away. Secondary metabolites are the most common way plants deter predators. Some examples of secondary metabolites are atropine, nicotine, THC and caffeine. Humans have found these secondary metabolite compounds a rich source of materials for medicines. It is estimated that 90% of the drugs in the modern pharmacy have their "roots" in these secondary metabolites.

First peoples herbal treatments revealed these secondary metabolites to the world. For example, Indigenous peoples have long used the bark of willow shrubs and alder trees for a tea, tonic or poultice to reduce inflammation.

Fig. 9: Pacific willow bark contains the compound salicin.

Both willow and alder bark contain the compound salicin. Most of us have this compound in our medicine cupboard in the form of salicylic acid or aspirin. Aspirin has been proved to reduce pain and inflammation, and once in our cells salicin converts to salicylic acid.

So how does it work? Salicin or aspirin acts as an enzyme inhibitor. In the inflammatory response two enzymes, COX_1 and COX_2 are key to this process. Salicin or aspirin specifically modifies an amino acid (serine) in the active site of these two related enzymes. This modification of the active sites does not allow the normal substrate to bind and so the inflammatory process is disrupted. As you have read in this chapter, this makes it competitive enzyme inhibitor.

8.9 PHARMACEUTICAL DRUG DEVELOPER

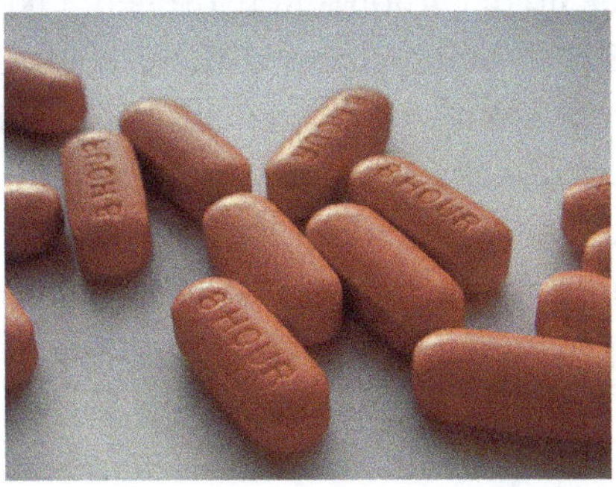

Fig. 10: Have you ever wondered how pharmaceutical drugs are developed?

Enzymes are key components of metabolic pathways. Understanding how enzymes work and how they can be regulated are key principles behind the development of many of the pharmaceutical drugs on the market today. Biologists working in this field collaborate with other scientists to design drugs (Figure 10).

Consider statins for example—statins is the name given to one class of drugs that can reduce cholesterol levels. These compounds are inhibitors of the enzyme HMG-CoA reductase, which is the enzyme that synthesizes cholesterol from lipids in the body. By inhibiting this enzyme, the level of cholesterol synthesized in the body can be reduced. Similarly, acetaminophen, popularly marketed under the brand name Tylenol, is an inhibitor of the enzyme cyclooxygenase. While it is used to provide relief from fever and inflammation (pain), its mechanism of action is still not completely understood.

How are drugs discovered? One of the biggest challenges in drug discovery is identifying a drug target. A drug target is a molecule that is literally the target of the drug. In the case of statins, HMG-CoA reductase is the drug target. Drug targets are identified through painstaking research in the laboratory. Identifying the target alone is not enough; scientists also need to know how the target acts inside the cell and which reactions go awry in the case of disease. Once the target and the pathway are identified, then the actual process of drug design begins. In this stage, chemists and biologists work together to design and synthesize molecules that can block or activate a particular reaction. However, this is only the beginning: If and when a drug prototype is successful in performing its function, then it is subjected to many tests from in vitro experiments to clinical trials before it can get approval from the U.S. Food and Drug Administration to be on the market.

Many enzymes do not work optimally, or even at all, unless bound to other specific non-protein helper molecules. They may bond either temporarily through ionic or hydrogen bonds, or permanently through stronger covalent bonds. Binding to these molecules promotes optimal shape and function of their respective enzymes. Two examples of these types of helper molecules are cofactors and coenzymes. Cofactors are inorganic ions such as ions of iron and magnesium. Coenzymes are organic helper molecules, those with a basic atomic structure made up of carbon and hydrogen. Like enzymes, these molecules participate in reactions without being changed themselves and are ultimately recycled and reused. Vitamins are the source of coenzymes. Some vitamins are the precursors of coenzymes and others act directly as coenzymes. Vitamin C is a direct coenzyme for multiple enzymes that take part in building the important connective tissue, collagen. Therefore, enzyme function is, in part, regulated by the abundance of various cofactors and coenzymes, which may be supplied by an organism's diet or, in some cases, produced by the organism.

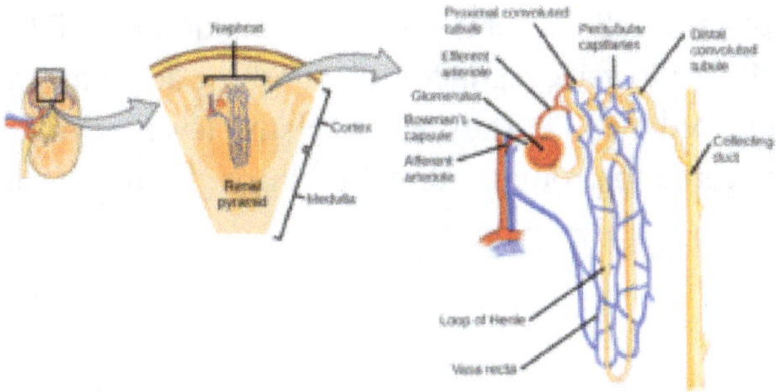

Fig. 11: Vitamins are important coenzymes or precursors of coenzymes, and are required for enzymes to function properly. Multivitamin capsules usually contain mixtures of all the vitamins at different percentages.

8.10 FEEDBACK INHIBITION IN METABOLIC PATHWAYS

Molecules can regulate enzyme function in many ways. The major question remains, however: What are these molecules and where do they come from? Some are cofactors and coenzymes, as you have learned. What other molecules in the cell provide enzymatic regulation such as allosteric modulation, and competitive and non-competitive inhibition? Perhaps the most relevant sources of regulatory molecules, with respect to enzymatic cellular metabolism, are the products of the cellular metabolic reactions themselves. In a most efficient and elegant way, cells have evolved to use the products of their own reactions for feedback inhibition of enzyme activity. Feedback inhibition involves the use of a reaction product to regulate its own further production. The cell responds to an abundance of the products by slowing down production during anabolic or catabolic reactions. Such reaction products may inhibit the enzymes that catalyzed their production through the mechanisms described above.

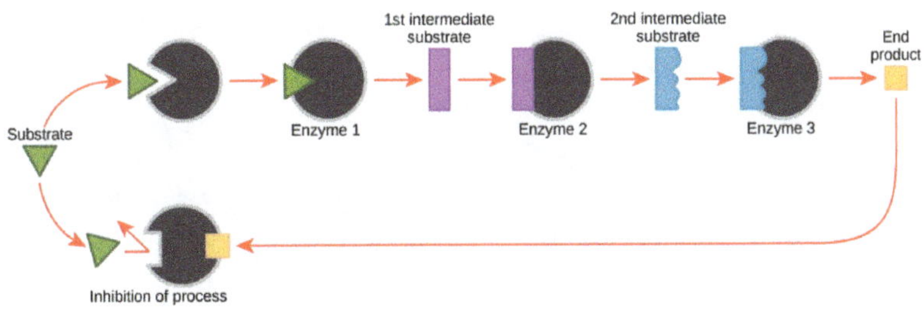

Fig. 12: Metabolic pathways are a series of reactions catalyzed by multiple enzymes. Feedback inhibition, where the end product of the pathway inhibits an upstream process, is an important regulatory mechanism in cells.

The production of both amino acids and nucleotides is controlled through feedback inhibition. Additionally, ATP is an allosteric regulator of some of the enzymes involved in the catabolic breakdown of sugar, the process that creates ATP. In this way, when ATP is in abundant supply, the cell can prevent the production of ATP. On the other hand, ADP serves as a positive allosteric regulator (an allosteric activator) for some of the same enzymes that are inhibited by ATP. Thus, when relative levels of ADP are high compared to ATP, the cell is triggered to produce more ATP through sugar catabolism.

9

BIOELECTRICAL AND ENGINEERING OF CELL BIOLOGY

9.1 INTRODUCTION: AN OVERVIEW

The last five decades of molecular and systems biology research have provided unprecedented insights into the molecular and genetic basis of many cellular processes. Despite these insights, however, it is arguable that there is still only limited predictive understanding of cell behaviours. In particular, the basis of heterogeneity in single-cell behaviour and the initiation of many different metabolic, transcriptional or mechanical responses to environmental stimuli remain largely unexplained. To go beyond the status quo, the understanding of cell behaviours emerging from molecular genetics must be complemented with physical and physiological ones, focusing on the intracellular and extracellular conditions within and around cells. Here, we argue that such a combination of genetics, physics and physiology can be grounded on a bioelectrical conceptualization of cells.

Biological electrical phenomena were recognized by Luigi Galvani and his contemporaries in the eighteenth century through the study of animal muscles and the nervous system. These early studies have led to the development of major fields, especially neuroscience and cardiology. Outside of these fields, however, studies of bioelectricity, i.e. electrical and electrochemical processes in cellular systems, have remained fragmented. While it was discussed as early as the 1970s that bioelectricity may be fundamental to understanding various cellular behaviours, the electrical investigations of cells were only focused on cellular bioenergetics.

- The bioelectrical view of cells as a more general concept has remained confined to the fringes of biological research for five decades, during which molecular biology has made astonishing advancements in our understanding of cells and our ability to manipulate genes.
- Interestingly, the resulting findings from molecular studies highlight once again the importance of bioelectricity, and now there is a revival of a bioelectrical view

of biological systems. Studies in diverse systems show that bioelectrical signals are at the heart of cell–cell communication in microorganisms, plants and animals (figure 1).

- Bioelectricity can underpin efficient growth and antibiotic resistance in bacterial biofilms and organization, morphogenesis and regeneration in mammalian and plant tissues. These findings, together with the realization that externally applied electrical fields can modulate multicellular processes such as regeneration in plant and vertebrate tissue, have resulted in the recent proposition that multicellular organization, and development more broadly, can, and should, be studied as a bioelectrical paradigm.

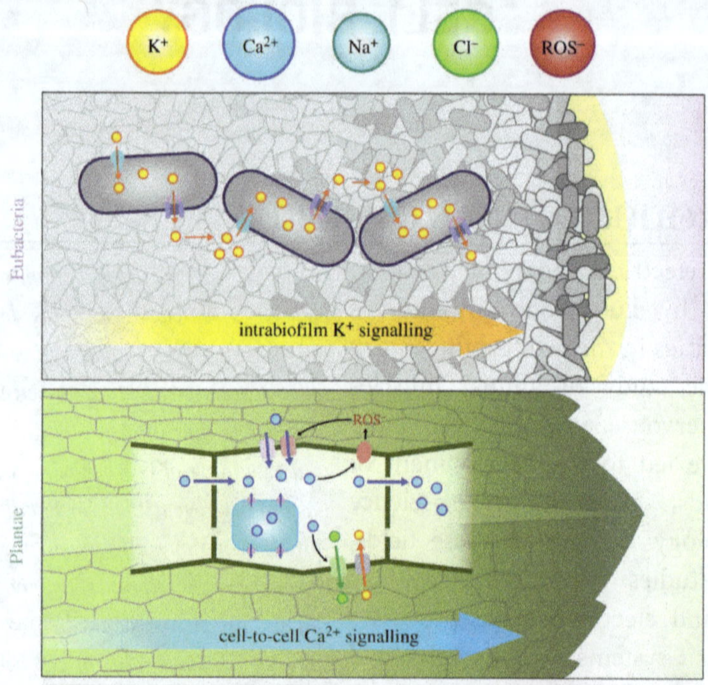

Fig. 1: Recent research shows that both prokaryotes and eukaryotes use ion- and redox-based electrochemical signals for communication. It has been shown that such communication enables the organization of growth and developmental processes across multiple length scales.

We argue here that bioelectricity can lead to a fundamental understanding of single-cell behaviours, beyond its roles in the multicellular context. By cell behaviours, we refer to high-level processes such as proliferation, dormancy and differentiation that are underpinned by dynamic changes in gene expression programmes, metabolic flux switching and mechanical cell properties. Notably, these changes are ultimately linked to the integrated physio-chemical properties of the electrochemically active cell–microenvironment interface. This motivates a bioelectrical view of the cell, the development of which can lead to a predictive understanding of cell behaviour.

9.2 BIOELECTRICAL NATURE OF THE CELL

- The bioelectrical conceptualization of cell behaviour can be illustrated with an analogy between a biological cell and a battery, both of which use redox reactions and ion movements (figure 2). In the case of the biological cell, its enclosed structure composed of multiple membranes partitions ions and charged and uncharged molecules (including macromolecules) across cellular compartments and within and outside the cell.
- The partitioning of charged molecules and ions gives rise to electrical and chemical potential differences across cellular membranes, and electrochemical gradients (ion motive forces, IMFs). A membrane potential (MP) arises from the combined electrical potential differences across a given membrane due to all charged molecules.

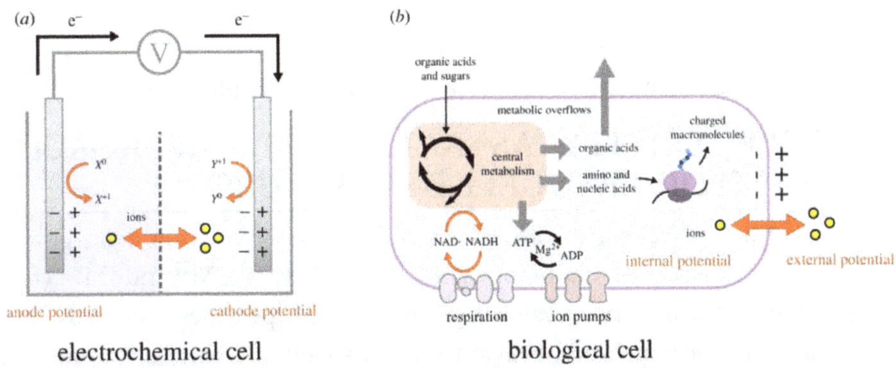

Fig. 1: The basis for a bioelectrical view of cells can be motivated by drawing an analogy between a battery (*a*) and a biological cell (*b*). Both systems rely on ion flows and redox reactions across interfaces.

- The maintenance of the MP and the coupling of IMFs to membrane-bound chemical reactions are key components of cellular bioenergetics and physiology, as recognized in the chemi-osmotic theory of respiration. While the specific mechanisms of MP and IMF can be different across different membranes (e.g. mitochondrial versus plasma membrane in mammalian cells or the inner versus outer membranes in a bacterial cell, their generation always involves electro-static and electro-dynamic processes. IMFs can arise from selective transport or differential permeability to charged molecules, or can be generated through membrane-bound redox reactions of the respiratory chain A key point to recognize is that IMF and MP are coupled to each other and further to cell volume and metabolism. The connection to volume arises from the distribution of ions and charged molecules (essential for the formation of MP), determining also the osmotic forces on cells. The connection to metabolism is achieved by four general mechanisms (figure 2b).

- First, the steady-state concentrations of key metabolic redox and energy carrier pairs (NADH/NAD^+ and ATP/ADP), which can determine metabolic pathway fluxes, can be altered by membrane-bound dehydrogenases and ATPases that can either use IMF to convert these pairs or use their conversions to sustain/generate it. Second, membrane transporters can couple the transport of metabolites, in particular organic acids and sugars, with the transport of ions, thereby linking this metabolically central process to IMF generation.
- Third, several 'master' compounds within central metabolism—such as glutamate, which is involved in nitrogen assimilation and the synthesis of many other amino acids—can also act as gating molecules to control the state of ion channels, thereby influencing IMFs via MP.
- Finally, the well-described excretions of metabolites and proteins from cells, as well as membrane-bound enzymatic processes, can influence the electrical and chemical potential of the cell microenvironment either directly or through redox reactions. In this context, it is worth noting that extracellular matrix polymers, such as collagen, chitin and cellulose, are shown to be piezoelectric.

9. 3 BIOELECTRICITY AS A HOLISTIC APPROACH TO UNDERSTAND DIVERSE CELL BEHAVIOURS

- The electrochemical nature of the cells and their microenvironments gives rise to a coupling between cell physiology and bioelectricity (i.e. MP and IMF) (figure 3). This bioelectrical conceptualization of the cell provides not only plausible explanations for many cell behaviours, but also a new framework to re-formulate much of the existing knowledge in cell physiology.
- To illustrate this point, we discuss here a few example cell behaviours in the bioelectrical context.

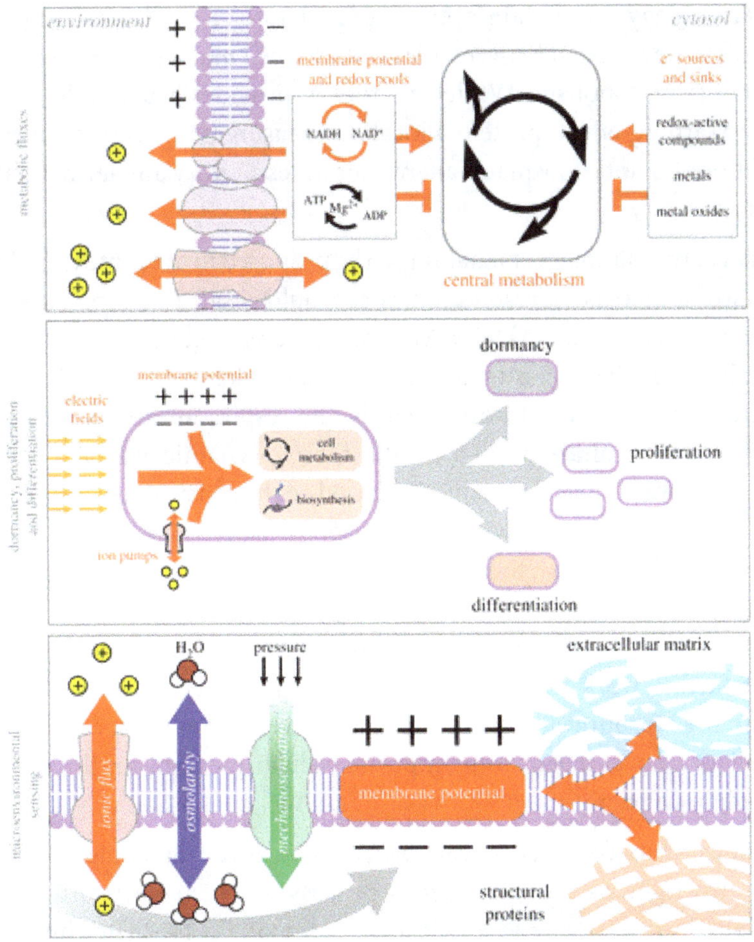

Fig. 3: Cartoon illustration of the coupling between the bioelectrical nature of the cell, in particular MP and IMF, and higher level cellular behaviours.

9.3.1 Metabolic Fluxes

- A key physiological response known as 'the Warburg effect' or 'overflow metabolism' is seen in many cell types including bacteria and is shown to underpin the behaviour of cancer cells and fermentative yeasts. This effect involves the metabolic switching of cells from respiration to respiration and fermentation (i.e. respiro-fermentation), despite the presence of oxygen. Respiration allows cells to generate mitochondrial IMFs through NADH oxidation/oxygen reduction and then harvest these for ATP synthesis. This process releases more energy per carbon source than fermentation, which involves the reduction of NAD^+. It is therefore puzzling that cells switch to respiro-fermentation in the presence of oxygen, when they could theoretically still respire and extract more energy per carbon. One explanation put forward to explain the Warburg effect is that it is a cellular strategy to maintain 'optimal' growth rates under limited enzyme capacity that has to be invested between biosynthesis, cell maintenance and fermentative/respiratory metabolism. While this theory provides a plausible rationalization of the Warburg effect, protein allocation measurements across different growth conditions do not show major alterations to respiratory and fermentative enzyme allocations.

- An alternative, simpler explanation can be proposed in the context of a bioelectrical framework: increasing respiration rates would result in an increased ATP/ADP ratio and a decreased NADH/NAD+ ratio, thereby reducing the thermodynamic feasibility of IMF generation and harvesting, which require NADH and ADP as substrates, respectively. In other words, the respiration to respiro-fermentation switch could be underpinned by the thermodynamic feasibility of these two processes under a given MP value and based on the constrained NADH/NAD+ and ATP/ADP ratios. In line with this bioelectrical view, it has been indicated that cancer cells, known for their overflow metabolism, have altered MP levels and that influencing the NADH/NAD+ ratio through engineered redox reactions directly alters the initiation point of the Warburg effect in bacteria and yeast. While further work will ascertain the validity of this bioelectrical viewpoint to explain the Warburg effect, we note that this viewpoint is experimentally testable and offers a novel means to manipulate cellular metabolism through redox and MP manipulations.
- Metabolism, more broadly, can be conceptualized as a bioelectrical process in its own right, as a coupled redox process in which electrons are transferred from an electron donor to an electron acceptor. To facilitate this process, cells use a variety of electron sources and sinks, including redox-active compounds, metals and their oxides. This opens up the possibility of using such compounds, or even electrode surfaces, to withdraw, or introduce, electrons into cellular metabolism. While this possibility has already been realized in biotechnological research (e.g. cell-mediated bioelectrosynthesis or waste-to-electricity generation using microorganisms), its true potential is in opening up new routes to study metabolism and its links to cell physiology through bioelectrical interfacing in mammalian and microbial cells. For example, redox-active, cell-permeable compounds have been used together with external electrodes to achieve an added, controllable redox cycling between cell metabolism and electrodes, thereby allowing the control of cellular physiological processes, such as the circadian clock and gene expression.

9.3.2 Dormancy, Proliferation and Differentiation

- The nature of the bioelectricity–metabolism coupling and its possible links to enabling different cell states are now being elucidated. In bacteria, it was recently shown that proliferating versus non-proliferating bacterial cells respond differently to electric fields. In addition, it has been shown that MP responses are different when bacterial cell metabolism or biosynthesis is perturbed by carbon starvation or antibiotic treatment, respectively. Other studies have also found that ionic fluxes can influence MP and physiological outputs such as the generation of metabolically dormant bacterial cells.

- In mammalian cells, proliferating cancer cells are suggested to have altered plasma or mitochondrial MP (reviewed in). Furthermore, MP is shown to be an indicator for stem-cell differentiation potential, alongside metabolic markers. Hyperpolarization of the plasma MP is sufficient to induce differentiation, and, conversely, depolarization of MP in a differentiated cell can induce reversal to a multi-potent progenitor. In this context, it is interesting that electrical fields can induce intracellular calcium oscillations in certain stem cells, possibly offering a route to control MP and differentiation in these cells.

9. 3.3 Microenvironmental Sensing

- A coupling between bioelectricity and physiology is also evident in cellular sensing of internal and external conditions across all cell types. Osmoregulation of cell size can be achieved through both the production and consumption of key metabolites (e.g. glutamate) or through the alteration of ionic fluxes across the membrane. Both processes can alter MP. For example, in bacteria, osmotic changes result in a significant motility response, possibly through changes in IMF causing flagellar rotation. Such a coupling between flagellar rotation and IMF has been characterized and recently been used to monitor changes in MP and cell metabolism through changes in flagellar rotation speeds.
- There is a large family of voltage-gated ion channels and non-specific porins that can be influenced by MP and, in turn, can influence MP and IMF by their actions. It is now increasingly clear that, in prokaryotic and eukaryotic cells, some ion channels can also respond to local mechanical forces within the membrane, thereby providing a direct link between MP, cell physiology and mechanical forces. The mechanical forces can come from the environment, including neighbouring cells within tissues or biofilms, or can be caused within the cell itself. Within the tissue context, for example, electrical and mechanical forces can act synergistically to re-model the extracellular matrix and further enable contractions of engineered muscle tissue. Within cells, (de)stabilization of microtubules is implicated to alter the mitochondrial MP in cancer cells, while, in bacteria, the alteration of MP can change the membrane distribution of structural proteins involved in cell division. It has also been suggested that action potentials in neurons are accompanied by a mechanical wave (soliton wave) across the membrane. These examples highlight the possibility that the link between MP and mechanical forces can provide a

key integrator of information from external and internal sources. Interestingly, one such process that potentially requires the integration of internal and external clues—the initiation of bacterial sporulation—was linked to bioelectrical changes in a recent study. Whether such bioelectrical changes during sporulation are caused by mechanical forces and/or involve other contributing factors remains an open research question to pursue.

- In plants, the best characterized electrical signalling responses are to mechanical stimuli caused by wounding and herbivory. A landmark study demonstrated that electrical potentials were rapidly initiated upon wounding, which propagated to systemic leaves, initiating the classical jasmonate-based wound response. Such wound-induced 'electrical' signals in plants are known as variation potentials (VPs) and show slower dynamics than action potentials seen in neurons. VPs appear to facilitate information transfer on a large scale, alerting all the intervening tissue of distal stimuli, and arise from the activation of ligand-dependent or mechano-sensitive calcium channels, facilitating Ca^{2+} influx. These wound VPs were shown to be dependent upon glutamate-like receptors (GLRs), which facilitate the generation of waves of Ca^{2+}. Consistent with this finding, glutamate triggers long-distance, Ca^{2+}-mediated plant defences against herbivory via GLRs. Notably, the application of a synthetic electrical stimulus of similar magnitude and duration to wound VPs activated transcriptional re-programming remarkably similar to that activated systemically following wounding. This study thus provides tantalizing evidence for using bioelectrical stimuli to engineer beneficial responses in plants.

9.4 BIOELECTRICAL ENGINEERING OF CELL BIOLOGY: POTENTIAL AND CHALLENGES

A bioelectrical view on cell behaviour, as advocated here, can help to establish a unifying framework that allows us to see cell behaviour arising from interfaces with its microenvironment through the fluxes of ions and redox-active compounds. This framework offers a new synthesis that can bridge molecular studies with a bioelectrical basis of cellular physiology. The resulting science can have a transformative effect on our understanding of cellular behaviour and pave the way to its direct control through predictive bioelectrical engineering, a process that is already advancing in the context of neuronal systems.

The above-highlighted couplings between cell physiology and bioelectricity can only be pursued experimentally through the development of integrated quantitative measurement techniques. To this end, electrochemistry offers a range of quantitative techniques that are capable of measuring the concentration of ions (e.g. H+ (i.e. pH), Ca2+ or Na+) or specific redox-active compounds (e.g. neurotransmitters, glucose, flavins, hydrogen peroxide and oxygen). The use and combination of such techniques have already resulted in the development of specific and powerful bio-electrochemical tools. Multi-electrode arrays allow the measurement of electrochemical events occurring at time scales of the order of microseconds and at the tissue level, while scanning ion conductance (SICM) and scanning electrochemical microscopy (SECM) allow mapping of ionic conductivity and redox reactions, respectively, at nano-scale to single-cell levels.

- Electrochemical approaches can also be used to modify the chemical composition of the cell microenvironment in a selective and controlled manner by producing or delivering reactant species that will trigger specific processes or by exposing cells to electrical fields and pulses. This opens up the possibility of 'dialling in' on cellular behaviour at the single-cell and tissue levels. It has been shown for example that electric fields can influence or stop mammalian cell division or trigger specific cellular responses and can be used to distinguish between metabolically active and dormant bacterial cells. Techniques such as SECM can directly deliver redox-active compounds to individual cells and measure their responses; for example, to interrogate the metabolism of cancerous versus normal cells.
- Regardless of the specifics of a given technique, using electrochemical approaches to study and engineer cell behaviour poses challenges. By its nature, an electrochemical measurement is complex, involving processes at the interface of an electrode and solution. In the case of cellular methods, there would even be multiple interfaces among cells, solution and electrodes. Thus, the characteristics of such interfaces must be considered to correctly evaluate the measurements. At single- and sub-cell scales, untangling the interface effects can be complicated and requires the application of modelling approaches, such as finite-element modelling to solve coupled mass transport–reactivity problems. The application of measurement and modelling in tandem provides a roadmap for elucidating cell/solution interface properties, as has been shown for understanding surface charge heterogeneities at the sub-cellular level. Another challenge for the application of electrochemical techniques is signal acquisition and processing.
- Recent efforts have focused on improving signal sensitivity (both signal/noise ratio and the specificity of signal) and temporal resolution. For example, it is possible to minimize electrode impedance by optimizing electrode size, morphology and materials. The use of such low-noise measuring systems revealed minute, yet regular, membrane capacitive current oscillations across large populations of mammalian cells, which have previously been considered electrically quiescent, such as C6 glioma cells.
- A number of technical challenges need to be considered when combining electrode measurements with cells and tissues, including sample preparation, interfaces of electrodes for in vitro or in vivo studies and maintenance of the integrity of the sensing/stimulating electrodes and the living system. For example, for in vitro

spatially resolved electrochemical measurements, cells must be immobilized, which may not reflect their natural state. Irrespective of the application (sensing or modifying cell behaviour), the electrode should not be cytotoxic to cells. It must be ensured that the physiological medium used neither is prohibitive for specific electrochemical techniques nor interferes with the measurement. By-products of a measurement (e.g. in the case of redox-based measurements) should not drastically disturb the cell microenvironment. This can be of particular concern, as by-products and coupled reactions, such as solvent breakdown, could impact and obscure experimental conclusions.

◉ These challenges are far outweighed by the potential of developing a bioelectrical basis of cell behaviour. As we argued above, this will not only open up completely new ways of understanding cells, but also allow the development of novel bioelectrical means to control and engineer cell behaviour. The latter is happening already with emerging applications of so-called electroceuticals to treat biofilms and cancer tumours. On the former premise, the ability to use nano- and micro-scale electrodes allows the elucidation of single-cell properties and responses in ways previously not possible. For example, recent applications of SICM have revealed cellular charge heterogeneities, which cannot be detected with bulk methods such as zeta-potential measurements of cells. SECM combined with SICM opens up new methods for stimulating cells with ions or redox-active compounds and following their bioelectrical responses in real time and space. The combination of such single-cell electrochemical measurements with other existing methods, such as fluorescence microscopy of ion-binding or MP-responsive dyes, and the tandem use of these measurements with modelling will provide a significantly improved theoretical basis for understanding the cell–ionic environment interface. The continued developments taking place in electrode manufacturing, electrochemical techniques and signal processing will enable bioelectrical control of cells and their behaviour at single-cell, tissue and perhaps even organ levels, using also implantable devices. These exciting prospects will require a true integration of cell biology, electrochemistry, physics and engineering in the years to come.

10

MOLECULAR BIOLOGY TECHNIQUES

10.1 NUCLEIC ACID EXTRACTION

The extraction of nucleic acid is a crucial step for biochemical and diagnostic processes. It is a series of steps to obtain pure nucleic acid samples that are suitable for different downstream applications. Different extraction methods result in the difference in yield, purity and extraction time.

10.2 DNA EXTRACTION

The choice of the DNA extraction method depends on the types of DNA (chromosomal and plasmid), source organism (bacteria, fungi, plant or animal), starting material (organ, tissue, cell, etc.) and desired results (molecular weight of the desired DNA, purity, extraction time required, etc.).

The extraction of DNA from biological material requires: cell lysis (cell disruption), inactivation of cellular nucleases, removal of biomolecules other than DNA, purification and quantification of the DNA. Common lysis procedures include mechanical disruption (for example, grinding, ultrasonication hypotonic lysis, osmotic lysis) and chemical lysis (for example, detergent lysis, alkali treatment). Lytic enzymes, chaotropic agents, and different types of detergents are the main components of chemical lysis.

10.3 COMMON DNA EXTRACTION METHODS

Nucleic acid extraction methods can be widely characterized into two different types: solutionbased methods and solid-phase based methods. In solution-based extraction methods, cell extracts are mixed with chemical solutions devised to purify nucleic acid. Solid-phase extraction methods work by causing nucleic acids to bind to solid supports, such as magnetic beads coated with silica or other materials. Different extraction methods result in different yields and purity of DNA.

10.3.1 Organic Extraction

A traditional method that can be used to obtain highly pure DNA. This method involves organic extraction (e.g. phenol:chloroform) followed by ethanol precipitation. In this method, cells are lysed and cell debris is usually removed by centrifugation. The remaining soluble material is then mixed with organic solvents such as phenol, or 1:1 mixture of phenol and chloroform. A phenol-chloroform organic solvent is used to concentrate DNA in a hydrophilic phase. Phenol dissociates proteins bound to DNA while chloroform denatures proteins and lipids. Three distinct phases will form: the *aqueous phase*, the *interphase* and the *organic phase*. Of these, the aqueous phase contains the DNA, whereas the proteins and lipids remain in the other two phases. The aqueous phase is then separated and treated with ethanol to precipitate the DNA (termed ethanol precipitation). The precipitated DNA can then be pelleted by centrifugation and dissolved in a buffer of choice for use in downstream reactions. If the protein content in cell extract is very high, then proteases such as pronase or proteinase K are used to degrade proteins before phenol extraction. RNA is digested from the preparation by treatment with RNase.

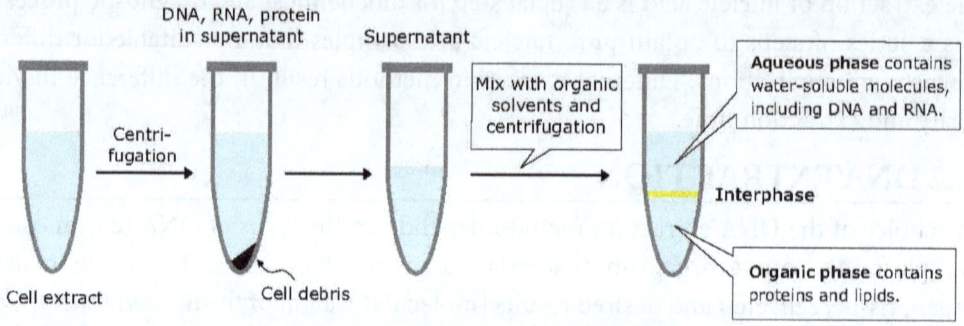

Fig. 1: Organic extraction method.

10.3.2 Silica-Based Method

It is a solid-phase DNA extraction method. Silicates have a high binding affinity for DNA under alkaline conditions and high salt concentration. The mechanism involved in this technique is the affinity between negatively charged DNA and positively charged silica material, resulting in selective binding of nucleic acids to the silica matrices (consist of silica material, in the form of either gel or glass particle). As a final step, DNA can be eluted from the silica matrix by any hypoosmotic solution, such as nuclease-free water or buffers such as alkaline Tris-EDTA.

10.3.4 Magnetic Bead-Based Method

Magnetic beads are paramagnetic materials with a uniform particle sizes that have been coated with a functional group that interacts with DNA. Under optimized conditions, DNA selectively binds to an appropriately-coated bead surface, leaving contaminants in solution. A magnetic field is used to separate magnetic beads from a suspension. Hence after DNA binding, beads are separated from other contaminating cellular components by placing a magnet outside of the tube to create a strong external magnetic field. This approach removes the need for vacuum or centrifugation, which minimizes stress or shearing forces on the target molecules.

10.3.5 Anion Exchange Method

Just like silica matrices, anion exchange resins are also widely used in DNA extraction. DNA xtraction by anion exchange chromatography is based on the specific interaction between negatively charged phosphates of the nucleic acid and positively charged surface molecules on the substrate. Unlike silicate negative charge, anion exchange resin makes use of the positively charged diethylaminoethyl (DEAE) cellulose to attract the negatively charged phosphate of nucleic acid. DNA binds specifically to the substrate in the presence of low salt, contaminants are removed by wash steps using a low or medium salt buffer, and purified DNA is eluted using .

10.4 QUALITATIVE AND QUANTITATIVE ANALYSIS OF NUCLEIC ACID

Nucleic acid quantitation is an important and necessary step prior to most nucleic acid analysis methods. Common methods used for determining quality and quantity of nucleic acid include absorbance, fluorescence and qPCR.

Absorbance based quantification method is performed at 260 nm (based on the fact that the heterocyclic rings of nucleotides absorb UV light with an absorption maximum at around 260 nm). Absorbance method is simple, quick and has a wide detection range. This method does not require any reagents. But this method is less sensitive than fluorescence or qPCR based methods and also cannot distinguish between ssDNA, dsDNA or RNA.

Nucleic acid can be quantified by measuring the absorption at 260 nm. For dsDNA, 1 absorbance unit is equivalent to 50 μg DNA (i.e. a solution of dsDNA with a concentration of 50 μg/mL will have an absorbance at 260 nm equal to 1); for ssDNA, it is equivalent to 33 μg DNA; for ssRNA, it is equivalent to 40 μg RNA. Absorbance is also used to measure purity of nucleic acids present in the sample. The ratio of absorbance at 260 nm and absorbance at 280 nm (A260/A280) and the ratio of absorbance at 260 nm and absorbance at 230 nm (A260/A230) are used to check nucleic acids purity. For a pure DNA sample, A260/A280 ratios is ~1.8 and for a pure RNA sample, the ratio of A260/A280 is ~2. A ratio of < 1.8 indicates the DNA sample is contaminated with protein or an organic solvent such as phenol. Typically, protein contamination can be detected by a reduction of this ratio; RNA contamination can be detected by an increase of this ratio.

The A260/A230 values for 'pure' nucleic acid are commonly in the range of 2.0–2.2. The A260/A230 is a sensitive indicator of contaminants that absorb at 230 nm. These contaminants are significantly more numerous than those absorbing at 280 nm and include chaotropic salts such as guanidine thiocyanate, EDTA, non-ionic detergents like Triton X-100 and phenol. Substances like polysaccharides also show absorbance at this wavelength, but will have a weaker effect. Fluorescence-based quantification methods use fluorescence dyes that preferentially bind a given species of nucleic acid (e.g. dsDNA, ssDNA or RNA). When a dsDNA binding dye, for example, is excited by a given wavelength of light, only dye in the dsDNA-bound state will fluoresce. As the dye binds to the target nucleic acid, fluorescent quantum yield increases as a function of shift in fluorophore molecular geometry. The intensity of fluorescent signal relates the amount of nucleic acid present. Fluorescence detection methods are comparatively more sensitive than absorbance.

10.5 POLYMERASE CHAIN REACTION

Polymerase chain reaction (PCR) is a rapid and versatile *in vitro* method for amplifying defined target DNA sequences present within the source of DNA. This technique was formulated by *Kary Mullis* in 1983. Usually, the method is designed to permit *selective amplification* of a specific target DNA sequence(s) within a heterogeneous collection of DNA molecules (e.g. total genomic DNA or a complex cDNA population). To permit such selective amplification, some prior DNA sequence information from the target sequences is required. This information is used to design two oligonucleotide primers (amplimers), which are specific to the target sequence and are often about 15–25 nucleotides long. After the primers are added to denatured template DNA, they bind specifically to complementary DNA sequences at the target site. In the presence of a suitably heat-stable DNA polymerase and DNA precursors (the four deoxynucleoside triphosphates, dATP, dCTP, dGTP and dTTP), primer initiates the synthesis of new DNA strands which are complementary to the individual DNA strands of the target DNA segment and will overlap each other.

10.5.1 Essential Components of PCRs:

- Thermostable DNA polymerase
- Primers
- dNTP (dATP, dCTP, dGTP and dTTP)
- Divalent cations (usually Mg2+) as a cofactor for activity of DNA polymerase
- Buffer, provides a suitable chemical environment for activity of DNA polymerase
- Template DNA

10.5.2 Primer Designing

In PCR, primer designing is the most important aspect for selective amplification. For primer design, some prior DNA sequence information from the target DNA is required. The information is used to design two primers, which are specific to sequences flanking the target DNA sequence. So, for most PCR reactions, it is very important to reduce the chance of the primers binding to other locations in the DNA than the desired one. Hence, certain rules for primer design are important to consider. These rules follow:

10.5.3 Primer Length

Length of the primers should not be very short or long. If the primers are too short, they might hybridize to non-target sites and give undesired amplification products. To illustrate this point, imagine a DNA molecule of 10 Mbp is used in a PCR experiment with a pair of primers eight nucleotides in length. The attachment sites for these primers are expected to occur, on average, once every 48 = 65,536 bp. This means that it would be very unlikely that a pair of 8 nucleotides long primers would give a single, specific amplification product. If the primer length will be 15 nucleotides, then the expected frequency of attachment site for the primer will be once every 415 = 1073741824 bp. This figure is higher than the length of the DNA molecule, so 15 nucleotides long primer would be expected to have just one attachment site. But we cannot take very long primer because the long primer influences the rate at which it hybridizes to the template DNA; long primers hybridize at a slower rate. It is generally accepted that the optimal length of primers is 18–20 nucleotides. This length is long enough for adequate specificity and short enough for primers to bind easily to the template at the annealing temperature.

Calculation of length of primers: The longer the primer, the higher its specificity for a particular target. The following equation can be used to calculate the probability that a sequence exactly complementary to the primer sequence will occur by chance in a DNA molecule that consists of a random sequence of nucleotides.

$$K = [g/2]^{G+C} \times [(1-g)/2]^{A+T}$$

where, K is the expected frequency of occurrence of target sequence in a DNA molecule, g is the relative G + C content of the DNA molecule and G, C, A and T are the number of specific nucleotides in the primer.

For a double-stranded genome of size N (in nucleotides), the expected number (n) of sites complementary to the primer is $n = 2NK$.

10.6 QUANTITATIVE REAL-TIME PCR

Quantitative Real-Time PCR is based on the general principle of PCR, which is used to amplify and simultaneously quantify a target DNA molecule. This is called Real-Time PCR because it allows the scientist to actually view the increase in the amount of DNA as it is amplified. Real-Time PCR systems rely upon the detection and quantitation of a fluorescent reporter, whose signal increases in direct proportion to the amount of PCR product in a reaction. These fluorescent reporter molecules include dyes that bind to the dsDNA (i.e. SYBR Green) or sequence-specific probes. SYBR Green binds to the minor groove of the dsDNA only. In the solution, the unbound dye exhibits very little fluorescence. This fluorescence is substantially enhanced when the dye is bound to dsDNA. SYBR Green has, however, a limitation that includes preferential binding to G.C rich sequences.

SYBR Green monitors the total amount of double-stranded DNA, but cannot distinguish between different sequences. To be sure that the correct target sequence is being amplified, a sequence-specific fluorescent probe is needed. An example is the TaqMan probe which is oligonucleotide designed to hybridize to an internal region of a PCR product.

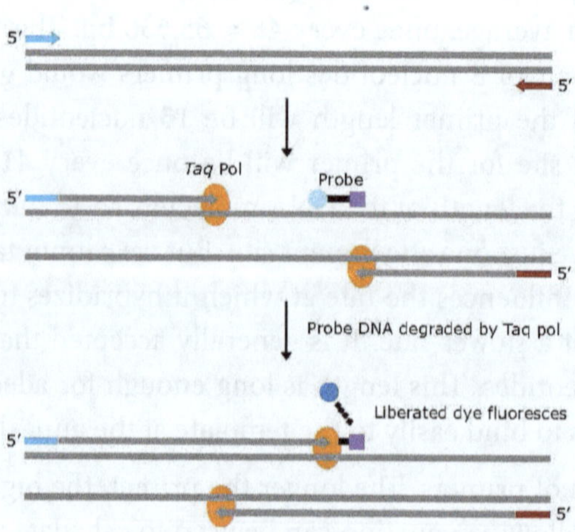

Fig. 2: Real-time PCR with TaqMan probe. The TaqMan probe binds to the target sequence after the denaturation step that separates the two DNA strands. As the *Taq* polymerase extends the primer during the next PCR cycle, it will eventually bump into the TaqMan probe. The *Taq* polymerase has a 5' to 3' exonuclease activity that degrades the DNA strand of the probe. This breaks the linkage between the fluorophore and quencher. The fluorophore is now free from quenching and its fluorescence increases.

The Taq Man probe has three elements: a short-wavelength fluorophore on one end, a sequence that is specific for the target DNA, and a quencher at the other end. As long as the fluorophore and the quencher are close to each other, fluorescence is quenched and no fluorescent light is emitted. This probe is designed to anneal to the center of the target DNA. When *Taq* polymerase elongates the second complementary strand during PCR, its 5' to 3' exonuclease activity cuts the probe into single nucleotides. This removes the close proximity between the fluorophore and the quencher and abolishes quenching. The short-wavelength fluorophore can now fluoresce and a signal will be detected that is proportional to the number of newly synthesized strands.

10.6.1 RT-PCR

RT-PCR (Reverse Transcription PCR) is a method used to amplify, isolate or identify a known sequence from a cellular or tissue RNA. The PCR reaction is preceded by a reaction using reverse transcriptase to convert RNA to cDNA. Some thermostable DNA polymerase can use RNA templates as substrate. An example of this is the recombinant form of Tth polymerase from *Thermus thermophilus*, which can catalyze high-temperature reverse transcription of RNA in the presence of MnCl2. It has both intrinsic reverse transcriptase and thermostable DNA-dependent DNA polymerase activities. This has led to the development of protocols for single-enzyme reverse transcription and PCR amplification.

10.6.2 Inverse PCR

Standard PCR is used to amplify a segment of DNA that lies between two inward-pointing primers. In contrast, inverse PCR (also known as inverted or inside-out PCR) is used to amplify unknown DNA sequences that flank one end of a known DNA sequence and for which no primers are available. The inverse PCR method involves a series of restriction digestion and ligation, resulting in a circular DNA that can be primed for PCR from a single section of known sequence. It involves isolating a restriction fragment that contains the known sequence plus flanking sequences. The restriction fragments circularize to form circular DNA under very low concentrations in the presence of DNA ligase. To perform PCR, two primers that bind specifically to the known sequence but they are oriented in opposite directions. Successful PCR with these primers produces a linear product in which central unknown region remains flanked by two short known sequences. RT-PCR RT-PCR (Reverse Transcription PCR) is a method used to amplify, isolate or identify a known sequence from a cellular or tissue RNA. The PCR reaction is preceded by a reaction using reverse transcriptase to convert RNA to cDNA. Some thermostable DNA polymerase can use RNA templates as substrate. An example of this is the recombinant form of Tth polymerase from *Thermus thermophilus*, which can catalyze high-temperature reverse transcription of RNA in the presence of MnCl2. It has both intrinsic reverse transcriptase and thermostable DNA-dependent DNA polymerase activities. This has led to the development of protocols for single-enzyme reverse transcription and PCR amplification.

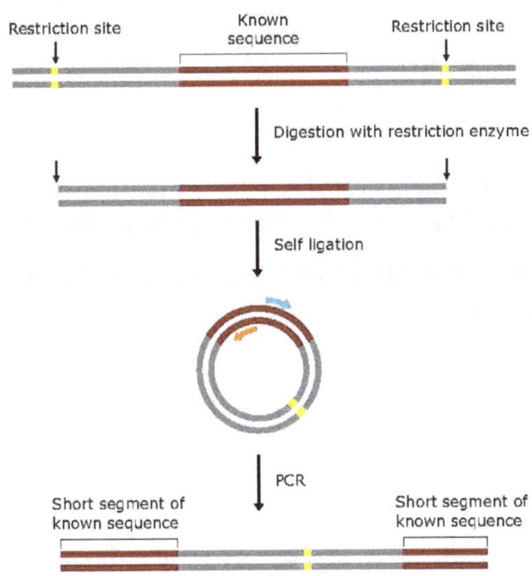

Fig. 3: Inverse PCR. A region of DNA in which part of the sequence is known. If the areas of interest lie outside the known region, inverse PCR can be used to amplify the

flanking region. It involves the digestion of genomic DNA with appropriate restriction endonucleases, intramolecular ligation to circularize the DNA fragments and PCR amplification. First, the DNA is cut with a restriction enzyme that does not cut within the region of known sequence. This generates a fragment of DNA containing the known sequence flanked by two regions of unknown sequence. Under low DNA concentrations, self-ligation is induced to give a circular DNA product. Finally, PCR is performed on the circular fragments of DNA. Two primers are used that face outwards from the known DNA sequence. PCR amplification gives linear products.

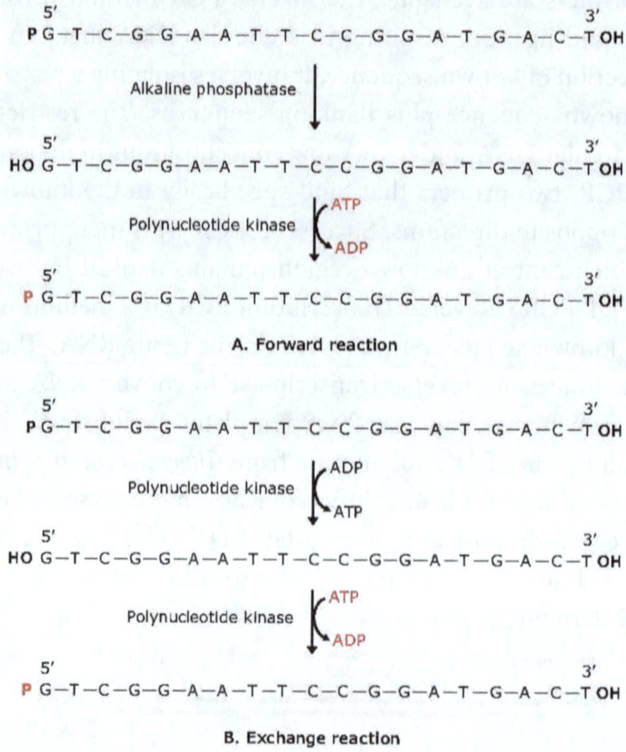

Fig. 4: 5'-end labeling. A. In the *forward reaction*, polynucleotide kinase transfers the gamma phosphate from ATP to the 5'-end of a polynucleotide (DNA or RNA). B.

In the *exchange reaction*, target DNA or RNA that has a 5' phosphate is incubated with an excess of ADP. Polynucleotide kinase first transfers the phosphate from the nucleic acid onto an ADP, forming ATP and leaving a dephosphorylated target. Polynucleotide kinase then performs a forward reaction and transfers a phosphate from ATP onto the target nucleic acid.

10.7 NONISOTOPIC LABELING

Nonisotopic labeling systems involve the use of non-radioactive probes. Two types of nonradioactive labeling are conducted – direct and indirect. Direct labeling strategies utilize probes that are directly conjugated to a dye or an enzyme, which generates the detection signal. Indirect labeling systems utilize probes that contain a hapten that will bind to a secondary agent generating the detection signal; the probe itself does not generate signal. Direct nonisotopic labeling, where a nucleotide which contains the label that will be detected is incorporated. Often such systems involve incorporation of modified nucleotides containing a fluorophore, a chemical group which can fluoresce when exposed to light of a certain wavelength. Most commonly used fluorophores for direct labeling are fluorescein, a pale green fluorescent dye and rhodamine, a red fluorescent dye.

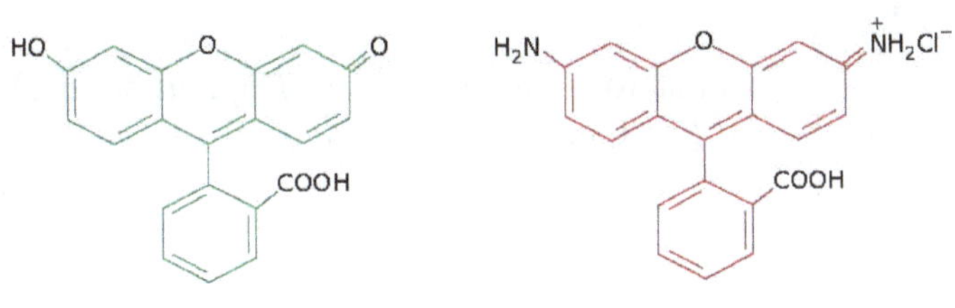

Fluorescein Rhodamine, core structure

Indirect nonisotopic labeling, usually featuring the chemical coupling of a modified reporter molecule to a nucleotide precursor. After incorporation into DNA, the reporter groups can be specifically bound by an affinity molecule, a protein or other ligand which has a very high affinity for the reporter group. Conjugated to the latter is a marker molecule or group which can be detected in a suitable assay. The reporter molecules on modified nucleotides need to protrude sufficiently far from the nucleic acid backbone to facilitate their detection by the affinity molecule and so a spacer of 4-16 carbon atoms long is required to separate the nucleotide from the reporter group.

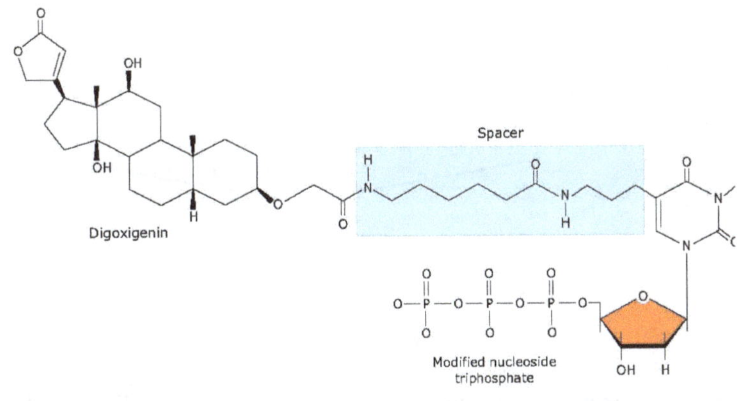

Fig. 5:The base of the nucleoside triphosphate depicted is an analog of thymine in which the methyl group has been replaced by a spacer arm linked to the plant steroid digoxigenin. The digoxigenin is detected by a specific antibody coupled to a visible marker such as a fluorescent dye.

10.7.2 Two Indirect Nonisotopic Labeling systems are Widely Used:

The biotin-streptavidin system utilizes the extremely high affinity of two ligands – biotin (vitamin H or B7) which acts as the *reporter*; and the bacterial protein streptavidin which is the *affinity molecule*. Biotin and streptavidin bind together extremely tightly with a dissociation constant in the order of 10–14 mol/litre, one of the strongest known in biology. Although both avidin and streptavidin bind to biotin with very high affinity, but avidin in some applications show high non-specific binding due to the presence of the sugars and high pI. The *streptavidinbiotin* system can be incorporated into virtually every immunoassay, whereby an antibody is conjugated to biotin (biotinylated antibodies) and then detected with streptavidin.

Digoxigenin is a plant steroid (obtained from *Digitalis* species) to which a specific antibody has been raised. The digoxigenin-specific antibody permits detection of nucleic acid molecules which have incorporated nucleotides containing the digoxigenin *reporter molecule*.

Both biotin and digoxigenin are linked to uracil, which is normally a component of RNA not DNA. Therefore, to label DNA, uracil must be incorporated into the DNA instead of thymine. If deoxyUTP labeled with biotin or digoxigenin is added to the polymerization reaction, DNA polymerase will incorporate the labeled uridine where thymidine would normally be inserted. The biotin or digoxigenin tags stick out from the DNA without disrupting its structure.

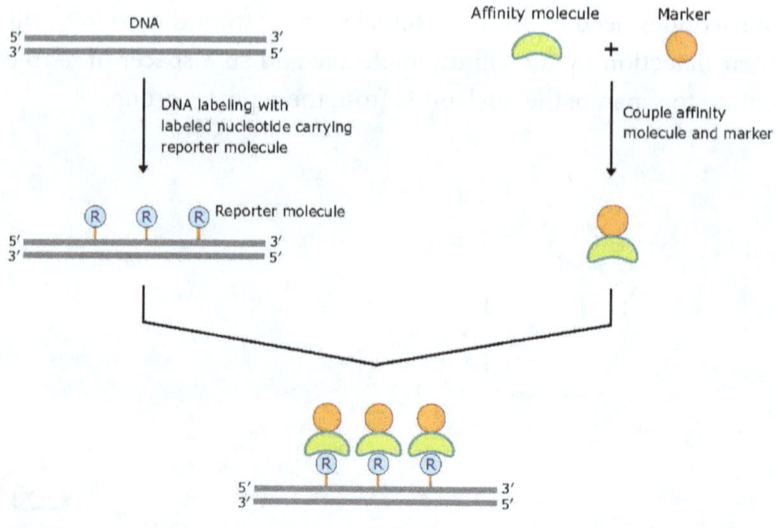

Fig. 6: Indirect non-isotopic labeling involves chemical linkage between a reporter molecule and a nucleotide. When this modified nucleotide is incorporated into DNA, then it binds specifically to an affinity molecule which has high affinity against the reporter molecule. A long spacer is introduced between nucleotide and reporter molecule so as to reduce steric hindrances for binding of affinity molecule.

10.7.3 Molecular Beacons

The *molecular beacon* is a structured fluorescent probe. A structured probe contains stem-loop structure regions that confer enhanced target specificity when compared with a traditional linear probe. Molecular beacons form a stem-loop structure, where the central-loop-sequence is complementary to the target of interest and the stem arms are complementary to each other.

It is an oligonucleotide (about 25 nucleotides long) that contains both a fluorophore and a quenching group at opposite ends. Its detection mechanism relies on the principle of FRET, in which a fluorophore in the excited state, can transfer energy to acceptor fluorophore, with subsequent emission of a fluorescent signal from the acceptor, or to a quencher dye, which dissipates the energy without emission of a detectable fluorescence signal. In order for this energy transfer to take place, the donor and acceptor molecules must be situated in close physical proximity.

10.7.4 Dot Blot Assay

In a dot blot assay, a specified volume of nucleic acid mixture is spotted onto a small area of a nylon membrane. In this technique, the nucleic acid molecules are not first separated by electrophoresis. Instead, a specimen containing nucleic acid mixture to be detected is applied directly on a membrane as a small dot. The dot is treated with an alkaline solution to denature DNA molecules. Finally, the dot-blot membrane is allowed to hybridize with a labeled probe.

The membranes are processed and exposed to film. If the dot of DNA contains a sequence similar to the probe, the film will turn dark in that area. If no dark spot appears on the film, it can be inferred that no DNA hybridization has occurred.

Slot blot is fundamentally similar to dot blot. The difference between dot and slot blot procedures is in the way that the nucleic acid mixture is blotted onto the membrane. In dot blot, the nucleic acids are blotted as circular blots, whereas in slot blot they are blotted in rectangular slots.

10.8 PHAGE DISPLAY

Phage display is a molecular technique that allows expression of foreign polypeptides or peptides on the surface of phage particles. This method was first described by George Smith in 1985.

In this technique, the DNA encoding the protein of interest is fused with a gene encoding one of the proteins that forms the viral coat. Genetic engineering techniques are used to insert foreign DNA fragments into a suitable phage coat protein gene. Phage display involved the use of filamentous phages such as fd, f1, M13, where the foreign gene was incorporated into a gene specifying a minor coat protein. Filamentous phage M13 is the most popular choice for phage display. M13 is a filamentous phage contains 6.4 kb single-stranded circular DNA. M13.

Enters *E. coli* through the bacterial sex pilus, a protein appendage that permits the transfer of DNA between bacteria. The single-stranded DNA in the virus particle [called the (+) strand] is replicated through an intermediate circular double-stranded replicative form containing (+) and (-) strands. Only the (+) strand is packaged into new virus particles.

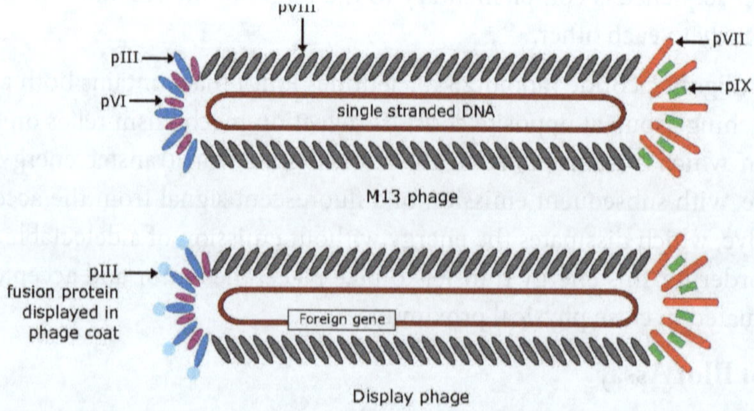

Fig. 7: Phage display. In order to display a peptide on the surface of a bacteriophage, the DNA sequence encoding the peptide must be fused to the gene for a bacteriophage coat protein (either to the pVIII gene or the pIII gene). The coat protein is expressed as a fusion protein on the virion surface without disturbing the infectivity of the phage.

10.9 DNA SEQUENCING

The term DNA sequencing encompasses biochemical methods for determining the order of the nucleotide bases, adenine, guanine, cytosine and thymine, in a DNA molecule. The methodologies for DNA sequencing are as follows:

10.9.1 Chain Termination Method

The enzymatic chain termination method (also dideoxy method) was developed by Frederick Sanger and coworkers in year 1977. The method is based on the DNA polymerase-dependent synthesis of a complementary DNA strand in the presence of natural deoxynucleotides (dNTPs) and dideoxynucleotides (ddNTPs) that serve as terminators. The DNA synthesis reaction is randomly terminated whenever a ddNTP is added to the growing oligonucleotide chain, resulting in truncated products of varying lengths with an appropriate ddNTP at their 3'-terminus.

The products are separated by size using polyacrylamide gel electrophoresis and the terminal ddNTPs are used to reveal the DNA sequence of the template strand.

10.9.2 Chemical Degradation Method

The chemical degradation method was introduced by Allan Maxam and Walter Gilbert in year 1977. In this method, the sequence of a dsDNA molecule is determined by treatment with chemicals that cut the molecule at specific nucleotide positions.

10.9.3 Pyrosequencing Method

The pyrosequencing method was developed by Mostafa Ronaghi and Pal Nyrén at the Royal Institute of Technology in Stockholm in 1996. In this method, the addition of a deoxynucleotide to the end of the growing strand is detectable because it is accompanied by the release of a flash of light.

10.9.4 Chain Termination Method

Chain termination method relies on the use of dideoxyribonucleoside triphosphates (ddNTP), derivatives of the normal deoxyribonucleoside triphosphates that lack the 3' hydroxyl group.

Purified DNA is synthesized in vitro in a mixture that contains single-stranded molecules of the DNA to be sequenced, the enzyme DNA polymerase, a short primer DNA to enable the polymerase to start DNA synthesis, and the four deoxyribonucleoside triphosphates (dATP, dCTP, dGTP, dTTP: A, C, G and T). If a dideoxyribonucleotide analog of one of these nucleotides is also present in the nucleotide mixture, it can become incorporated into a growing DNA chain. Because this chain now lacks a 3' OH group, the addition of the next nucleotide is blocked, and the DNA chain terminates at that point.

To determine the complete sequence of a DNA fragment, the double-stranded DNA is first separated into its single strands and one of the strands is used as the template for sequencing. Four different chain-terminating dideoxyribonucleoside triphosphates (ddATP, ddCTP, ddGTP and ddTTP) are used in four separate DNA synthesis reactions on copies of the same single-stranded DNA template.

Reaction produces a set of DNA copies that terminate at different points in the sequence. The products of these four reactions are separated by electrophoresis in four parallel lanes of a polyacrylamide sequencing gel. Sequencing gels are run in the presence of denaturing agents, urea and formamide. They routinely contain 6–20% polyacrylamide and 7 mol/l urea.

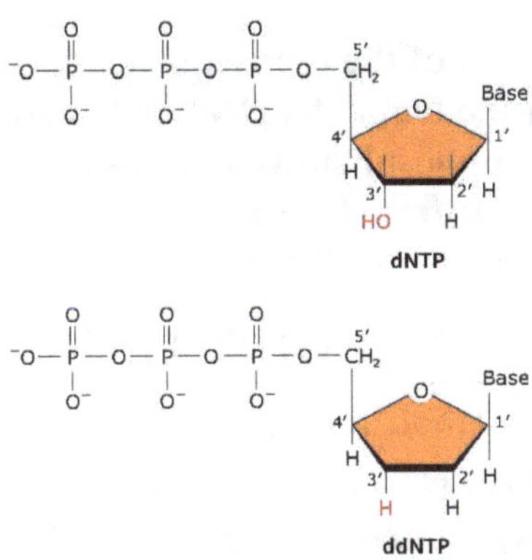

Fig. 8: The structures of dNTP and ddNTP differ in the number and location of hydroxyl groups on the 2' and 3' carbons. DNA polymerase cannot add another nucleotide to a chain ending in dideoxyribose because its 3' carbon does not have a hydroxyl group.

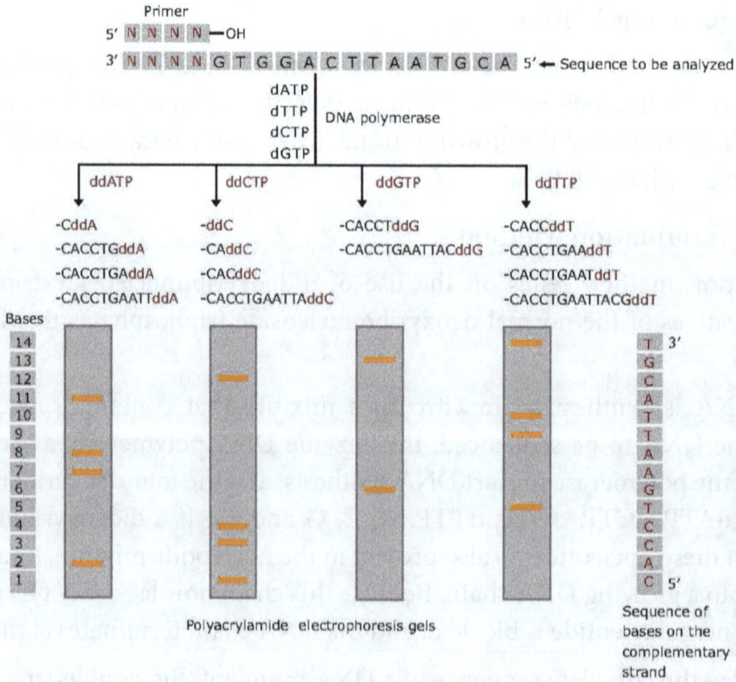

Fig. 9: DNA sequencing by chain termination method (Sanger method).

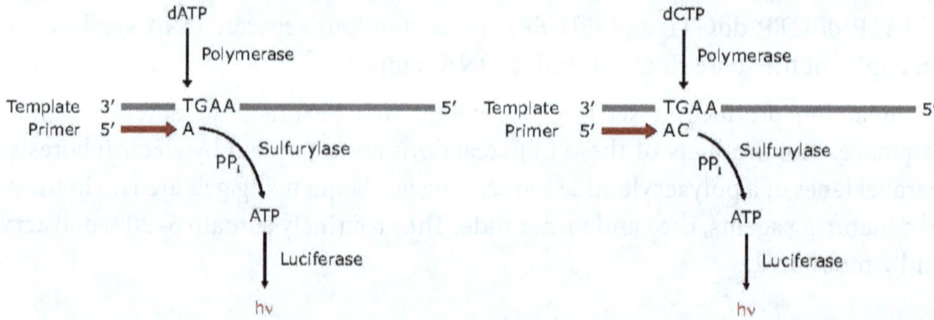

Fig. 10: Schematic representation of the pyrosequencing enzyme system. If the added dNTP forms a base pair with the template, DNA polymerase incorporates it into the growing DNA strand and pyrophosphate is released. ATP sulfurylase converts the PPi into ATP which serves as substrate for the light producing enzyme luciferase. The produced light is detected as evidence of that nucleotide incorporation has taken place.

10.10 NEXT-GENERATION SEQUENCING TECHNOLOGIES

Nucleic acid sequencing is a method for determining the exact order of nucleotides present in a given DNA or RNA molecule. The chain-termination method (also commonly referred to as Sanger or dideoxy sequencing method), published in 1977, has remained the most commonly used DNA sequencing technique to date. The Human Genome Project, led by the International Human Genome Sequencing Consortium and Celera

Genomics, was accomplished with *firstgeneration* Sanger sequencing. Since completion of the first human genome sequence, demand for cheaper and faster sequencing methods has increased greatly. This demand has driven the development of *second-generation* sequencing methods (or next-generation sequencing).

Next-generation sequencing platforms perform massively parallel sequencing, during which millions of fragments of DNA from a single sample are sequenced in unison. Massively parallel sequencing technology facilitates high-throughput sequencing, which allows an entire genome to be sequenced in less than one day. In the past decade, several next-generation sequencing platforms have been developed.

To illustrate how next-generation sequencing process works, consider a single genomic DNA (gDNA) sample. The gDNA is first fragmented into a library of small segments that can be uniformly and accurately sequenced in millions of parallel reactions. The newly identified strings of bases, called *reads*, are then reassembled using a known reference genome as a scaffold (resequencing), or in the absence of a reference genome (*de novo* sequencing). The full set of aligned reads reveals the entire sequence of each chromosome in the gDNA sample.

Illumina (Solexa) sequencing

This system was initially developed in 2007 by Solexa and was subsequently acquired by Illumina, Inc. Illumina next-generation sequencing utilizes a fundamentally different approach from the classic Sanger chain-termination method. It is based on sequencing by synthesis (SBS) technology – tracking the addition of labeled nucleotides as the DNA chain is copied – in a massively parallel fashion. Illumina sequencing systems can deliver data output ranging from 300 kilobases up to 1 terabase in a single run, depending on instrument type and configuration.

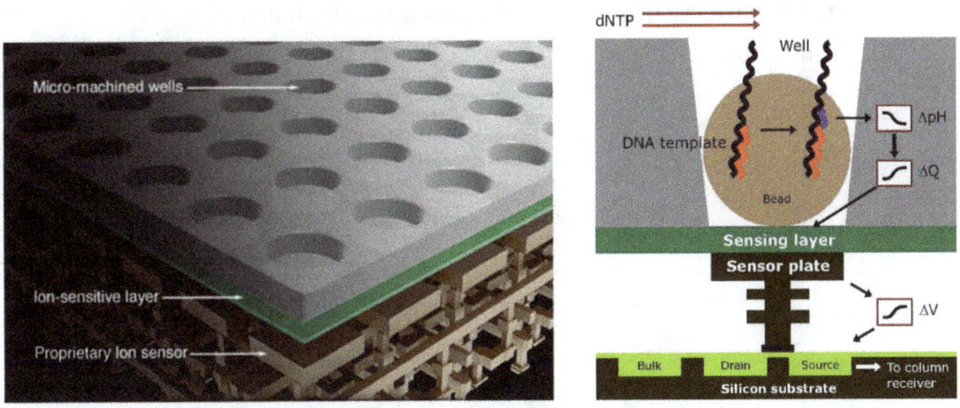

Fig. 11: Ion Torrent technology- Sensor, well and chip architecture. An image of a well, a bead containing DNA template, and the underlying sensor and electronics. Cross section view of a single well that houses ion sphere particles with a clonal amplified DNA template. A hydrogen ion (proton) is released when a nucleotide is incorporated by DNA polymerase. The proton is then detected by the sensing layer due to the change of pH, therefore translating the chemical signal to a digital input.

10.11 BIOSENSORS

A *biosensor* is a measuring device which is used to detect chemical compounds by converting a biological response into an electrical signal. The electrical signal it produces carries the necessary information about the process under investigation. The most widely accepted definition of a biosensors is: '*a self-contained analytical device that incorporates a biologically active material in contact with an appropriate transducer for the purpose of detecting the concentration or activity of chemical species in any type of sample.*'

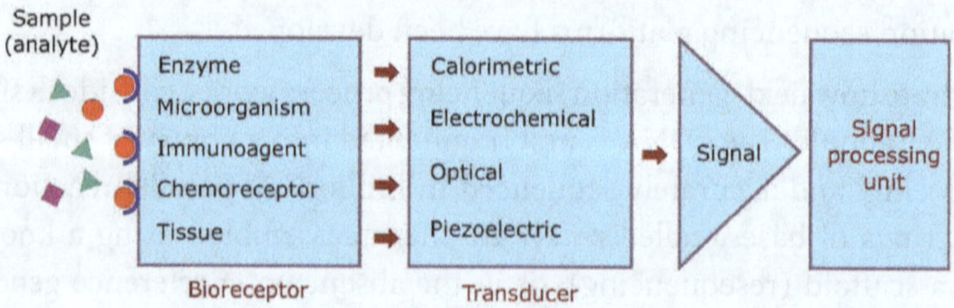

Fig. 12: Schematic diagram showing the main components of a biosensor. The bioreceptor - interacts with the specific analyte of interest to produce signal; the transducer - transforms the signal resulting from the interaction of the analyte with the bioreceptor into another signal (i.e. transduces) that can be more easily measured and quantified; and signal processing unit - converts the signal into a workable form (amplified, processed and displayed).

There are three main parts of a biosensor:

i. Biological detection system, a *bioreceptor*. Bioreceptor may be an enzyme, an antibody or similar binding molecule, a living cell or organelles.
ii. A *transducer*, which transforms the signal resulting from the interaction of the analyte with the bioreceptor into a measurable signal.
iii. A signal processing system, which converts the signal into a workable form.

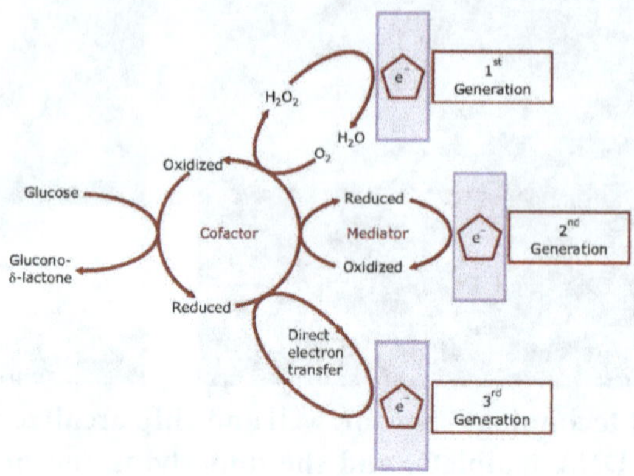

Figure 13. The evolution from 1st to 3rd generation electrochemical biosensors. The figure highlights modifications in the biosensor layout with each generation using glucose sensors as an example.

10.12 CRISPR/CAS SYSTEMS

CRISPR/Cas systems are highly diverse adaptive microbial immune systems used by most archaea (~90%) and many eubacteria (~40%) to protect themselves from invading viruses and plasmids. These systems allow the cell to recognize and distinguish incoming 'foreign' DNA from 'self' DNA. CRISPR/Cas systems consist of two general parts: CRISPRs (clustered regularly interspaced short palindromic repeats) and Cas (CRISPR-associated) proteins.

CRISPRs consist of highly conserved short repeated sequences separated by similarly sized short spacers sequences. The size of CRISPR repeats and spacers varies between 23 to 47 bp and 21 to 72 bp, respectively. The bacterial genome may contain more than one CRISPR locus. The CRISPR loci have highly diverse and hypervariable spacer sequences, even between closely related strains. Spacers are unique sequences originating from viral or plasmid DNA. These sequences are used as recognition elements for invading viral or plasmid DNA. By adding new spacers in their own genome, bacteria able to recognize new matching viral or plasmid genomes.

Another feature associated with CRISPR loci is the presence of a conserved sequence, called leader, located upstream of the CRISPR with respect to the direction of transcription. CRISPR activity requires the presence of a set of CRISPR-associated (cas) genes, usually found adjacent to the CRISPR, that code for Cas proteins essential to the immune response. The Cas proteins with perform a variety of functions such as nucleases, helicases and polymerases. The CRISPR immune system works through the cooperation of many diverse Cas proteins, resulting in CRISPR-Cas systems currently being grouped into two classes, six types, and over 30 subtypes.

Emmanuelle Charpentier and Jennifer Doudna are awarded the Nobel Prize in Chemistry 2020 for discovering the CRISPR/Cas9 system, a method for genome editing.

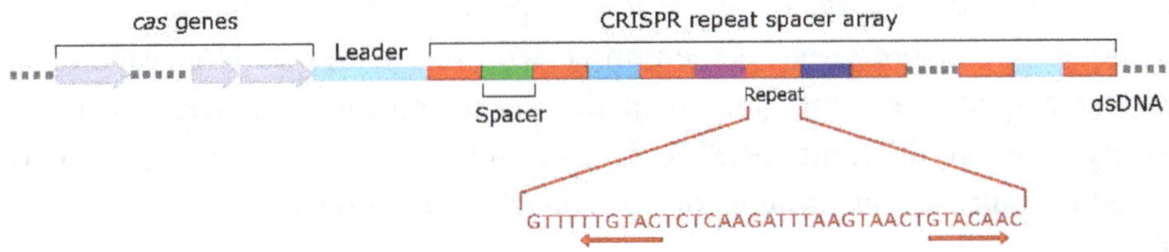

Figure 14. Locus organization: CRISPR loci consist of 23 to 47 bp palindromic repeat sequences (in red), separated by 21 to 72 bp spacer sequences. The spacers have no common features in their sequences. The *cas* genes are located close to the CRISPR loci. The number of CRISPR loci, their sequences, and the type of *cas* gene vary among species.

10.13 GENERAL MECHANISM

A simple analogy to understand the CRISPR mechanism is to imagine the design of a personalized weapon. Suppose, earth is attacked by an alien species and none of the earthly weapons are working on it, so someone would go and pick up the body parts of alien and based on that a new weapon specific for that alien would be developed. Also, some of the body parts will be stored for future memory. Similar to this, a bacterium, when attacked by bacteriophage, is not aware of any defence mechanism. Hence, a small Cas protein picks the part of DNA of phage and bring it close to the CRISPR loci, where it is integrated into the CRISPR site.

A crRNA (CRISPR RNA) is synthesized from that loci which along with other proteins then recognize the complementary sequence present in the invading DNA of phage and degrade it.

The CRISPR/Cas system works in a similar manner.

The general mechanism of CRISPR works in three distinct phases: adaptation (or *acquisition*), expression and maturation (or *biogenesis*) and interference (or *targeting*).

10.13.1 Adaptation Phase

Adaptation is the first step in the process of developing a weapon against foreign invading DNA (viral or plasmid DNA). During adaptation, Cas proteins capture short segments of foreign nucleic acids (termed protospacers) and integrate them as 'spacers' into the CRISPR locus in order to record a molecular memory of past invaders. The invading DNAs are directionally integrated, as new CRISPR spacers, into a CRISPR array that is separated by repeat sequences. Some CRISPR/Cas systems employ an alternative mechanism of adaptation — namely, spacer acquisition from RNA, via reverse transcription by a reverse transcriptase encoded at the CRISPR/Cas locus.

10.13.2 Expression and Maturation Phase

Once the desired sequence (i.e. protospacer) from the invading DNA is integrated into the loci of CRISPR, the CRISPR locus is transcribed into a pre-CRISPR RNA (pre-crRNA). CRISPR loci are transcribed from an upstream promoter located in the AT-rich leader sequence. The precrRNA is then processed into mature crRNAs, each containing a transcribed spacer sequence joined to the partial repeat sequence. The pre-crRNA processing differs in different CRISPR/ Cas variants. It may be mediated by a complex composed of multiple Cas proteins or a single multidomain Cas protein or by non-Cas host RNases.

10.13.3 Interference Phase

The crRNA forms a complex with Cas proteins. The crRNA of the crRNA-Cas complex makes base pairing with the protospacer of the invading DNA. Finally, crRNA-directed cleavage of invading DNA occurs by Cas proteins at protospacer, a site complementary to the crRNA spacer sequence.

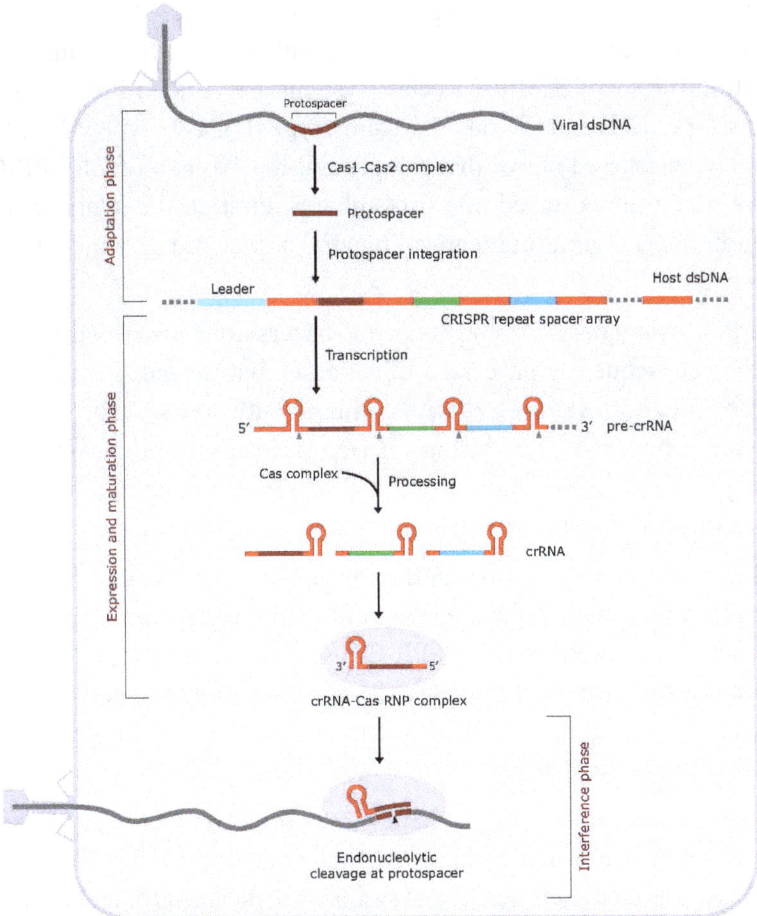

Fig. 15: The CRISPR/Cas system has three phases. 1.

In the adaptation stage, the Cas1–Cas2 complex, which comprises two Cas1 dimers and a single Cas2 dimer, acquires a protospacer from the invading viral DNA and integrates it as a new spacer into the CRISPR array. 2. In the expression and maturation stage, the CRISPR array is transcribed and then processed into mature crRNAs, each containing a transcribed spacer and part of the repeat sequence. These crRNAs form ribonucleoprotein (RNP) complexes with Cas proteins. 3. In the interference stage, the crRNA–Cas RNP complex identifies the invading target DNA through complementary basepairing. Cleavage of the target sequence causes the destruction of the invading viral genome and therefore protects the bacteria from the viral infection.

10.14 TYPES OF CRISPR/CAS SYSTEMS

The diversity of Cas proteins and the presence of multiple CRISPR loci make classification a complex task. According to the current classification scheme, all CRISPR/Cas systems are divided into two distinct classes, based on the differences between the interference complexes – *class 1 systems* and *class 2 systems*. The class 1 systems involve multi-subunit effector complexes (composed of multiple Cas proteins— Cas3, Cas5–Cas8, Cas10 and Cas11). By contrast, class 2 systems involve single subunit effectors (composed of Cas9, Cas12 or Cas13). The two CRISPR-Cas class systems are divided into three types each, Class 1 systems: types I, III and IV Class 2 systems: type II (Cas9), type V (Cas12) and type VI (Cas13) As we discussed above that the general mechanism of CRISPR/Cas system based adaptive immunity is staged into three phases. First, in the *adaptation phase*, Cas proteins integrate short regions of the invading viral or plasmid genome into the CRISPR array as new spacers.

During crRNA biogenesis, the repeats and spacers are transcribed into a long pre-crRNA, which is subsequently processed to generate mature crRNAs that each contain a single spacer sequence. Mature crRNAs combine with one or more Cas proteins to form an effector complex, which during *interference*, senses and cleaves protospacers, complementary to the cr- RNA. Although all CRISPR/Cas systems adhere to this general pathway, they exhibit striking diversity in their processing and interference steps.

Type I, II and V systems recognize and cleave DNA, type VI can edit RNA, and type III edits both DNA and RNA. How the effect of type IV system on DNA or RNA is still unknown. Nearly all characterized CRISPR-Cas systems (except type III) mediate both adaptation and interference to the foreign genetic elements by specifically recognizing PAM (protospacer-adjacent motif). The PAM is a short sequence located adjacent to the protospacer on the invading DNA.

It is essential for recognition (by Cas1 and Cas2 proteins) and the distinction between self and non-self DNA. The absence of the PAM sequence at the CRISPR locus in the host genome protects it from self-cleavage. Type I systems are the most frequent and widespread systems, which target invading DNA in a PAM dependent manner. This system employs cascade (CRISPR-associated complex for antiviral defence) - a multiprotein complex containing mature crRNA and Cas proteins– Cas6, Cas5, Cas7, Cas8, and Cas11, to target DNA recognition and cleavage. Cascade first recognizes the PAM sequence on the target DNA, and promotes the pairing of crRNA. During the interference stage, triggers the recruitment of the endonuclease Cas3, which initiates degradation of the invading target DNA.

Similar to type I systems, type III systems employ cascade for antiviral defense and target both DNA and RNA substrates. Type III systems work in PAM-independent manner and are categorized into four subtypes named A–D. The interference target of type III-A is mRNA, while the interference target of type III-B is the same as that of type I and II CRISPR-Cas systems, which is DNA. However, the interference targets of types III-C and D are unclear. Type II systems (also type V and VI systems) only employing a single protein and crRNA in he RNP complex against invaders. The type II systems comprise of Cas9 protein, crRNA and tracrRNA (trans-activating crRNA) and work in a PAM dependent manner. TracrRNA is a hairpin RNA transcribed from a repeat region. Cas9 binds and stabilizes the tracrRNA-crRNA duplex. tracrRNA is responsible for activating RNase III to promote the pre-crRNA processing. Mature crRNA combines with tracrRNA and Cas9 activates cleavage. Type II systems include three subtypes, II-A, II-B, and II-C.

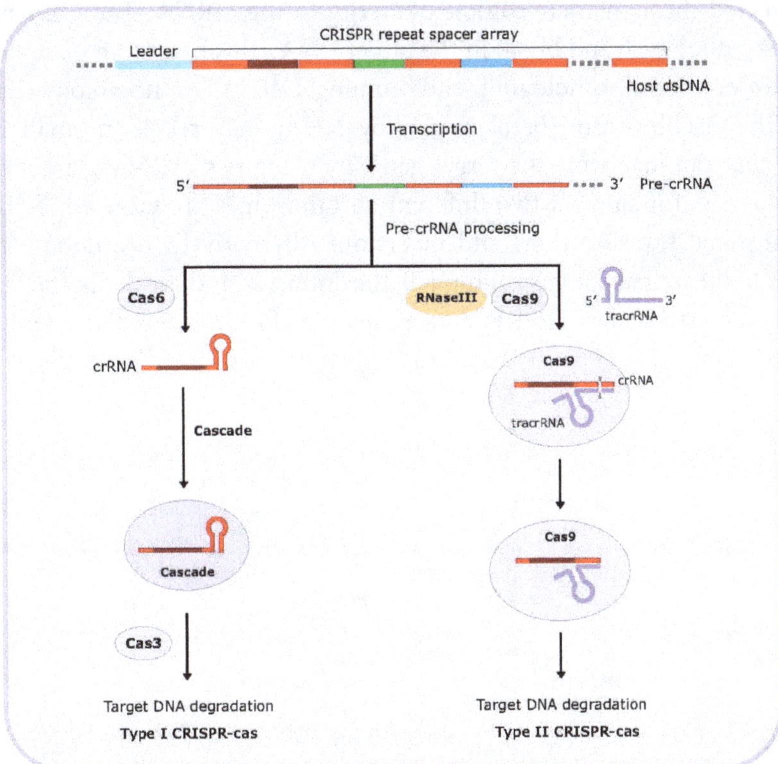

Fig. 16: Comparison of mode of action of types I and II CRISPR-Cas systems.

These two systems differ in their processing and interference steps. During the expression stage, CRISPR loci are transcribed and the resulting pre-crRNA is processed into short crRNAs by cleavage in the repeat sequences. In type I system, pre-crRNA cleavage is carried out by Cas6 endoribonucleases. In type II systems, this process involves the expression of a tracrRNA, which base-pairs with the repeats in the pre-crRNA transcript. The resulting duplexes are cleaved in the repeat sequences by RNase III in a Cas9-dependent reaction. Thus, in both CRISPR–Cas systems, cleavage of the pre-crRNA occurs in the repeat sequences, hence, mature crRNA consists of a spacer flanked by partial repeats. In type II systems, the tracrRNA remains bound to the crRNA and the mature crRNA–tracrRNA duplexes are complexed with Cas9. In type I systems, mature crRNA is bound by a Cas protein complex (i.e. Cascade). During the interference stage crRNAs function as guides for the Cas proteins, as they recognize and bind to complementary sequences in invading genome.

10.15 CRISPR/CAS9 AND TARGETED GENOME EDITING

The CRISPR/Cas bacterial immune system is used as a simple, RNA-guided platform for highly efficient and targeted genome editing. CRISPR genome editing technology takes advantage of DNA repair. crRNAs guide Cas9 proteins to target sequences, which are recognized through their complementarity to the crRNA. The Cas9 nuclease then introduces a double-strand break in the target DNA. This break can be repaired through two pathways— non-homologous end- joining (NHEJ) or homology-directed repair (HDR). The double-strand break repair by NHEJ may result in small insertions or deletions that can inactivate the target gene. By using two crRNAs, larger deletions can be generated, and if sites on two different chromosomes are targeted, NHEJ can result in chromosomal translocations and inversion. Alternatively, if a donor template with homology to the targeted locus is supplied, the double strand break may be repaired by the homology-directed repair (HDR) pathway allowing for precise replacement mutations to be made.

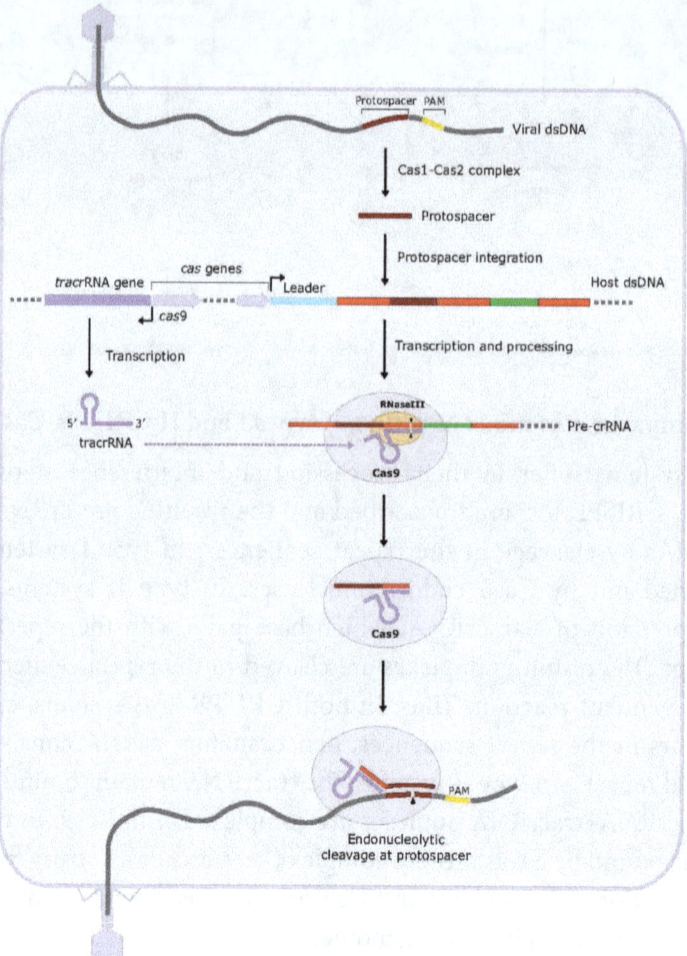

Fig. 17: Type II CRISPR/Cas system.

In the acquisition step, invading DNA is directionally integrated, as new CRISPR spacer, into a CRISPR array that is separated by repeat sequences, thus creating a memory of invading genetic element. In the expression and maturation step, the CRISPR locus is transcribed into a precrRNA, which is then processed into mature crRNAs that contain spacer sequences joined to partial repeats. The CRISPR locus also encodes a tracrRNA that has complementarity to the repeat regions of crRNA transcripts. In the interference stage, a crRNA–tracrRNA hybrid is formed through binding of the complementary repeat region sequences, and this RNA hybrid guides the Cas nuclease towards complementary DNA sequences, which leads to the targeting and cleaving of invading genetic elements.

10.16 TOOLS FOR PROGRAMMABLE GENOME EDITING

Scientists can probe the function of a gene or open reading frame by mutating or deleting a locus of interest and observing the resulting phenotype. In the past decade, researchers hypothesized that by exploiting endogenous, cellular DNA repair pathways, one could create mutations or precise edits at a desired location in the genome, termed *genome editing*. Double-strand breaks are lethal to cells, thus organisms evolved mechanisms to repair these lesions. Scientists proposed that by generating a targeted, double-strand break at a site of interest, then during the repair process errors may occur, resulting in a mutation at a desired site. Additionally, endogenous double-strand break repair pathways could also stimulate the incorporation of exogenous DNA, creating very specific researcher-designed edits. Thus, nuclease-mediated approaches have been extensively explored for site-specific gene editing.

Meganucleases (or *homing nucleases*), are among the first classes of nucleases that were engineered to target specific genomic sites for gene editing purposes. Meganucleases are a group of nucleases that recognize long nucleotide sequences and induce a double-stranded break at their targeted site. The long recognition sequence of meganucleases may occur only once within a genome, thereby facilitating its use for site-specific genome editing. Meganucleases can be reengineered to target novel sequences through strategies such as protein engineering, structure- based design, and molecular evolution, although the procedure is usually labor-intensive.

A. ZFN enzyme comprise a DNA-binding domain formed by zinc finger modules (ZF), each recognizing a unique 3-base pair sequence on the target DNA, and a DNA-cleaving domain composed of FokI. Two ZFN are designed to recognize DNA sequences that flank the desired cleavage site. In the example, each ZFN comprises four ZF which recognize 12 bp on opposite strands. Upon interaction of ZF with the target site, a FokI dimer catalyzes a targeted double-strand break. B. Targeted DNA cleavage is achieved by a pair of TALEN molecules. Each TALEN is comprised by a TALE and a FokI. C. CRISPR/Cas9 systems.

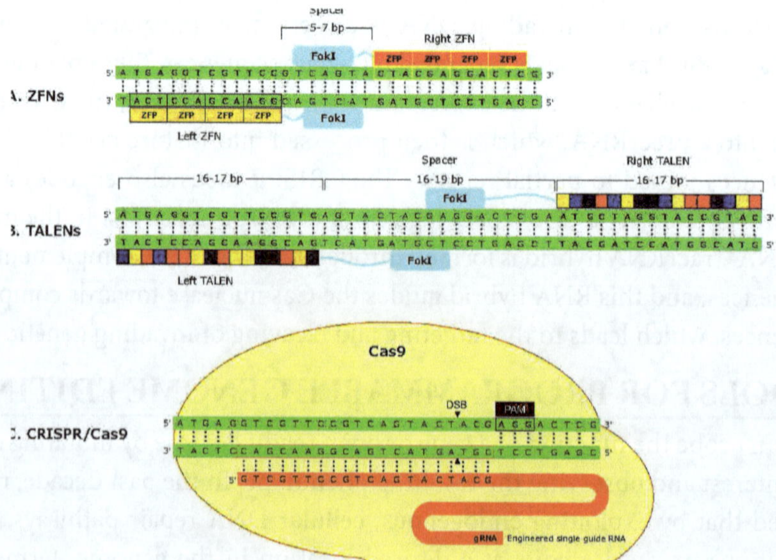

Fig. 18: Schematic representation of components of ZFNs, TALENs and CRISPR/Cas9 systems.

Other examples of programmable genome editing machines include *zinc-finger nucleases* (ZFNs) and *transcription activator-like effector nucleases* (TALENs), in which the DNA-binding domains of transcription factors have been fused with the nuclease domain of the restriction enzyme FokI, an obligate dimer. When targeted to paired adjacent sequences, the FokI domains of these programmable, site-specific nucleases form a dimer that activates the nuclease activity, thus creating a double-stranded break near their binding sites. Researchers can exploit the cell's endogenous DNA repair pathways to create mutations at the desired doublestranded break sites. However, because these tools function through protein–DNA interactions, targeting to a new site requires engineering and cloning a new protein, which precludes ZFNs and TALENs from being used for high-throughput applications. In contrast to most known DNA-binding proteins, Cas9 is an RNA-guided nuclease whose sequence specificity largely arises from Watson–Crick base pairing between its guide RNA and the target DNA site, in addition to a direct interaction between Cas9 and a short protospaceradjacent motif (PAM) of DNA. Thus, Cas9 can be programmed to target new sites simply by changing its guide RNA sequence, making it an ideal platform for high-throughput sequencespecific gene editing, as well as other applications. Its natural endonuclease activity has been co-opted for sequence-specific editing of the genome in a wide range of organisms, including bacteria, fungi, plants and animals. To enable sequence-specific genomic regulation, ucleasedeactivated Cas9 (dCas9) has been engineered and can be fused to a variety of effectors, such as transcriptional activators, repressors, and epigenetic modifiers. In addition to applications in genome editing and regulation, DNA-binding proteins, such as ZFs, TALEs, and dCas9, have been fused to fluorescent proteins (FPs) to allow direct imaging of genomic loci in living cells.

11

BIOFLUID MECHANICS OF SPECIAL ORGANS

11.1 INTRODUCTION: AN OVERVIEW

The lymphatic system performs many crucial functions in health and is involved in diverse disease states. Lymphatics are responsible for fluid and protein balance, gathering approximately 4 liters/day of interstitial fluid and returning it to the venous system. As this fluid is filtered, undesirable elements such as tumor cells and foreign pathogens are normally destroyed in lymph nodes. This system is also part of the primary transport mechanism for the immune system.

Despite the importance of the lymphatic system in health and disease, comparatively little is known about its function, certainly relative to arterial and venous circuits. The fact that the lymphatic system performs its pumping functions against pressure gradients and gravity makes it an interesting topic for engineering analysis. Clearly, the integration of the system components from the cellular to the systemic level is crucial in this function. Our future modeling efforts will focus on capturing physiologic features of clinically important disease states (edema, cancer, etc.), and predicting the likelihood of success of various interventions. Currently, there is no cure for edema, and prognoses of metastasized cancers remain grim.

- Lymphedema, a debilitating disease for which there is no known cure, affects a large number of cancer patients who have undergone lymph node dissection as well as trauma victims. The lymphatic system is also the major transport route for metastases of the most deadly cancers.
- Understanding and modeling the transport of lymph remains a challenge. Tissue pressure is very low, and cannot account for even a small percentage of total lymph flow. Given its return to the venous system in the upper torso, much of the body's lymph must be pumped "uphill."

- There are extrinsic mechanisms such as muscle pumping that aid this task, but much of the work comes from the contraction of lymphatic vessel smooth muscle, with valves preventing backflow. Lymphatic vessels alter pumping characteristics in response to changes in diameter and flow, with the degree of response to these signals being regionally dependent. This implies that lymphatic endothelial cells respond to shear stress and circumferential stretch with substances that trigger lymphatic smooth muscle cell action. Unfortunately, little is known about lymphatic endothelial mechanotransduction, communication with smooth muscle, and how this integrates into a functioning system overall.
- In the field of fluid flow within the human body, blood flow in the systemic circulation has been the main focus since the recognition by William Harvey that blood was in fact in continuous circulation and carried by a network of blood vessels. But beyond the systemic circulation, other fluids and other fluid flow phenomena pervade the body so totally that it would be hard to imagine a bodily function of any kind that does not involve fluids or fluid flow.
- In fact, the study of the systemic circulation is limited to not only the type of fluid involved but also to the type of service which the systemic circulation provides - the first being blood, of course, the second is transport - the principal function of the systemic circulation is to provide a means of transporting blood continuously from the heart to tissue cells and back again. Some of the most fascinating fluid flow phenomena within the human body involve fluids other than blood and a service other than transport- the lymphatic and pulmonary systems provide two striking examples. In this chapter describes the special fluid mechanics of these two systems. While transport is still involved in both cases, this is not the only service which they provide and blood is not the only fluid involved. In both systems, filtration, extraction, enrichment, and in general some "treatment" of the fluid itself is the primary function. In the pulmonary system, the liquid lining of the lungs plays a pivotal role, and the mechanical interaction between tissue and liquid is of key importance to lung viability. In disease states such as respiratory distress syndrome, the lining fluid can become dysfunctional, leading to reduced gas exchange and damage to sensitive pulmonary tissues.
- The study of the systemic circulation has also been conventionally limited to treating the system as if it were an open-loop system in which the main hemodynamic variables such as pressure and flow are governed by the laws of fluid mechanics independently from the physiological controls and regulations that govern these same variables. This implies that any failure of the system can be fully explained in terms of the laws of fluid mechanics, which of course is not the case. While a system failure due to a physical obstruction in a blood vessel can be readily explained in terms of the laws of fluid mechanics, a system failure due

to arrhythmia cannot. In this paper we examine the clinical implications of these issues and of the special biofluid mechanics issues that arise in the lymphatic and pulmonary systems.

11.3 IMPORTANCE OF LYMPHATIC SYSTEM IN HEALTH

- The human lymphatic system plays several crucial roles in maintaining health, including fluid and protein balance, immune cell transport, and nutrient uptake. In approximate terms, out of 8000 liters of cardiac output per day, 20 liters of fluid flows out the capillaries, 16–18 liters of which is reabsorbed by the blood vessels. That leaves 2–4 liters to be returned via the lymphatic system. Returning interstitial proteins to the blood circulation is also of crucial importance, since we lose approximately 50% of circulating proteins through the capillaries every day. This movement of fluid, cells and solutes from the blood to the interstitial spaces, into the lymphatic compartment and back to the blood is essential to many crucial body functions.
- The act of pumping this fluid back into the venous system at a point near the heart presents significant challenges. Interstitial tissue pressures hover around atmospheric. Thus, lymphatic system flow cannot rely on pressure-driven flow as in blood vessels. Furthermore, the location of the return point to the venous system in the upper torso means that the system must pump "uphill."
- The lymphatic system begins with terminal lymphatics that collect fluid and constituents from the interstitial space. There are approximately 40,000 terminal lymphatics per kg of tissue. These rather porous structures consist of endothelial cells, intercellular primary valves, and a loosely organized substrate. Their porosity is such that the concentration of proteins is practically identical in interstitial fluid and lymph, eliminating osmotic pressure gradients that might hinder fluid movement. The pressure gradient from the interstitium to the inside of terminal lymphatics is variable in magnitude and direction, but is typically less than 1 mmHg].
- Terminal lymphatics outflow to small collecting vessels with the aid of valve structures in between endothelial cells that prevent backflow. Valves are in fact ubiquitous in the lymphatic system. The vessels between valves are referred to as lymphangions. The lymphatic vessels eventually flow into lymph nodes. Lymph from afferent vessels flows in and is slowed within the subcapsular sinus. This facilitates interaction with the immune cells located in lymph node compartments, including the high-endothelial venous sinuses. Filtered lymph then flows out through efferent vessels, eventually collecting in the thoracic duct that flow into the venous system.

- Movement of fluid through the lymphatic system facilitates immune response. The delivery of foreign bodies to the lymph nodes is enabled by lymph flow, as is the crucial trafficking of monocytes, dendritic cells and lymphocytes. Finally, mesenteric lymphatic vessels play an important role in lipid uptake, which flow into the lymph and then into blood vessels.

11.4 IMPORTANCE OF LYMPHATIC SYSTEM IN DISEASE

- Given the variety of challenges presented to the lymphatic system in health, it should not be surprising that there are numerous complexities in disease conditions. Local tissue swelling, or edema, occurs when the lymphatic system fails to remove fluid efficiently from the interstitium. There is no known cure for this critical condition, which can eventually lead to immune dysregulation and malnutrition. A functional definition of edema is a 50% increase in interstitital volume.
- Some have suggested that presence of positive tissue pressure indicates edema, but some tissue pressures are positive in health. Edema affects 300 million people worldwide, including approximately 10 million patients who contracted the condition secondary to breast or pelvic cancer therapy, recurrent infections, trauma or vascular surgery. Congenital lymphedema affects one of every 6000 newborns, but may not appear until puberty.
- Acquired lymphedema caused by the filaria parasite (elephantiasis) affects 3–5 million people in Western countries, and is a common cause of disability in developing countries. In the US, the majority of cases appear due to malignancy or its therapy. The onset of lymphedema in these cases can appear within days, or years later. It affects approximately half of breast cancer survivors undergoing axillary lymph node dissection; in some cases taking years to develop. The risk following groin dissections is approximately the same. Lymphedema is almost always accompanied by immune system disorders.
- The lymphatic system plays a tremendously important role in the development and spread of cancer. The formation of new lymphatic vessels (lymphangiogenesis) in tumors is quite aggressive, and is an avenue for many cancers. Furthermore, when cancerous tumors metastasize, offending cells travel through the lymphatic system. While it is a normal function of macrophages and lymphocytes to eliminate cancerous cells, this defense system can become overwhelmed or subverted. A large portion of cancer deaths are attributable to metastasis through the lymphatic system. Pathologies may also develop in the presence of injuries due to lymphatic system dysfunction. In serious injuries involving open wounds, the lymphatic system is crucial in preventing infections through immune response.

- An important distinction between lymphatics and blood vessels is that the latter quickly shut down through vascular smooth muscle contraction, whereas lymph vessels remain open to aid in wound drainage.

11.5 LYMPHATIC SYSTEM PUMPING CHARACTERISTICS

- As discussed above, the lymphatic system is presented with the challenge of pumping several liters per day of fluid, proteins, cells, and foreign bodies from various interstitial spaces back to the venous system. It performs this function often against both an unfavorable pressure gradient and against gravity. While there are a few key mechanisms for overcoming this challenge, the degree to which each of these mechanisms functions varies tremendously around the body.
- Lymphatic pumping in the neighborhood of skeletal muscle benefits from extrinsic pumping mechanisms, similar to those that aid venous return of blood from the lower limbs. External compression of lymphatic vessels due to muscle contraction or other means results in propulsion of lymph forward, with backward flow prevented by competent valves. Another possible extrinsic aid to pumping is positive tissue pressure. This pressure varies among different tissues. However, the estimated flow rate due to this pressure is a small percentage of overall lymph flow.
- Much of the work associated with lymphatic pumping is performed via intrinsic mechanisms, i.e., those associated with direct action of the lymph vessels. The lymph vessels are exquisitely designed for this purpose. The intramural smooth muscle cells of lymphatic vessels possess far more mitochondria than smooth muscle of any other type. Additionally, lymphatic vessels have proportionally large numbers of blood vessels feeding their walls, compared to the *vasa vasora* of comparably sized blood vessels. Some of this extra vasculature may be due to lack of internal blood supply. Regardless, it is clear that lymphatic vessel walls contain significant contractile machinery.
- There is tremendous regional variability in pumping capability. In general, peripheral vessels contract more, presumably to overcome the significant downstream resistance to flow. It has been shown that mesenteric lymphatics pump rather strongly, compared to cervical and femoral lymphatics, as well as the thoracic duct. There appear to be two primary intrinsic regulators of lymphatic smooth muscle. The first is a response to stretch, which has been shown to increase the frequency of the pacemaker activity. Presumably, filling and thus stretching a lymphatic vessel is an indication to propel fluid forward to avoid edema. The second regulator is flow-related, and appears to be due to shear stress sensitivity of lymphatic endothelium. Increasing flow causes increases in both diastolic and systolic diameter (borrowing terminology from the heart), and decreases

ejection fraction as well as contraction frequency . Once again, it is important to note the regional variability in sensitivity to these regulating mechanisms. The thoracic duct is more sensitive to flow-induced smooth muscle relaxation than the mesenteric lymphatics . If extrinsic mechanisms were successful at propelling flow, then smooth muscle cell contraction would actually increase flow resistance as well as being metabolically costly. One possible regional exception to this was noted in a study of rat iliac lymphatics, with an inhibition of the intrinsic pump at high inlet pressures. Removal of the endothelium blocked any response to changing inlet pressure. Similar endothelium-dependent contractile inhibition was noted by von der **Weid et al.,** who denuded the endothelium in guinea pig mesenteric lymphatic vessels, and inhibited the response to acetylcholine. Thus, lymphatic endothelium-expressed nitric oxide (NO) is likely a key factor in contractile control, as is endothelin-1 (ET-1). The production of both substances by blood vessel endothelium is known to be shear stress dependent.

- Shear stress estimations, or indeed flow measurements of any kind, are difficult to perform in lymphatic vessels because the relatively low flow rates and delicate nature of these small vessels renders traditional blood flow measuring techniques useless. Based on high-speed (500 frames/s) in situ microscopic imaging of rat mesenteric lymphatic vessels, we have developed automated vessel and lymphocyte tracking algorithms to estimate lymph flow rate and wall shear stress, and then correlated those with the phasic pumping activity. Average flow rates and wall shear stresses through individual vessels were 13.95 mliter/hour and 0.64 dyne/cm^2, respectively, with peak wall shear stresses ranging 4–12 dyne/cm^2. Diameter changes in these nominally 100 mm diameter vessels were typically 40% or greater over the contractile cycle. We have observed some vessels contracting down to nearly zero diameter in other experiments.
- Lymphatic endothelial cells share some characteristics with blood vessel endothelial cells, such as alignment and elongation with the direction of flow. Crucial differences, however, include the expression of VEGFR-3 and Prox-1, both lymphangiogenic factors. Prox-1 knockout mice do not develop lymphatic vessels. Application of shear stress to cells resulted in inhibition of pan-cadherins only with lymphatic endothelial cells, and was noted at shear stresses as low as 2 dynes/cm^2 . There are no reports of molecular-level changes in lymphatic endothelial behavior related to mechanical stimuli.
- The lymphatic system is equipped with several extrinsic and intrinsic pumping mechanisms. When extrinsic mechanisms fail to produce sufficient flow, endothelial-based regulatory mechanisms kick in. However, the relative importance of the different pumping mechanisms varies throughout the body. Because of this regional variability, lymph pumping occurs in the absence of

synchronized activity. While it appears that the intrinsic pumping is coordinated over short distances, there is no guarantee that the unsteady propulsion of lymph from one region occurs in phase with, or even at the same frequency as, that from adjacent tissue.

- Scientist developed a lymphatic flow network model that captured some of the essential flow characteristics. His network consisted of at most seven generations of branches; still far away from representing terminal lymphatics and the small lymphangions that perform the important pumping work. Vessel contraction was assumed to occur in response to wall strain at a single frequency and phase, regardless of tissue location. This model was constructed well before much of the basics of lymphatic function were revealed, however, such as the effects of flow on smooth muscle contraction inhibition.

11.6 BIOFLUID MECHANICS OF THE PULMONARY SYSTEM

11.6.1 Physiological Significance

- The delicate structure of the lung epithelium makes it susceptible to surface tension induced injury. In cases of sepsis leading to multi-organ failure, large regions of the lung can fill with fluid (become atelectic), eliminating gas-exchange to a significant fraction of the lung. In addition, liquid fills the lungs of all neonates at the time of birth. Those born prior to 25 weeks gestation are likely to have an immature surfactant system, resulting in elevated interfacial surface tension. As a result, large regions of the lung may remain atelectic or close due to interfacial instabilities. This disease is referred to as respiratory distress syndrome (RDS). Subsequent lung collapse after inflation can occur in surfactant deficient neonates as well as adults with "wet" lungs.
- The deformation response of biological cells exposed to different flow conditions is a problem with important clinical implications. For example, patients suffering from acute lung injury (ALI) cannot breathe on their own due to the collapse and fluid occlusion of small pulmonary airways. These patients must be placed on a mechanical ventilator in order to survive. However, the microbubble flows generated during ventilation can exacerbate the existing lung injury and the mortality rate for ALI is very high (30–40%). Even for those patients who survive ALI, the tissues of the lung may be damaged due to the mechanical environment, which exposes the sensitive epithelium to abnormal physical forces that can initiate or exacerbate lung injury. This may occur with mechanical ventilation, leading to ventilator-induced lung injury (VILI).
- Clearly the damage to the lung related to VILI and ALI is associated with liquid obstructions that accompany atelectasis. In order to understand this relationship, models of air-liquid interfacial flows and fluid-structure interactions in single

airway and a network of airways that represent the bifurcating structure of the lung should be investigated. The liquid film consists of a mucus layer on top of a periciliary fluid layer. Mucus is a non-Newtonian fluid possessing viscoelastic characteristics. Surface tension on an air-liquid interface induces liquid flows, which may cause the lung's airways to close due to the formation of a liquid plug as a result of drainage of the liquid lining coating the airways if the liquid film is sufficiently thick. The stability of the liquid layer is also influenced by the air core flow as well as the rheological properties of the liquid. Once occluded, either by a short plug or an extended collapsed region, the airway must be reopened to maintain ventilation to distal regions of the lung.

- It is hypothesized that these events that may damage the lung – experimental and theoretical fluid mechanics models are essential to provide insight towards this hypothesis and may lead to clinical protocols that could reduce the morbidity and mortality of those suffering from ALI or RDS. In the sections that follow, we present predictions and observations related to interfacial flow-induced damage to the lung, and propose future studies that may enhance our understanding of these multi-scale problems.

11.6.2 Predictions of Fluid Mechanical Events Related to Atlectrauma

- Closure of airways can occur by a Rayleigh instability of the liquid lining, or an instability of the elastic support for the airway as the surface tension of the air – liquid interface pulls the tube shut, or both. Airway closure may occur in normal lungs at low lung volumes resulting in liquid plugging]. The thickness of liquid film is 2–4% of the airway diameter under normal conditions, but may be as large as 20% in disease.
- The effect of viscoelasticity of lung mucus on airway closure is still ongoing research. While to date most investigations use Newtonian models, the simplest viscoelastic model is that of a Maxwell fluid, where there are both viscous and elastic elements. When this viscoelastic system is oscillated, the dynamic response is frequency dependent and force *vs* displacement has both in-phase (elastic) and out-of-phase (viscous) components. Compared with a Newtonian fluid of the same base viscosity, the viscoelastic fluid layer grows faster and closure times are shorter because of a smaller effective viscosity. Due to the same mechanism as that shown, oscillatory core flows can saturate the instability. Since the effective viscosity of the viscoelastic fluid depends on the frequency of the force applied, there may exist a series of optimum frequencies of the oscillatory flow, that would increase the longitudinal displacement of the liquid bulge and hence maintain airway patency.

- Since the film flow is induced by the pressure gradient due to the surface tension forces, surfactant can retard the closure process. Several studies have investigated the effect of surfactant on airway closure.
- The process of opening collapsed atelectic region involves complex fluid-structure interactions that depend on airway geometry, fluid properties and surfactant biophysical processes. Theoretical and computational models have been developed. These models predict that a 'two-branch' response can exist with distinctly different low-velocity and high-velocity behavior.
- Recent studies show that the low-velocity branch is unstable and this unstable solution is unfeasible under the transverse gravitational force. These investigations also provide insight into mechanical stresses that can damage the lung. However, in general these are estimates based upon steady, fully-developed flow. Investigations of unsteady and transient flows in more complex geometries would improve our understanding of regions of the lung that might be highly sensitive to interfacial flows.

11.6.3 Predictions of Micro-Mechanical Stresses That May Damage the Lung

- Airway epithelial cells experience significant mechanical stresses during the reopening event. In extended regions of airway closure, computational analyses have predicted the existence of spatial/temporal gradients in shear and normal stress that may cause a local "pinching" effect on airway epithelial cells. , short regions of closure can occur wherein a liquid plug obstructs the airway. This liquid plug propagation results in a significant change in mechanical stress in the vicinity of the front meniscus surface.
- In both situations, the behavior of these stresses and the system at macro-scale is modulated by surfactant behavior at the molecular scale, which demonstrates the importance of multi-scale interactions in this system.

11.6.4 Observations of Damage Due To Fluid Mechanical Events

- Computational studies indicate that microbubble or liquid plug flow imparts a complex combination of normal and shear stress to the epithelial cells on airway walls. Experiments in *in vitro* systems clearly demonstrate that these flows can impart injurious mechanical stresses on airway and alveolar epithelial cells.
- It should be noted that the calculations of applied stresses are based upon simplifications of a flat cell monolayer, fully developed flow and a Newtonian lining fluid. In reality the surface geometry of the monolayer is complex and topological effects can greatly amplify the magnitudes of the hydrodynamic stresses. These hydrodynamic stresses may result in cell deformation, membrane rupture and/or cell detachment.

- Recent study shows that sub-confluent cells experience more damages and cell detachment than confluent cells. Hydrodynamic stresses may also be transduced into injurious biological responses including the up-regulation of inflammatory pathways and altered surfactant secretion.

11.7 PHYSICO-CHEMICAL INTERACTIONS

- Pulmonary mucus possesses viscoelastic and shear thinning characteristics with a yield stress. It is still unclear which property of the mucus is most important in the context of atelectrauma. Experimental evidence shows that the Maxwell fluid is inadequate to model the viscoelasticity of the mucus. In fact, the lining fluid properties vary substantially from region to region within the lung. Therefore, a reliable rheological model of lung liquid is required to make progress in estimations of the importance of viscoelasticity in normal and pathophysiological conditions.
- Surfactant physicochemical effects have been investigated under ideal conditions. However, surfactant biophysical properties are extraordinarily complex. For example, the surfactant on air-liquid interface is prone to buckle and collapse when its concentration exceeds the maximum dynamic concentration. Multilayer models of pulmonary surfactant are essential for enhancing our understanding of surfactant physicochemical hydrodynamics. Finally, surfactant is released in the lung in lamellar bodies and can create multi-molecular micelles.
- Accurate models of micelle transport, rheological characteristics and micelle dynamics at the air-liquid interface will substantially improve our understanding of these complex interactions. These models may be useful in developing strategies for pulmonary ventilation that could expand upon the "surfinacatant" work that could lead to scenarios that could enhance surfactant uptake.

11.8 CELL MODELS

- Under hydrodynamic stresses applied, cellular deformation and its injury may be a function of various biophysical parameters including cytoskeletal structure, micro-mechanical properties, cell morphologies and cell-cell contact mechanics. Dynamic response of cells to a rapid change in surface force and the response of cells on a flexible membrane are necessary to study experimentally.
- New computational models of this system may be necessary to be developed in order to accurately evaluate experimental conditions. Pharmacological intervention may lead to protective strategies for VILI through modification of cell stiffness.

11.9 AIRWAY INTERACTIONS

- While most models to date have relied on single or small networks of airways, in reality pulmonary airways are surrounded by parenchyma that consists of numerous alveoli, all of which are connected to distal airways. Therefore, the dynamics of each airway and alveolus is interdependent. As such, the behavior of one component may affect all others through parenchymal tethering. For this reason, the collapse and reopening process of an airway may affect other reopening processes.

- To understand these mechanical interactions, total lung lobe models need to be developed. Since the pressure in liquid-coated airways and alveoli are functions of their geometry and surface tension, this will naturally lead to multi-scale modeling. This modeling will allow us to identify differences in inflation/deflation mechanisms of atelectic regions of the lung, and may yield improved technologies for reducing the morbidity and mortality associated with RDS, ARDS and VILI.
- Though simplified models of airway closure and reopening phenomena in single airways have led to advances in our understanding of atelectrauma, future investigations should explore the importance of physicochemical interactions, cell mechanics, and airway interactions.

11.10 SYSTEM DYNAMICS VERSUS HEMODYNAMICS IN THE CONTROL OF BLOOD FLOW

- From a historical perspective, work on the systemic circulation has gone through a number of distinct eras: from the ancient era where blood was thought to be manufactured in the heart and then dispatched to the rest of the body, to the next era where it was recognized that blood is actually in continuous circulation through the body, then the era of steady Poiseuille flow where the focus was on the details of flow within the blood vessels, to the era of pulsatile flow, first in a rigid tube, then in an elastic tube, which led to the era of wave propagation and wave reflections and, finally, to the modern era where computer power coupled with the Navier-Stokes equations has made it possible to simulate blood flow in every possible configuration. It is interesting that every era was marked by an unwavering commitment to the subject of the day, all else being regarded as secondary, just as the pulsatile nature of blood flow was regarded as secondary in the era of Poiseuille flow, or as wave reflections were regarded as secondary in the early days of pulsatile flow.
- Secondary issues were actually seen as "aberrations" in the design of the cardiovascular system, even with some negative connotation. Indeed, to this day we continue to think of wave reflections in this negative way, as an added "cost" to the system, or an "afterthought" in the design of the system. Wave reflections have yet to be embraced in our minds as an intrinsic part of the system, not secondary or negative in any way, but simply an integral part of the system, essential as any other in the efficient working of the system, and as such they may actually be used by the system to some advantage.

- In the modern era our commitment is to the Navier-Stokes equations and to our computer power to solve them under every conceivable condition, thereby making it possible to simulate most of the blood flow phenomena which we observe within the human body. As in previous eras we feel that we have the subject essentially in hand, though, again, with the exception perhaps of some "secondary" issues. In what follows we consider one such issue, namely that of control and regulation of fluid flow within the human body, more specifically the control and regulation of blood flow.
- While blood flow is certainly governed by the Navier-Stokes equations, it is at no time governed by the Navier-Stokes equations alone. Blood flow is part of a closed-loop control system, or more precisely it is under a hierarchy of local and global control loops in which the primary object is control of pressure and flow rate. While the Navier-Stokes equations govern the relation between pressure and flow, it is an open-loop relation which they govern. In that relation, a change in pressure produces a change in flow and there the matter ends. There is no provision for any feedback from the change in flow back to the change in pressure, yet in the physiological system feedback loops are ever present.
- Feedback loops in the cardiovascular system are mediated by a number of neural and humoral control mechanisms. As in a man-made control system, each of these loops is associated with a controlled variable, a sensor, and a controller that is able to affect the value of the controlled variable. While a great deal has been learned about certain elements of these control systems, much is still unknown. Typically, a controlled variable such as blood pressure or heart rate is identified and some of the sensors associated with it are located, but the control system loop governing that variable can rarely be sketched out or isolated in complete form. This is not only because some of the sensors or pathways are unknown but because in most if not all cases the loops involved, rather than being distinct closed loops, may typically be interlaced with each other. The control systems for heart rate and stroke volume, for example, may have both local and central loops and these loops may interlace with each other as well as with others associated perhaps with the control of blood pressure and pCO_2.
- To deal with blood flow in its full breadth, therefore, it is necessary to integrate hemodynamics with the full dynamics of the system because the integrity of blood flow in any organ or system depends not only on hemodynamics within the blood vessels of that organ or system but also on the full dynamics of the system as a whole. The first of these is governed by the Navier-Stokes equations, the second by the control loops and mechanisms associated with blood flow in that particular organ. Put another way, while satisfying the Navier-Stokes equations is a necessary condition for an accurate description of blood flow within the

vascular bed of an organ, it is not a sufficient condition for a description of the overall dynamics of that flow in the context of control and regulation, or in the context of what is the ultimate objective of blood flow in that organ, namely cell perfusion.

⊙ Arrhythmia, in all its forms, is a familiar example of this dichotomy. It affects the harmony between the pumping rhythm of the heart and the compliance of the aorta, thereby affecting cardiac output and the dynamics of the systemic circulation. In the heart, similarly, a disruption in the harmony between the pulsatile rhythm of the input pressure driving the flow into the coronary circulation, the pulsatile rhythm of the contracting cardiac muscle and its effect on coronary vessels imbedded within it, and the combination of wave propagation and wave reflections within the coronary vascular tree, will lead to a failure of the dynamics of the coronary circulation while the flow within individual vessels is continuing to satisfy the Navier-Stokes equations. A much simplified version of this scenario is shown in Fig. 1 where the dynamics of the flow in a vascular tree is disturbed by vasodilatation or vasoconstriction at the peripheral levels of the tree. The result is a change in the distribution of admittance along the tree and therefore a change in the dynamics of blood flow through the tree. The flow in individual vessel segments remains orderly in the sense that it continues to satisfy the Navier-Stokes equations. Thus, while the overall dynamics of the flow have been affected, hemodynamics within the tree have not.

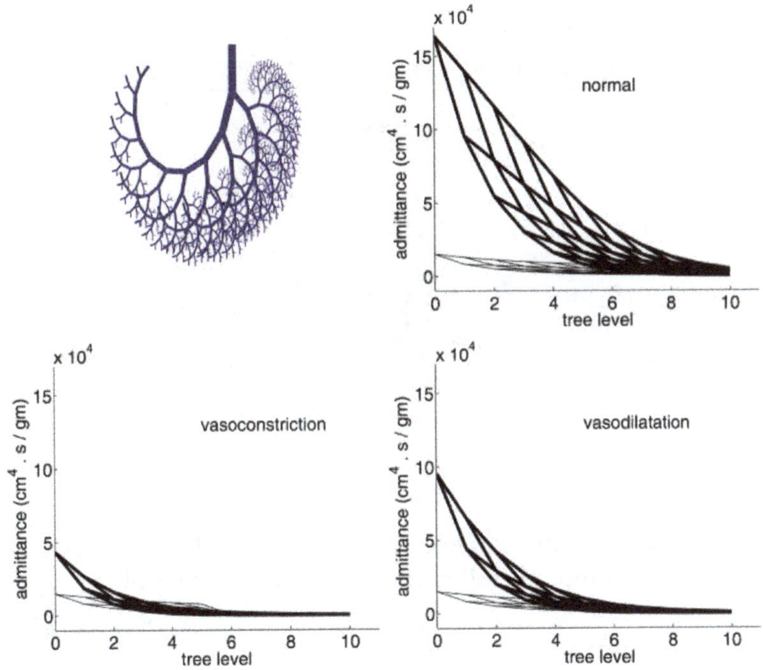

Fig. 1: Admittance distribution in an 11-level tree model (top) and how it is affected by vasodilatation or vasoconstriction in the peripheral vessels (bottom).

- In sudden cardiac death there is a precipitous fall in cardiac output to levels that can no longer sustain cerebral or cardiac function. By definition, the timing of this failure cannot be attributed to any prevailing coronary artery disease, so the disease if present is seen only as a participant in but not the immediate cause of the event. According to currently held view, coronary artery disease is seen as a precursor to conditions of myocardial ischemia and infarction, leading to a disruption in heart rhythm, fall in cardiac output, fall in coronary blood supply and finally heart failure, as illustrated in Fig. 2. A difficulty with this course of events is that it proposes that a fall in cardiac output is followed by a fall in coronary blood flow.
- The mechanism for this sequence is not clear if indeed possible, because if the fall in cardiac output leads to a drop in aortic pressure, it is well known that autoregulation of coronary blood flow ensures that this does not lead immediately to a fall in coronary blood flow. Because of autoregulation, a fall in cardiac output is more likely to be the *consequence* rather than the *cause* of a fall in coronary blood flow, the latter being precipitated by a disruption in the dynamics of the coronary circulation produced by coronary artery disease, cardiac damage, or a disruption in neural activity. This alternative course of events is illustrated in Fig. 3.

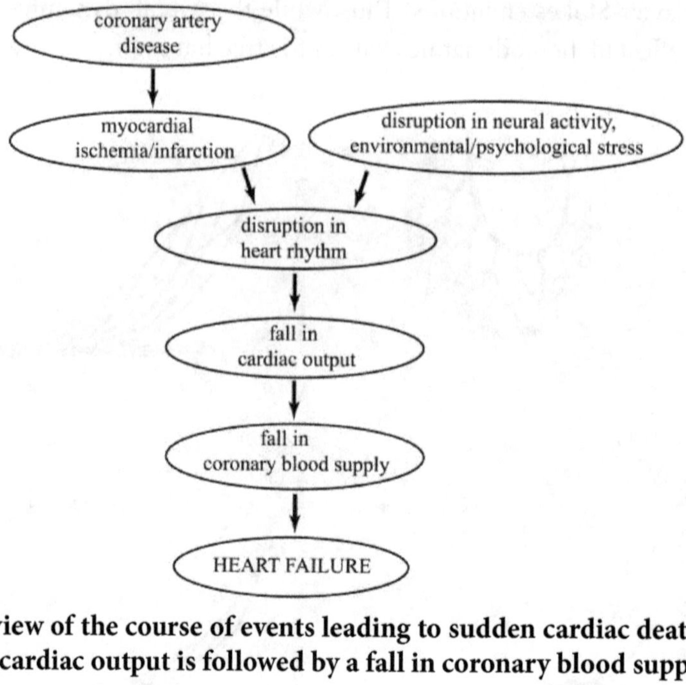

Fig. 2: Current view of the course of events leading to sudden cardiac death in which a fall in cardiac output is followed by a fall in coronary blood supply.

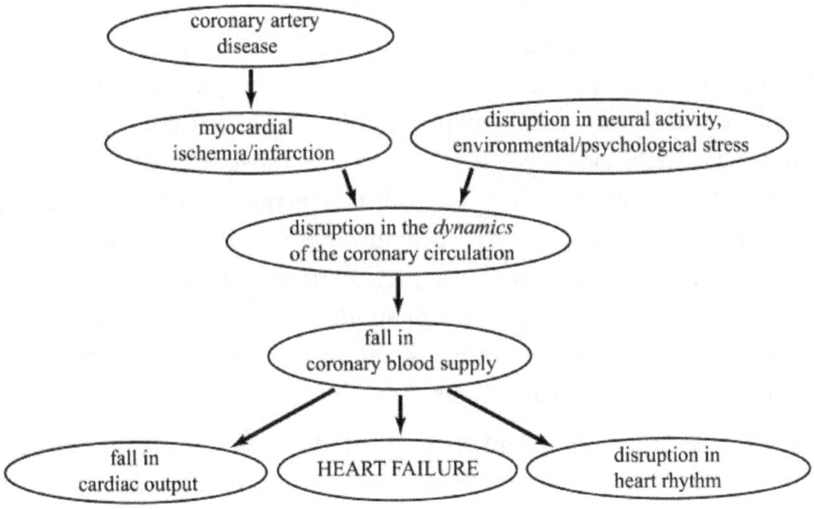

Fig. 3: A more likely course of events leading to sudden cardiac death in which the fall in cardiac output is the consequence rather than the cause of a fall in coronary blood supply, and the primary trigger is a disruption in the dynamics of the coronary circulation.

- Thus, the integrity of blood flow in its widest sense depends in the first place on hemodynamics within the vascular bed but ultimately on the dynamics of the system as a whole, and in the same way that the first of these may be disrupted by a vascular pathology, the second may be disrupted by what may be termed a "dynamic pathology". Arrhythmia is a clear example of a dynamic pathology while sudden cardiac death is, ultimately, the result of a dynamic pathology.
- A dynamic pathology is not unlike a structural or functional pathology in a tissue or organ, but it deals specifically with *dynamics*. Furthermore, it deals with the dynamics of an integrated system as a whole, such as the dynamics of the systemic or of the coronary circulation, including all the feedback loops involved in the control and regulation of the system in each case. Thus, an atherosclerotic lesion in a coronary artery is a vascular pathology, but the ultimate significance of this lesion lies only in the dynamic pathology which it produces, more specifically in the way the lesion affects the dynamics of the coronary circulation as a whole, and in particular in the way it affects the ability of the coronary circulation to provide the heart with sufficient blood supply. If the occlusive progression of the lesion is sufficiently slow so that the occlusion is compensated for by vascular restructuring, for example, the dynamic pathology may be minimal or nonexistent. Thus, the presence of vascular pathology does not necessarily imply the presence of dynamic pathology, but the reverse is not necessarily true, namely the absence of vascular pathology does not necessarily imply the absence of dynamic pathology.

- Perhaps the most celebrated example of a dynamic pathology is the phenomenon of a "broken heart" which has recently made the transition from the realm of folklore to that of medicine where it is now accepted as a legitimate clinical syndrome. Following a multi-patient clinical study, **Wittstein et al** reported that "Emotional stress can precipitate severe, reversible left ventricular dysfunction in patients without coronary disease. Exaggerated sympathetic stimulation is probably central to the cause of this syndrome." Because of the absence of coronary artery disease in these patient, the precipitating pathology in this case is again a dynamic pathology as in the case of sudden cardiac death, namely a disruption in the dynamics of the coronary circulation followed by a fall in coronary blood supply and a fall in cardiac output.
- One of the most important aspects of dynamic pathologies is that unlike structural pathologies they do not leave a footprint after they have resolved or produced their damage. Following sudden cardiac death, indeed following any death attributed to heart disease, only structural pathologies can usually be found and they become the focus of attention. Any dynamic pathologies involved are not in evidence because they are no longer at play. This is the reason for which the broken heart syndrome has taken so long to be accepted by the medical community.
- The concept of a dynamic pathology, which is a failure in the overall dynamics of a dynamic system, is well familiar in science and engineering but is rather foreign in the clinical setting where the focus is on the more tangible hemodynamics. The broken heart syndrome, with the distinction of being transient and reversible, offers the opportunity to make the dynamic pathology more tangible, that is it makes it possible to detect the dynamic pathology while it is at play and in pure form, because the syndrome is not associated with coronary artery disease.

12

SPATIAL ORGANIZATION AND BIOCHEMISTRY IN BUILDING BLOCKS OF CELLULAR SIGNALLING PATHWAYS

12. 1. INTRODUCTION: AN OVERVIEW

Biochemical pathways and networks are central to cellular information processing. While a broad range of studies have dissected multiple aspects of information processing in biochemical pathways, the effect of spatial organization remains much less understood. It is clear that space is central to intracellular organization, plays important roles in cellular information processing and has been exploited in evolution; additionally, it is being increasingly exploited in synthetic biology through the development of artificial compartments, in a variety of ways. In this paper, we dissect different aspects of the interplay between spatial organization and biochemical pathways, by focusing on basic building blocks of these pathways: covalent modification cycles and two-component systems, with enzymes which may be monofunctional or bifunctional. Our analysis of spatial organization is performed by examining a range of 'spatial designs': patterns of localization or non-localization of enzymes/substrates, theoretically and computationally..

- A hallmark of all cells is their ability to perform a range of complex information processing tasks. This is a key feature which differentiates them from non-living objects of a similar scale. This is critical for the maintenance of life and conditions in a cell and for a flexible response to the environment. This is achieved via a combination of biochemical modification pathways and networks, as well as gene regulation.
- A broad range of research has focused on elucidating information processing in biochemical networks, involving a combination of approaches. One basic approach involves using experiments, mathematical

modelling or a combination thereof to understand the behaviour of signalling pathways in specific contexts. Associated with this is also the isolation of basic themes in signalling and information processing, which transcend the individual context. Another approach which has emerged from these studies is a broader multifaceted approach to understanding multiple aspects of the behaviour and functioning of these networks. This includes non-trivial information processing characteristics (adaptation, oscillations, multistability), the roles of ingredients such as feedback, stochasticity, complex post-translational modification (e.g. multisite phosphorylation), and the robustness of networks to achieve certain tasks (*e.g.*). Taken together these approaches span an entire range from the detailed quantitative investigation of specific systems to the investigation of basic elements encountered in multiple systems to broader theoretical investigations of basic underpinning features, and have significantly advanced our understanding of these systems.

- Biochemical signalling networks in both bacteria and eukaryotes consist of basic building blocks. In this regard, bacteria present key distinguishing ingredients, one feature being the presence of two-component systems (TCSs). They typically consist of a membrane-bound histidine kinase, which is an environmental sensor, and a response regulator, which serves as the output. Their widespread presence is borne out by the fact that they (histidine kinase, response regulator) are among the largest gene families in bacteria. One of the main mechanisms of output regulation (phosphorylation) is via a phosphorelay mechanism. TCSs and phosphorelays form building blocks of a wide range of regulatory processes (such as osmotic regulation, nitrogen fixation, virulence and adaptive responses) that allow bacterial cells to function and survive. One of the characteristics of many enzymes in bacteria (in particular histidine kinases) is that they are bifunctional, functioning as kinases or phosphatases. Interestingly, both TCSs and bifunctional enzymes are also encountered in eukaryotic systems: TCSs in fungi and bifunctional enzymes in metabolic and biosynthesis pathways. Bifunctional enzymes play important roles in multiple contexts and possess distinct information processing characteristics, and for both reasons they have been the focus of a range of experimental and theoretical investigations.
- The reversible modification of substrates by kinases and phosphatases (covalent modification cycle; CMC) is a basic building block for signal transduction in eukaryotes. This basic unit is built upon in many ways; for instance, by having a cascade of modifications where the modified protein at one stage acts as an enzyme for the next stage, or by having successive modifications of substrate (e.g. multisite phosphorylation) for regulating substrate function. Other elements which are naturally incorporated within this set-up include feedback regulation

(both positive and negative). In general, there is a much more intricate and sophisticated organization of biochemical networks in eukaryotes.

- One of the fundamental aspects of cellular information processing is the presence of spatial organization (e.g.). Indeed, the presence of basic spatial 'compartments' in the form of membranes, nucleus, the cytoplasm and organelles is fundamental to eukaryotes. Furthermore, it is evident that the spatial organization is intertwined with the biochemical modification pathway, at multiple levels. This is clearly seen in many cases and stems from the fact that signalling responses occur in locations different from where they were initiated. Additionally, spatial organization at specific levels (e.g. the membrane, through the presence of membrane regions such as lipid rafts) can serve as a basic way to route signals and organize information flow. Furthermore the presence of compartmentalization of various types, and the fact that these have been actively exploited in evolution, is testament to this.
- Spatial proteomics is emerging as a key large-scale investigative tool in this regard. Bacterial systems (previously long assumed to be essentially well mixed) themselves exhibit clear hallmarks of spatial organization, even though the more elaborate machinery associated with spatial organization in eukaryotes is absent. This is seen both in localization patterns of species, which may change dynamically in the course of the life of a cell, as well as in spatiotemporal self-organized patterns, such as the Min waves in Escherichia coli. These spatial organization patterns may act at the level of enzymes and substrates, and may involve monofunctional or bifunctional enzymes. An example of this is seen in the model system Caulobacter crescentus. Spatial control mechanisms regulating fundamental processes such as the cell cycle, cellular differentiation and polarity have been studied in this organism. Caulobacter cells use multiple spatial control mechanisms: for example, the dynamic localization of proteins at cell poles and the spatial distribution of signalling proteins play an important role during its asymmetric development. Furthermore the choreographed temporal and spatial control of multiple bifunctional enzyme modules (enzymes, substrates) is at the heart of cell-cycle regulation and the transition between different phases. Disrupting enzyme activity as well as the localization (or spatial activity) disrupts a cell's development (it fails to form an organelle at one of its poles). This is an example of the combination of spatial organization and biochemical building blocks serving as the basic modes of regulating a central biological process.
- An additional strong impetus to investigating spatial organization comes from synthetic biology. Here bottom-up approaches have been used to engineer compartmentalization and localization, and this has been used in the context of genetic circuits, biochemical pathways, as well as hybrid communicating systems

(e.g. cell population communicating with a circuit). Furthermore elaborate spatial compartment networks are being constructed, with locations of compartments capable of being dynamically regulated. Multiple aspects of the compartments, including the transport and localization of individual components within them, as well as their shape, can be manipulated chemically. Spatial organization thus implemented is a vital component of artificial cells, with a number of potential synthetic biology applications, that also offers insights into the functioning of natural systems.

- While some aspects of spatial effects in biochemical pathways have been studied experimentally and (to a lesser extent) theoretically, a broad understanding and synthesis of different facets of the biochemistry space interplay, for use in systems and synthetic biology, starting from basic building blocks, is largely absent.
- In this paper, we investigate multiple basic aspects of the interplay between chemical modification pathways and spatial organization. To do this, we focus on basic building blocks of chemical modification—TCSs and CMCs—and examine both monofunctional and bifunctional enzymes. This encompasses a range of broadly encountered basic biochemical building blocks. We develop a systems framework to examine the role of spatial organization by focusing on multiple spatial 'designs' (organization, involving either diffusion or localization of species). In all cases, a basic reference point is all species localized in the same location, in which case the modification can be described in purely kinetic terms. Incorporating different spatial designs allows us to explore the impact of spatial organization as a perturbation of the co-localized case. We focus on basic information processing characteristics of the chemical modification modules (ultrasensitivity, threshold behaviour, concentration robustness, bistability) and how they are altered by spatial organization. We also examine how spatial organization may be used to directly manipulate the enzyme activity. We examine further aspects of the chemistry–space interplay by dissecting alterations in spatial organization (either during a cell life cycle or through evolution), and the effect of alterations of biochemistry, when spatial organization is present.
- There are many reasons for the relevance of this study and the way it is approached. Understanding the effects of spatial organization requires a synthesis of the basic effects of space impacting different modification pathways. By focusing on the basic building blocks of chemical pathways, on one hand, and basic spatial designs, on the other, we can create a broad platform to obtain a range of insights applicable in multiple contexts. Importantly, many instances of non-trivial organization of pathways can be broken down into spatial organization of particular modification stages, which is exactly what our systems analysis addresses. This analysis is also relevant to 'spatial mutations' (experimentally

studied) resulting from mislocalization of species in natural systems. The effect on particular information processing behaviour is relevant because these have all been postulated or demonstrated to be of relevance in cells, transcending any individual context. Many aspects of cellular information processing networks (behaviour, robustness, evolution) have been studied in temporal terms with a synthesis which neglects a basic aspect of cellular organization. Revealing the effect of the spatial dimension to cellular information processing can highlight when it may be important and change our understanding of these aspects: some of this is already seen in a study of basic building blocks. Finally, there is considerable interest in compartmentalizing pathways and manipulating the localization and transport of species in synthetic biology, and these are also vital aspects of artificial cells. The broad growth of systems and synthetic biology has spawned an area which aims to bridge the living and the non-living. Spatial organization of chemistry is one of the important themes at the junction of the living and the non-living, and our approach is relevant here as well.

12. 2 MODELS

Our goal is to dissect the interplay between basic biochemical modification cycles and their spatial organization. Consequently, our approach is to employ basic and fairly generic descriptions of the biochemistry (and additionally examine relevant augmentations) as well as spatial organization to isolate key patterns which emerge from the most basic aspects of the interplay

2.1. Kinetic description of the enzymatic modification modules

Our study of biochemical modification cycles involves variations along two independent axes: CMCs and TCS biochemistry, on one hand, and monofunctional and bifunctional enzymes, on the other (figure 1). Taken together, these represent some of the most basic biochemical modification scenarios which are encountered. We study the effect of spatial organization for all these modules. The kinetic models of these modification cycles employ widely used descriptions of basic enzymatic modification steps. A single covalent modification of a substrate by an enzyme is modelled in a standard way with the enzyme reversibly binding to the substrate to create a complex leading to an irreversible substrate modification and release of enzyme. For reversible modification of substrates by kinases and phosphatases, this basic description results in a standard model of the CMC (see figure 1a; in all cases below X and X* denote the unmodified and modified forms of the substrate; kinase and phosphatase are denoted by K and P, respectively, for the case of monofunctional enzymes). A TCS on the other hand involves phosphotransfer in the forward direction (X to X* conversion). For a TCS (with monofunctional enzymes), the main difference in the model (figure 1b) is that the modification of a substrate is associated with a transfer of a phosphate group of the enzyme, which then becomes inactivated. Thus, here the kinase is present in two (reversibly interconverting) forms, inactive (K) and active (K*).

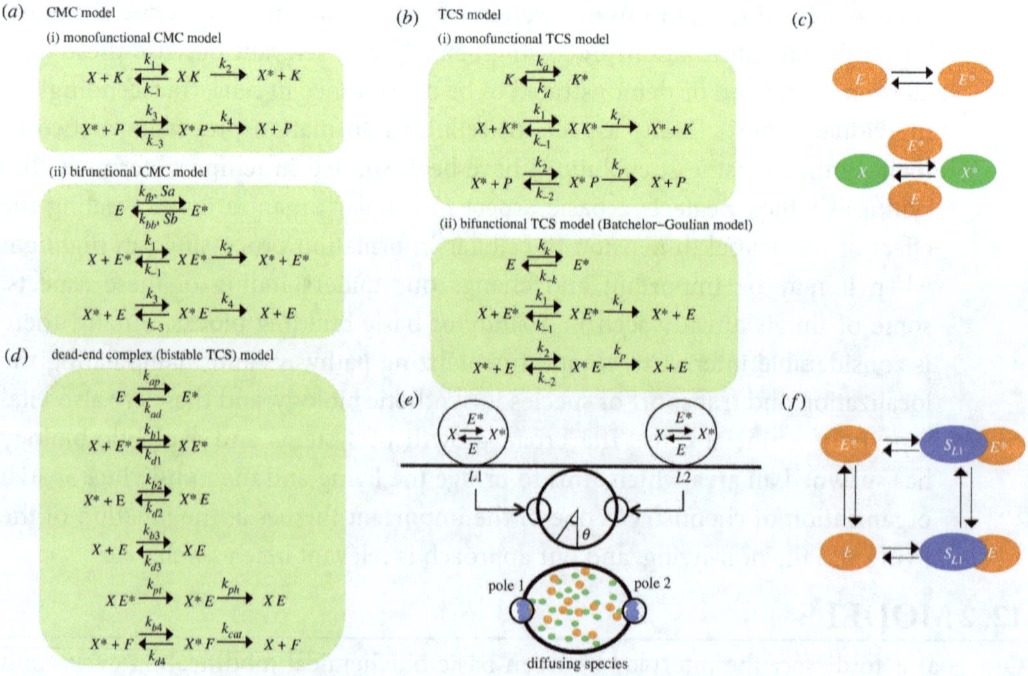

Fig. 1: Schematic of models employed in the paper.

X and X* represent unphosphorylated and phosphorylated substrates, respectively. (a) Covalent modification cycle (CMC), mediated by monofunctional (kinase K and phosphatase P) and bifunctional enzymes (kinase form E* and phosphatase form E). (b) Two-component system (TCS) models with monofunctional and bifunctional enzymes (same notation as before, K and K* represent inactive and active forms of the kinase; see text). (c) A schematic of a bifunctional enzyme mediating forward and reverse modifications of a substrate. The interconversion between the enzyme forms in general may be mediated by signals (e.g. Sa and Sb in the bifunctional model in (a)). (d) A dead-end complex model of a bifunctional TCS which gives rise to bistability: this model is an augmentation of the basic bifunctional TCS, incorporating an additional phosphatase F and a dead-end complex XE (based on Igoshin et al. [66]). (e) A schematic of a sample spatial organization: bifunctional enzyme localized at two regions (substrates are non-localized and diffusing). θ denotes the spatial co-ordinate here and in subsequent figures (see text). (f) A schematic model of active localization of two enzyme forms (E, E*) at a specific location (see text for details), incorporating the interconversion between the free forms (E*, E), and the localized forms (SL1E*, SL1E). SL1 denotes the localizing molecule, and its attachment to the enzyme form localizes it.

An important basic variation of biochemistry is the presence of bifunctional enzymes, which can act as both kinase and phosphatase. In all cases we assume two distinct enzymatic forms (kinase form E* and phosphatase form E), which are associated with some basal interconversion. For bifunctional enzymes as part of TCSs (figure 1b), we use the description of the TCS modification cycle, with the only differences being that there is a pair of interconversion reactions between the enzyme forms reflecting this and that there is no inactive form of kinase explicitly described. This is the model employed by Batchelor & Goulian. We also study bifunctional enzymes within a CMC (figure 1a). Here again, we augment the basic CMC model to incorporate interconversion of enzyme forms (discussed further below). This model, which again incorporates most aspects of bifunctionality, with covalent modification, allows us to establish correspondences with the bifunctional TCS model .

12.2. 1 Model Variations and Augmentations

We examine some basic augmentations to our models, because they lead to distinct patterns of information processing of interest. In the case of bifunctional enzymes in a CMC, we examine the possibility of feedback (both linear and nonlinear) of a modified substrate (directly or indirectly) affecting the balance of enzyme forms (such feedback is consistent with the biochemistry of modification). The nonlinear feedback mechanism results in bistability, a feature which is also seen in the linear feedback case with extra enzyme–substrate interactions. Another variation of the basic model is the one studied, which uses a signal to tune the balance of kinase and phosphatase, explicitly incorporating signal binding and uptake. This allows for ultrasensitive switch-like behaviour. The possibility of a 'dead-end complex' (an augmentation to the basic model; see figure 1d) generating bistability has been studied in both bifunctional and monofunctional TCSs. We perform a focused analysis of these cases. The kinetic (ODE) models are presented in electronic supplementary material, §§2.1 and 2.2.

12.2.2. Comments on Kinetic Models

A few points regarding the bifunctional enzyme models we employ are worth emphasizing. Firstly, in these models, the enzyme forms are all active (models with inactive enzyme forms have also been studied but are not presented here). Further, we ignore the possibility of ternary or higher complexes, which may be associated with substrates binding to enzyme forms at different locations, or dimerization. Some of these aspects have been studied in the literature. Given our focus on spatial perturbations, we kept the models as simple as possible to ascertain a fundamental understanding (which arises from the most basic aspects of bifunctionality) without invoking such additional features. Secondly, there are models of bifunctional enzymes where multiple kinetic steps are described in great detail. Finally, there could be mechanistic variations in specific contexts and these are outside the scope of the current study. The models employed (which use concise descriptions of the key steps) are sufficient for the nature of the questions we address, and the analysis and insights presented here (which transparently emerge from the underlying assumptions, sometimes seen by contrasting with a closely related model with a basic difference) can serve as a platform for examining these auxiliary factors in a subsequent study.

12.2.3 Spatial Organization

The ODE models depict kinetic modification cycles, with enzyme and substrate in the same location and no species diffusing. Our focus in spatial organization is on the effect of localization of specific entities in certain regions and how that impacts the overall modification chemistry. Our modelling framework to achieve this involves an explicit spatial description, using PDEs. At the outset, this brings a set of new ingredients alongside the kinetics: the description of localization and diffusive transport of species, in a spatial domain with specified boundaries. The models involve the localization of entities (enzymes or substrates) to certain regions of the spatial domain, which we refer to as compartments. These regions are representative of compartments which may or may not be bounded by compartment boundaries. While our model incorporates these regions of localization (and not any compartment boundaries explicitly), many essential insights carry through to the case of compartments with boundaries (discussed in the electronic supplementary material).

The basic effects of localization are studied by the imposition of different patterns of localization (discussed in detail below) of enzymes and/or substrates in one or two regions (representing poles of a cell, for instance) within the spatial domain in one dimension (figure 1e). Other species diffuse freely in the intervening space, and this is described by Fickian diffusion. In all cases, the localization is realized by a combination of the initial conditions (ensuring that a species is in a given location) and non-diffusibility. As such, our depiction of the spatial localization of entities is relevant both in natural biology (for example bacteria, through localization at specific regions such as cell poles, or eukaryotes, where the model could depict a cross section of a cell, or the membrane of a cell, with localization in compartments) and in synthetic cell-free systems, with compartmentalization of reactions.

The study in a one-dimensional spatial domain is sufficient to understand the basic effects of spatial organization (which ultimately stems from different patterns of localization/non-localization of species, coupled to chemical reactions). This serves as a basis for examining specific spatial aspects (e.g. geometry) in subsequent models in two or three dimensions, while still providing important insights relevant to those cases. We have considered both no-flux boundary conditions (to describe the cross section of a cell for instance) and periodic boundary conditions (to describe the membrane of a cell in two dimensions for instance, as depicted in figure 1e) in our PDE models. The results from these choices of boundary conditions are commensurate: periodic boundary conditions involving either a single compartment (pole) or two compartments (poles) symmetrically localized is exactly equivalent to no-flux boundary conditions in half the domain size. Further, the essential qualitative conclusions, which are the focal point throughout the paper, do not depend on this choice of boundary conditions (as can be seen by an analytical study of the models). θ denotes a (dimensionless) distance co-ordinate in all cases and is scaled from 0 to 2π to serve as a common reference, and our computational results focus on the concentration of the modified and unmodified substrates X^* and X as a function of the spatial co-ordinate or other parameters.

12.2.4 Active Localization of Species

While localization of species can be described by localizing the initial condition and having it non-diffusible, we also need to consider ways in which localization is imposed on diffusing species. This is the key ingredient for cells to impose localization dynamically. To describe this, the localization is realized by having the species binding to an immobile species at a location (reversibly) but the bound species being non-diffusible. In the case of bifunctional enzymes, we can consider (i) equal localization (equal binding/unbinding rates of each enzyme form at the location) and (ii) a bias in localization, where one form of the enzyme is localized more strongly than the other owing to differences in binding (or unbinding) rates. These are easily described in the basic model (see electronic supplementary material, §2.3) and transparently analysed analytically. Within this model one can examine variations such as the presence/absence of (i) interconversion of localized forms and (ii) enzymatic activity of localized enzymes.

12.2.5 Parameters

We now discuss the role of parameters in relation to the goal of our study. The parameters belong to two classes: kinetic parameters and spatial parameters (diffusivity, domain size), which we discuss in turn. It is clear that even the simple models have a number of kinetic parameters. We aim to understand basic qualitative aspects of the effects of spatial organization on biochemical modification. This is because the qualitative characteristics are the ones most easy to discern experimentally, and in general have the greatest impact in terms of biological outcome (and are consequently the first point of interest, from a modelling viewpoint). We further note the following. (i) Many of the basic qualitative effects of spatial organization stem from basic considerations (e.g. dilution, nature of localization patterns) and are seen across multiple kinetic parameter sets. Furthermore, in many cases this is directly understood through analytical work. As an example, for the case of species localized in one location, with a single species diffusing throughout the domain, the impact of the spatial organization on the modification kinetics, at steady state, is only through the dilution owing to species spreading outside the region of localization. This can be characterized analytically and computationally, revealing a basic underlying trend. While the quantitative effect depends on both kinetic parameters and the size of the domain, the underlying qualitative insight follows from basic analysis. (ii) The kinetic parameter regime is studied inasmuch as it represents a distinct basal information processing behaviour, which is another explicit vantage point of analysis. This allows us to examine the effect of spatial organization on this type of reference behaviour, and pin down the reasons for any qualitative changes. Taken together we identify both basic effects of spatial organization as well as those on particular reference kinetic characteristics. (iii)

Finally, in many cases (where relevant) we observe similar trends across different models of biochemical modification, attesting to the essential robustness of the conclusions. With regard to spatial parameters, we use a sample compartment size (about 10–20% of the entire domain) and have studied a range of diffusivities of species. With regard to kinetic parameters, we employ parameter sets drawn from the literature, along with parameter sets in specific cases corresponding to particular information processing characteristics (based on kinetic analysis in the literature). For instance, for basic analysis of both CMCs and TCSs with monofunctional and bifunctional enzymes, we use modification parameters with comparable ranges of binding, unbinding and catalytic constants (ensuring that the system is not operating in a limiting regime), similar to parameters and in ranges used in the literature (and have also examined other parameter sets underpinning the study). Special parametric regimes are analysed separately (analytically) if needed.

Many aspects of qualitative behaviour/trends introduced by spatial organization are established analytically. In other cases, computational results directly support conclusions which we draw (e.g. the possibility of a certain behaviour occurring). All in all, we employ generic kinetic models, along with focused augmentations (enabling specific behaviour), and isolate basic aspects of their interplay with spatial organization. These arise from basic ingredients of the kinetic modification (e.g. the bifunctionality of enzymes) and the most basic aspects of spatial organization and can be transparently revealed.

The dimensionless equations are analysed computationally via simulations (solver ode15s) in Matlab; the kinetic schemes in Matlab are cross-validated by comparing with simulations in COPASI (where models for the kinetic schemes are automatically generated). In the case of spatial organization, a standard finite difference discretization of the PDE is employed (and results checked by doubling discretization points), which converts the PDE into a set of ODEs. This is complemented by bifurcation analysis of the steady states of ODEs for the kinetics (and, in certain cases, bifurcation analysis of the steady state of the PDE, which can be mapped onto the steady state of the ODE with a modified conservation condition: in these cases, as discussed analytically in electronic supplementary material, §3, the steady state of the PDE satisfies the kinetic ODE equations, and the only effect of spatial organization is a dilution effect which manifests itself in the conservation condition). Further details are presented in electronic supplementary material, §2.4.

12.3 TRANSITIONS IN BIOCHEMISTRY AND SPATIAL ORGANIZATION

We discuss basic aspects of spatial organization of monofunctional and bifunctional enzymes. The extent to which enzyme activity can be manipulated spatially (by localization in particular) is examined next, and finally we examine how transitions in biochemistry and spatial organization impact pathway behaviour.

12.3.1 Monofunctional Enzymes with Spatial Organization

We first examine a monofunctional enzyme pair (kinase, phosphatase) reversibly modifying a substrate: this modification being described by a CMC. A structured approach to the analysis of spatial organization facilitates the dissection of the interplay between space and biochemistry.

When the enzymes and substrates are co-localized in one location (which we refer to as a compartment), the behaviour of the modification system can be described via its kinetics. Spatial organization enters through altered localization patterns of enzymes and/or substrates, which we henceforth refer to as spatial designs (e.g. figure 2). We use the co-localized scenario as a reference point and examine the effect of spatial designs relative to this. Relative to this basic co-localized design, there are multiple possible alterations of localization of enzyme and substrate. The basic substrate localization alterations we examine are as follows.

- X being diffusible and exiting out of the compartment while X^* is localized (this substrate localization pattern is denoted by S_1).
- X^* doing the same, while X remains localized (denoted S_2).
- Both X and X^* being able to exit (denoted S_3).

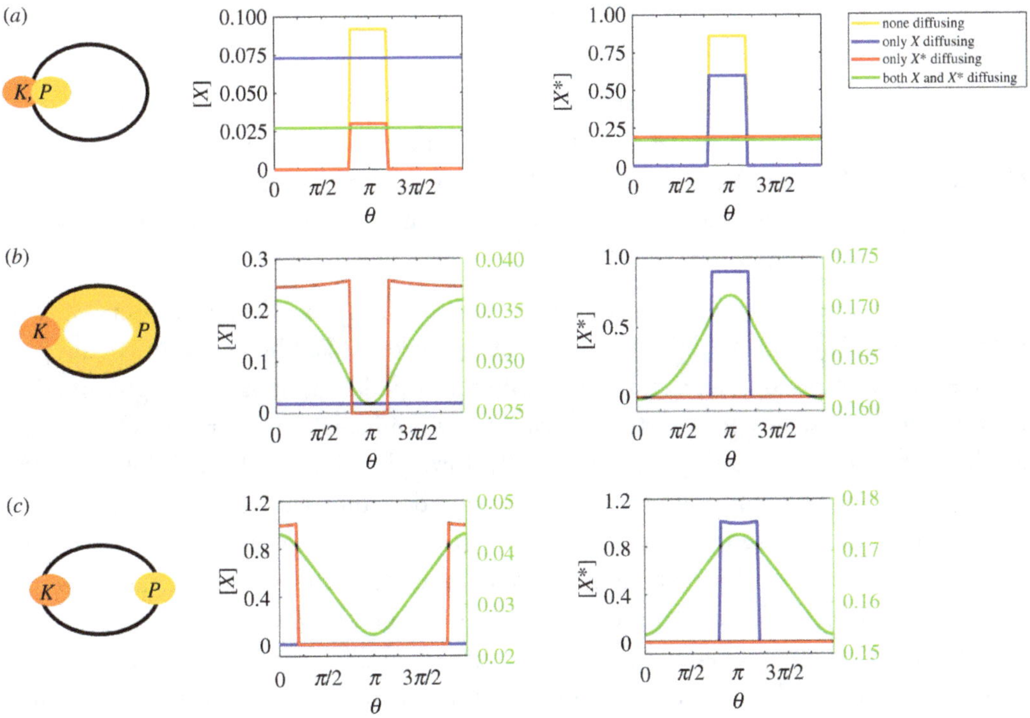

Fig. 2: Spatial designs in monofunctional CMCs.

(a–c) Three different kinds of localization of enzymes, depicted schematically in the left column: (a) K and P are localized in the same location (design E0) , (b) K is localized in one location and P is everywhere (design E2) and (c) K and P are localized in two distinct locations (design E1). In each subplot in the middle and right column, curves describe different substrate localization patterns: all species localized (in the same location as the kinase) (design S0), only X diffusing (design S1), only X* diffusing (design S2), both X and X* diffusing (design S3), respectively. θ is the spatial coordinate. This figure demonstrates that enzyme localization can help generate non-zero uniform (red and green curves in (a)) and graded (green curves in (b) and (c)) X* profiles; combinations of substrate and enzyme localization patterns (blue curve in (c) for X) can induce zero responses (also red curves in (b) and (c) for X*). Note that the green curves in (b) and (c) are associated with a different scale, also depicted in green.

The other important aspect is the localization of enzymes. Here we have multiple possibilities. (i) Kinase localized at one location while phosphatase is localized at another location (this enzyme localization pattern is denoted by E1). (ii) Kinase localized at one location while phosphatase is present everywhere (denoted by E2). (iii) Both enzymes present everywhere (denoted by E3).

Any of the associated spatial designs can be denoted by a pair of indices depicting the enzyme and substrate localization patterns. D00 denotes the original localization pattern, whereas Di,j denotes the localization pattern corresponding to Ei and Sj, with i, j = 0, 1, 2, 3. E0 corresponds to completely localized enzyme and S0 corresponds to completely localized substrate.

The first point to note is that transitions in localization are associated with changes in indices; furthermore, a single alteration in localization pattern is associated with the change of a single index. Transitions in localization patterns may emerge both in evolution and during the life cycle of a cell, with dynamic alterations in localization. They may also be associated with mutations.

Figure 2 depicts a snapshot of results for each of the spatial designs associated with monofunctional enzymes. All kinetic parameters are kept fixed here, and the qualitative changes introduced by (alterations of) different spatial designs are representative. Examining all these designs together reveals a few points. (i) An alteration of substrate localization affects the steady state primarily through dilution effects. Note that the extent of dilution depends on the size of the overall domain (relative to the region where other species are localized) and also which substrate is diffusing (e.g. figure 2a). (ii) The alteration of enzyme location can introduce qualitative differences. (a) One example is the transition between uniform and graded behaviour (e.g. green curves in figure 2a and 2b,c, the latter being graded). Graded behaviour occurs when the enzymes are not in the same location and both substrates diffuse. Spatial uniform (non-zero) concentration profile of the substrates (which are not localized) is seen in all other cases, with the exception of the following scenario. (b) The possibility of either modified or unmodified substrate attaining a zero steady-state concentration (e.g. red curve in figure 2b for X* and blue curve in figure 2c for X) is another example of a distinct qualitative behaviour observed. This can happen when the overall cycle of interconversion is disrupted owing to placement of enzymes, so that a particular substrate is not (or does not remain) in contact with the enzyme which acts on it. This can happen if a substrate form is modified at a particular location, but this modified form cannot diffuse/reach a location containing its associated enzyme. An echo of this is observed in compartmentalization of multisite modification and phosphotransfer cascades, but can be traced back to this basic feature.

The essential qualitative insights stem from basic aspects of the interplay of biochemistry and spatial organization as seen above. Quantitative aspects depend on parameters such as the domain size. Substrate diffusivity affects the steady-state behaviour only in the case of separated enzymes with both substrates diffusing. In this case, when the diffusion coefficient of both equally diffusing species is high (in dimensionless terms), the substrate concentration profiles are essentially uniform with levels explicitly determined by the kinetics, total enzyme/substrate amounts and domain size. In other cases, a gradient is obtained, and the profiles of species can be determined analytically in certain regimes. This can be consolidated further by mathematically mapping steady-state concentrations onto a compartmental model, highlighting the limitations of thinking about such systems in simplified compartmentalized terms.

A complementary effect of spatial organization of dilution owing to delocalization is that of elevation by localization. To illustrate this, we consider the case where kinase and phosphatase are localized in one location, with the modified substrate X^* also localized there. The presence of a non-localized unmodified substrate everywhere in the domain and diffusible now allows for an elevated X^* output (relative to unmodified species also localized, at the same total local concentration), because the substrate modification system can draw on substrate from outside the region of modification (not shown). We note here that the ODE kinetic model applies at steady state both to the case where all species are co-localized in one location and also to the case of no localization of any component, with enzymes present uniformly: in the latter case there are no gradients of any species, and the diffusion plays no role at steady state. Thus the above result could be interpreted by saying that localizing the enzymes and X^* at a given region can elevate X^* concentration levels. The effect of this elevation is seen strikingly in a case of product inhibition (figure 3c), where X^* inhibits or deactivates the active form of kinase. In this instance, we find that, by having the spatial design just discussed, the resulting elevation of output (from the red to green curve) can even completely compensate for the effect of product inhibition. This localized placement of enzymes (along with localization of X^*) leads to a higher output than a uniform global placement of enzymes and substrates. Analytical results in the electronic supplementary material, §3.1, reveal conditions under which this behaviour is observed. This is another illustration of how a spatial design (which can involve localizing X^* and enzymes) can achieve this counterintuitive result.

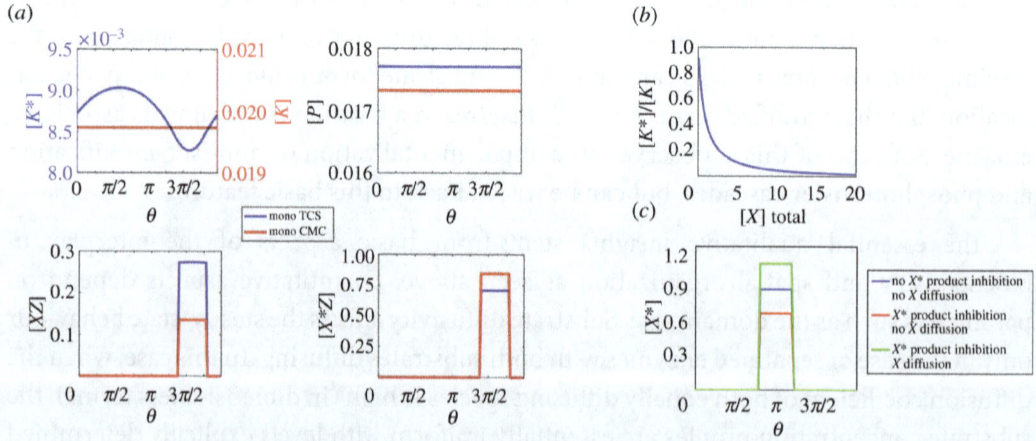

Fig. 3: Spatial retroactivity in a monofunctional TCS.

With localization of substrate and enzymes diffusing/non-localized, the concentration profiles of (active) kinase and phosphatase are different in a monofunctional TCS (blue curves in (*a*)) and CMC (red curves in (*a*)): in the TCS, the concentration profile of (active) kinase is graded (K^* denotes active kinase in the TCS) but that of phosphatase is uniform, while they are both uniform in the CMC (K is the active kinase in the CMC, where the enzyme is assumed to be constitutively active). This shows how spatial information of the substrate can be backpropagated to the (active) enzyme at steady state, in a TCS. Also shown in (*a*) for reference are the localized substrate profiles (denoted by XZ and X^*Z, which involve active substrate localization in a particular region). (*b*) How the balance of active and inactive kinase concentrations depends on the substrate in the TCS, in the kinetic (ODE) model, which is the source of the behaviour observed. (*c*) An illustration of how localizing enzymes and X^*, while allowing for X to diffuse (in the case of the monofunctional CMC), can cause an elevation of output, which can even compensate for the effect of product inhibition (green curve compared with red curve, with the green curve being above the blue curve).

12.3.2 Distinct Features Associated with Two-Component Biochemistry

The effects of spatial design studied above are relevant for modification through both CMCs and TCSs. This is because the interplay of factors responsible for the observed behaviour is similar in both cases. We now examine distinct aspects which emerge when the biochemical modification cycle is representative of a TCS: the key difference in TCSs is that the forward modification occurs via phosphotransfer and the reverse reaction by covalent modification.

The fact that substrates can affect upstream enzymes by sequestration, in a cascade (e.g. enzymatic modification cascade), is well appreciated, and is an example of what is termed 'retroactivity'. However, the spatial dimension to retroactivity is not well appreciated, and we examine one basic instance of this: the case of a TCS where the substrates are localized, but the enzymes are diffusible. In this case (figure 3a, blue curve), at steady state, the (free) phosphatase concentration profile is spatially uniform while that of the (active) kinase is not. The dichotomy in this behaviour is understood by noting that, owing to covalent modification, the net effect of substrate on the phosphatase at steady state is only a degree of sequestration. The free enzyme profiles at steady state are determined purely by diffusion, since the net effect of the substrate in consuming the enzyme at steady state is zero (see electronic supplementary material, §3.1). By contrast, because of the phosphotransfer mechanism, the free (active) kinase profile is not uniform, but instead is determined by the balance of kinase diffusion and the balance between interconversion of inactive and active kinase (which has a dependence on the substrate at steady state). This is further consolidated by a comparison with the CMC in the same setting (where the kinase is constitutively active): there, at steady state both enzyme profiles are spatially uniform (figure 3a, red curve). This illustrates when and how retroactivity can result in the backpropagation of spatial information, and how this depends on the nature of the chemical modification mechanism.

12.4 BIFUNCTIONAL ENZYMES WITH SPATIAL ORGANIZATION

We now examine bifunctional enzymes. As discussed in the Introduction, bifunctional enzymes and their substrates may be subject to both temporal and spatial regulation. As an example, in the cell cycle in *C. crescentus*, multiple scenarios involving the localization of enzyme and substrate are observed (this is true for each of two interacting bifunctional enzyme modules). Similar scenarios could also emerge through mutations which affect the localization of enzyme or substrate.

Having studied the basic aspects of spatial organization in monofunctional enzyme modification cycles, here we primarily focus on those aspects which are particularly related, directly or indirectly, to bifunctional enzymes: both distinct characteristic behaviours of modification through bifunctional enzymes and other interactions which may naturally be relevant here. We examine both TCSs and CMCs, identifying both similarities and subtle differences between the two.

In terms of characteristic behaviour, we focus on the fact that bifunctional enzymes are associated with some form of absolute concentration robustness (ACR), which directly stems from the bifunctional nature of the enzyme. This has been studied both experimentally and theoretically. We also note that bifunctional enzymes may be associated with monostable or bistable switch-like behaviour, as discussed in the literature, both through general considerations and in specific cellular contexts. Finally, bifunctional enzymes may be subject to feedback from the substrate, which alters the balance between enzyme forms. These characteristic aspects serve as reference points from which to evaluate the effect of spatial organization. We note here that feedback and bistability are also observed in monofunctional enzyme systems, and many of the associated insights we draw are also relevant to monofunctional enzymes. This emerges transparently from our analysis and is due to the fact that, in these instances, key dominant factors have similar consequences in both these systems.

12.4.1 Spatial Designs

We first comment on the effect of different spatial designs of bifunctional enzymes and their substrates (figure 4). In contrast to monofunctional enzymes (depicted in figure 2), we note that localization of bifunctional enzymes at two locations does not result in a graded response (figure 4*b*), unless the balance of enzymes is different at those locations (figure 4c). Furthermore, as seen in figure 4 (in contrast to figure 2), there is no scenario (with a non-trivial balance of kinase/phosphatase forms) where the X^* concentration is zero everywhere: this is because the bifunctional nature of the enzyme guarantees local reversibility, in contrast to scenarios involving monofunctional enzymes, where that may not be guaranteed. This means that, even through spatial mutations, a scenario of zero/complete substrate modification will not be generated as long as the enzyme is not completely separated from both substrates. We now examine the effect of substrate spatial organization on different characteristic behaviours of bifunctional enzymes.

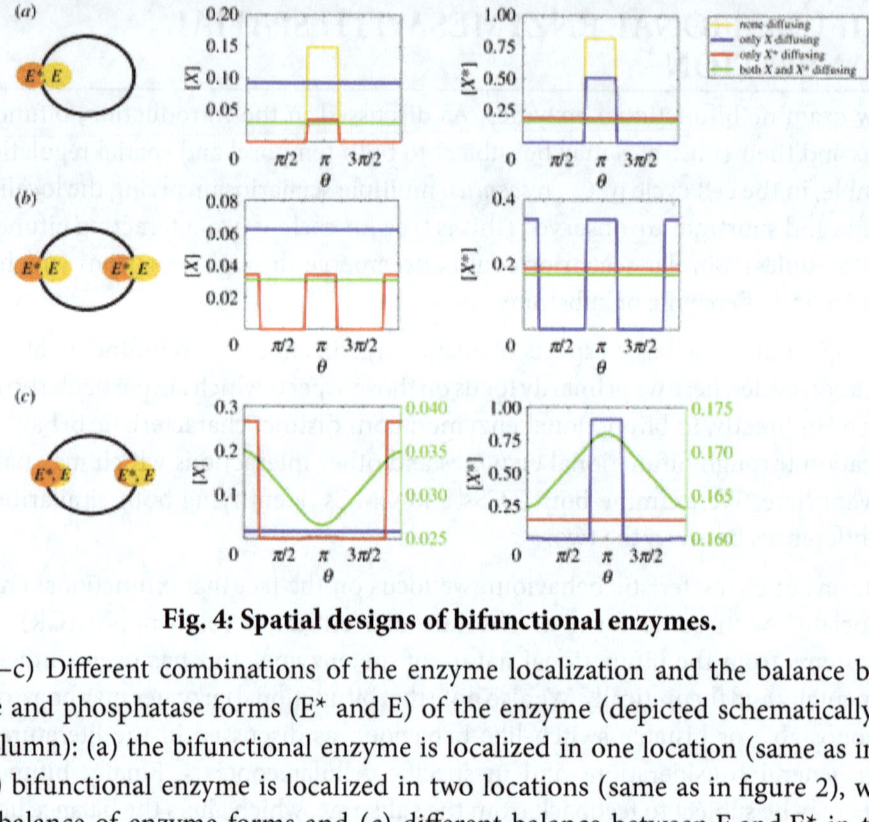

Fig. 4: Spatial designs of bifunctional enzymes.

(a–c) Different combinations of the enzyme localization and the balance between kinase and phosphatase forms (E* and E) of the enzyme (depicted schematically in the left column): (a) the bifunctional enzyme is localized in one location (same as in figure 2), (b) bifunctional enzyme is localized in two locations (same as in figure 2), with the same balance of enzyme forms and (c) different balance between E and E* in the two locations. A spatially varying balance of enzyme forms is a necessary factor for producing a graded response (green curve in (c): associated green scale). Moreover, because of the local reversibility of the biochemical cycle, resulting from the nature of the bifunctional enzyme, there is no case of zero concentration of substrate everywhere. Similar results are also found in the TCS (electronic supplementary material, figure S2).

12.4 2. Absolute Concentration Robustness

The same enzyme performing opposite modifications is at the core of different kinds of concentration robustness (in particular a weak dependence of certain outputs on total enzyme amounts). We discuss both the covalent modification and the TCS.

12.4.3 Covalent Modification

Here, for enzymatic modification of substrate in the unsaturated regime, both X^* and X^*/X are constant over a wide range of total enzyme concentrations (figure 5a). In this case, the total amount of enzyme can essentially be factored out of the steady-state kinetics. Note that, when enzymatic kinetics is not in the unsaturated regime, X^*/X remains constant, while the X^* dose-response curve is essentially flat, followed by a change (electronic supplementary material, figure S1(A)). This ratio remains constant because, at steady state, the substrate modification does not have any net impact on the balance of enzyme forms (the only impact is a sequestration of enzyme). An increased amount of enzyme, however, results in a decreased amount of available substrate $X + X^*$ (because of sequestration in complexes), which is the source of the variation of X^*. A sensitivity of X^*/X can be obtained by having feedback from substrate X^* mediating the conversion from E to E^*, but this also needs the substrate modification kinetics to be far from the unsaturated regime (electronic supplementary material, figure S1(B,C)). These aspects are all explained analytically in electronic supplementary material, §3.2.

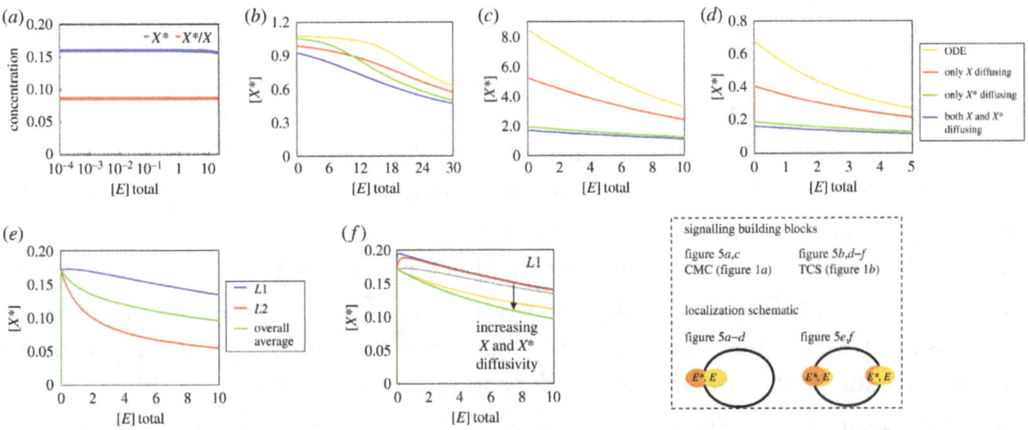

Fig. 5: The effects of spatial organization on ACR in a bifunctional CMC and TCS.

(a) A regime of concentration robustness in the CMC ODE model (unsaturated substrate modification), broadly mirroring behaviour studied in the TCS model. (b–d) When enzyme is localized in one location (same as figure 2), the outputs X* show different degrees of variation with respect to the total concentration of enzyme for different spatial designs of substrate in a bifunctional CMC (c) and TCS (b,d). Note that the kinetic regime of the CMC in (c) is different from that in (a), with no ACR in the kinetics. Panel (b) shows (depicted over a broader range) that the non-localization of substrate significantly impacts the ACR in the TCS. In the CMC (c), substrate diffusion can create a regime of ACR, arising from the interplay of spatial organization and the bifunctional nature of the enzyme (green and blue curves in (c) and electronic supplementary material, figure S3(A)). Similar behaviour can also be seen in a TCS for lower total substrate amounts (d). (e,f) In the TCS, with enzymes in multiple locations (denoted by L1 and L2), if the balance of enzyme forms is different at the two locations, the ACR (for local or average substrate concentrations) is not maintained if the total enzyme concentration is varied simultaneously in the two locations (contrast with figure 5d where the robustness behaviour is observed for the case of X, X* diffusing, for the same total substrate amount).

12.4.4 Two-Component Systems

The source of ACR has been studied in detail and others, both theoretically and experimentally. The analysis indicates that, when the total amount of substrate is (substantially) greater than that of the enzyme, a regime of concentration robustness exists. This is seen in simulations from the ODE model. This behaviour occurs when the enzyme concentration is less than or comparable to the substrate concentration.

12.4.5 Effect of Spatial Organization on Absolute Concentration Robustness

Spatial organization, as before, is introduced via an altered localization pattern of enzyme or substrate, relative to the case where all species are co-localized. We discuss this in turn. The exiting of enzyme from the compartment results in a reduced amount of enzyme available for modification. For both the CMC and the TCS, this generally enhances the effect of ACR based on the discussion above. In the case of substrate diffusion, there are three different possibilities of diffusing species: only X, only X* and both X and X*. As seen in figure 5b (for TCS), the particular zone of concentration robustness seen in the kinetics is now essentially destroyed. Furthermore, when both species diffuse, the system shows no echo of ACR. This is because the condition of the substrate concentration being much larger than that of the enzyme is now violated, owing to dilution. Now if we examine the CMC (in a kinetic regime where ACR for X* did not hold), the maintenance of the X*/X ratio continues to hold good, irrespective of which species diffuses. On the other hand, if we examine the output X* (figure 5c), we find that spatial organization in fact introduces ACR of the output, depending on which species is diffusing, and furthermore this is strongest when either X* or both X and X* diffuse, in the current instance. This is also seen in the TCS for a reduced level of substrate (figure 5d; also see electronic supplementary material, figures S4 and S5 for further demonstration of the introduction of these effects and dependence on the spatial design). A detailed analysis in the electronic supplementary material, §3.2, reveals the interplay between spatial organization (and the specific pattern of spatial localization) and bifunctionality, in introducing a new regime of concentration robustness (even when it did not exist in the kinetics). This is explained analytically in the electronic supplementary material. The essential insight is that the combination of bifunctionality of the enzyme (which imposes specific kinetic restrictions) along with dominant dilution of certain combinations of substrate species (which crucially relies on the spatial design) gives rise to this behaviour. In these cases, the dominant factors in the substrate conservation arise from the substrates present everywhere in the domain (i.e. non-localized), which may completely trump any effects of substrate sequestration in enzyme complexes. This explains why this absolute robustness effect is operational even if locally total enzyme amounts are greater than total substrate amounts (which from a purely kinetic study would preclude the behaviour from occurring), and also indicates why such a robustness could be created even when the completely co-localized scenario (with the same total amount of substrate) did not show this behaviour. This also demonstrates the importance of the identities of species spreading.

We examine enzymes localized in two distinct locations with substrates shuttling between them. If the balance of enzymes is the same at both locations the steady-state substrate concentration profiles are spatially uniform (seen in figure 4b). Here, the role of spatial organization is qualitatively analogous to the single location case (with two species diffusing) and can be quantitatively mapped onto that. When the balance of enzymes is different at the two ends, then in general the ACR property is lost. Simulations in both CMCs and TCSs show this. Variation of the enzyme amount (equally in the two locations) shows an absence of concentration robustness of the local as well as the global average of X^* in computational analysis (figure 5e,f, electronic supplementary material, figure S6(A–F); also see electronic supplementary material, figure S6(G,H), which shows two cases where the figure 5d setting is augmented by enzyme at another location, revealing this). Can ACR be observed? If the substrate diffusivity is high (in dimensionless terms, so that substrate profiles are uniform), then if enzymatic modification of substrates is in the unsaturated limit, we see that X^*/X is independent of the total enzyme concentration (at each location, varied equally). This along with substrate conservation shows X^* to be independent of total enzyme amounts. Here, even with substrate modification not in an unsaturated regime, for large enough domains, the domination of $X + X^*$ in the substrate conservation condition allows for the behaviour, similar to the cases above (see electronic supplementary material).

When enzymes are localized at two (or more) locations, it is worth asking if the system can exhibit local concentration robustness to variation of enzyme amount at a specific location (while the others are fixed). This can be seen if (i) substrate diffusion is very weak, and total substrate amounts are sufficiently high (here substrate concentrations at this location remain essentially determined by the kinetics), or (ii) the balance of enzyme forms is the same at the locations (this is independent of diffusivity).

ACR is experimentally seen in the OmpR/EnvZ TCS system. We can infer from the above analysis for such systems where this behaviour is observed experimentally that either (i) there is no localization of enzyme/substrate or enzymes and substrates are both co-localized, so that the kinetic description is valid, or (ii) there can be localization of enzyme in one location or possibly more than one location, in which case the balance of enzymes is the same in all locations (though the possibility of a different balance of enzyme forms may also be possible, if substrates are highly diffusible). Here there are multiple possibilities. (a) The origins of this behaviour can be essentially traced to the enzyme kinetics, along with the dilution effect of the substrate. (b) The spatial organization of substrates (which species is (non)localized) plays a key role, along with bifunctional enzyme kinetics in establishing this effect, which may not be seen even if all substrate were co-localized with enzyme. In both cases (a) and (b), our analysis predicts that experiments performed with cells of different sizes (e.g. at different growth stages) can reveal different quantitative characteristics and margins of concentration robustness, something not seen in case (i), if all species are co-localized. (a) and (b) can be distinguished by an experiment which co-localizes substrate with enzyme: the robustness property and margin is enhanced in (a) but can be destroyed in (b). Increasing domain size can diminish the robustness margin in (a) but enhance it in (b) (see electronic supplementary material).

12.4.6. Bistability

Bistability is of natural interest in bifunctional enzymes as it allows for opposite enzyme activities to be demonstrated under the same conditions and at the same time. We focus on two different ways in which bistability may be achieved in bifunctional enzymes. One way is through co-operative allosteric feedback from substrate affecting the balance of enzyme forms. This is shown in the CMC system, clearly demonstrating that this results in a regime of bistability. Another way in which bistability can be achieved is through the presence of additional enzyme–substrate complexes, called dead-end complexes. The role of additional enzyme–substrate complexes has been studied in. The effect of dead-end complexes in generating bistability in bifunctional enzymes in TCSs has been studied in. It is worth pointing out that the latter mechanism does not require an explicit feedback mechanism between enzyme and substrate, and this works within the constraints of TCS biochemistry.

12.4.7 Effect of Spatial Organization on Bistability

When the enzyme is localized in one location, then substrate diffusion can result in a destruction of bistability (see red curve in figure 6a demonstrating this for the CMC, with X* diffusing). This is because the dilution of modified substrate weakens the feedback and hence weakens the propensity for bistability. Similar insights hold good if X exits the compartment. A very similar trend is also seen in the TCS bifunctional system as well (figure 6b–d), where bistability was realized through the presence of dead-end complexes, rather than co-operative feedback. The dilution effect weakens the implicit feedback, eventually destroying bistability (these results are supported by the analytical work in electronic supplementary material, §3.2). Can bistability be preserved in such a spatial design? We show that, by starting with a greater initial substrate, we indeed get a bistable dose–response curve though its range is altered (green curve, figure 6a—this curve was obtained by incorporating the effects of dilution modifying the conservation condition, and using bifurcation analysis). Another way of maintaining bistability is if the initial compartment is a large fraction of the domain size, simply because of minimal distortion due to dilution. Both these aspects are seen in the TCS system as well. Similar insights hold good when both substrate species diffuse. Here, when the substrate modification kinetics is in the unsaturated regime and substrate diffusivities are equal, we can explicitly account for the dilution effect essentially independent of enzyme concentrations in the CMC system (analysis in electronic supplementary material). Figure 6e,f also shows how bistability can allow for enzymes localized in different regions to realize the different steady states, but this happens only when the substrate diffusivity is low (in dimensionless terms).

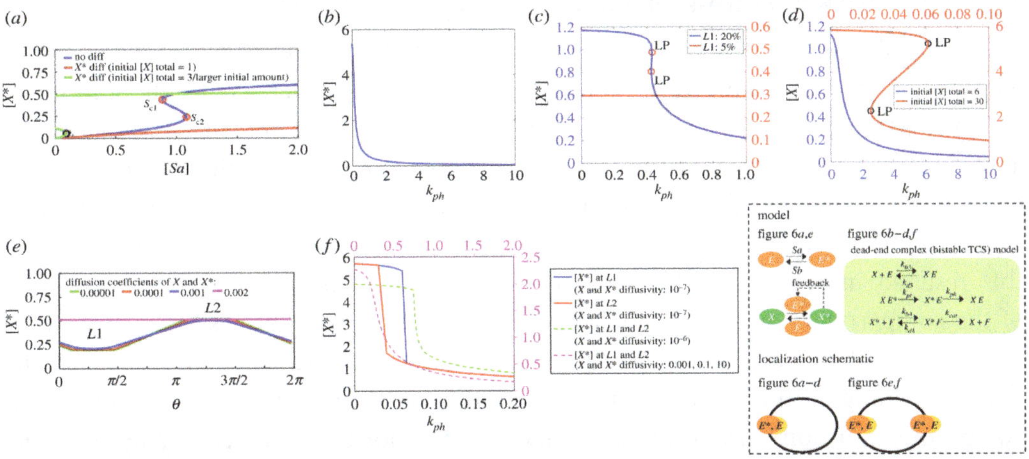

Fig. 6: The effects of spatial organization on bistability in a bifunctional CMC and TCS.

The perturbation of bistability arising from (a) a CMC with feedback (when X* is diffusing), showing how the associated dilution effects can distort and even destroy bistability (red curve). (b–d) Different cases of a TCS with a dead-end complex model (corresponding to X diffusing, X* diffusing or both diffusing, respectively) all show a similar result (blue curves in (b) and (d), red curve in (c)). kph is a kinetic parameter in the dead-end complex model (see electronic supplementary material). However, a larger initial amount of substrate ([X]total) and relative compartment size can facilitate recovery of the destroyed bistability (green curve in (a), blue curve in (c) and red curve in (d)). (e,f) Bistability can be leveraged to obtain distinct steady states at different locations both in the CMC (green, blue and red curves in (e), showing distinct steady states at the two locations: the opposite combination is not shown) and TCS cases (red and blue curves in (f) depicting concentrations at the two locations), but only when the substrate diffusion is weak. A schematic of models used is depicted.

The above insights are relevant to cases where a localized enzyme and substrate may result in bistability, at a certain stage of cellular growth: it is suggested that bistability in the localized PleC bifunctional histidine kinase activity ensures irreversibility during the transitions between phases of the cell cycle and robustness to fluctuating nutrient levels in *Caulobacter*. There is a subsequent alteration to the localization pattern of substrate, as there is a progression to other phases of the cell cycle. We infer that, through a sufficiently high concentration of localized substrate, a cell can allow for the maintenance of bistability, over natural growth sizes of the cell. In such a case, the breakdown of bistability would require abnormally large cells (which may arise from mutation). If the effect of bistability is not meant to be propagated to other stages of the cellular growth, this dilution must destroy bistability, which then constrains the balance of initial substrate concentration and size of the cell. In particular, a cell not growing sufficiently large cannot remove the effect of this bistability.

12.4.8 Feedback Resulting In Threshold-Type Behaviour

Our discussion of bistability in CMC bifunctional systems stemmed from co-operative feedback. It is worth examining the effect of a simpler non-cooperative feedback in such a system, and we see that it is possible to obtain a threshold effect. In such a case (figure 7a), above a certain value of signal, the output is zero (see electronic supplementary material, §3.2). We focus on a signal level below this threshold, so that a non-zero steady state is seen in the kinetics. Here we find (figure 7b) that, if X* alone leaked out of the compartment, this resulted in a reduced (but non-zero) steady state. If both X* and X leaked out of the compartment, a zero steady state for X* was obtained. Interestingly, even when only X diffused out of the compartment a zero steady state resulted. This indicates that, depending on the extent of dilution, and the particular species leaking out of the compartment, a non-zero steady state may or may not be sustained. Analytical work (electronic supplementary material) shows this.

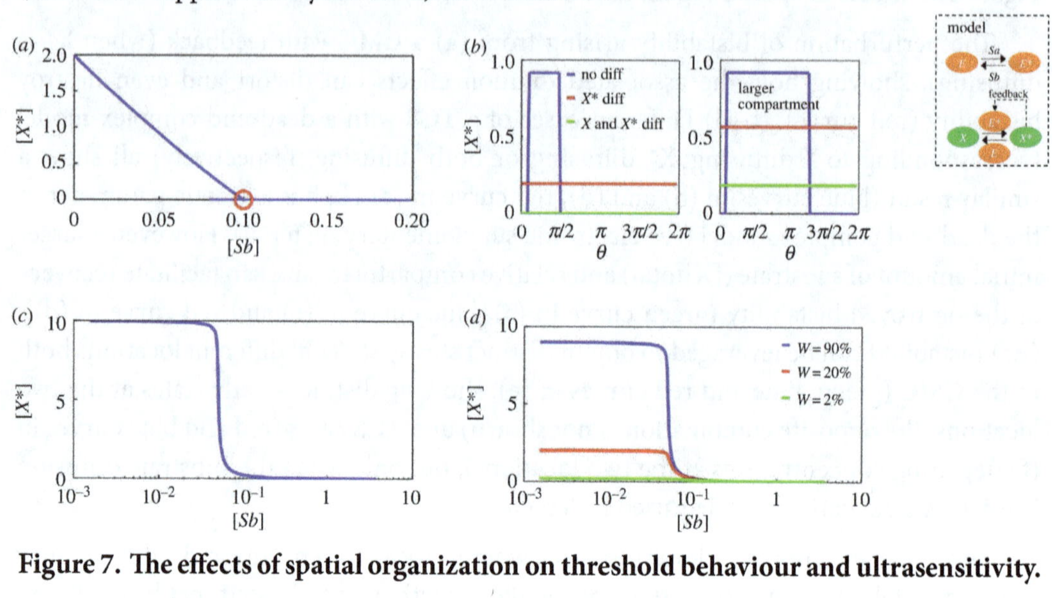

Figure 7. The effects of spatial organization on threshold behaviour and ultrasensitivity.

(*a*) In the case of a bifunctional CMC with feedback (see schematic), a threshold effect can result from the kinetics (*Sb* is a kinetic parameter reflecting the E^* to E basal conversion: beyond a certain level zero output results). (*b*) Depending on which substrate species is localized/non-localized, zero or non-zero concentrations for X^* may be obtained, though this also depends on the size of the region where enzymes are localized, relative to the ambient region. (*c,d*) The perturbation of ultrasensitive behaviour in the kinetics (no feedback) (*c*), through dilution, and the dependence on compartment size (*d*) (see text for details).

While our analysis of both bistability and feedback above has focused on bifunctional enzyme systems, we point out that similar behaviour can also be observed in monofunctional enzymes, and be traced back to the same fundamental ingredients (co-operative/linear feedback, dead-end complexes). Consequently, the trends associated with spatial organization above are relevant to these cases as well.

12.4.9 Ultrasensitive Switch-Like Behaviour

Figure 7c shows how an ultrasensitive switching behaviour is obtained in the kinetics, similar to. This behaviour can be eroded because of the diffusion of the substrate species out of the compartment (figure 7d). As the size of the compartment relative to the surrounding domain is reduced, the switch amplitude is reduced, and the sharpness of switching is altered. This has been studied in other cases of enzymatic modification.

12.4.10. New Capabilities

In addition to placing constraints on kinetic and spatial parameters for the realization of different behaviour, spatial organization can also provide new capabilities. Kinetically, a bifunctional enzyme (for given conditions, total amounts of species and signals which affect the balance of enzyme forms) exhibits either a kinase-dominant or phosphatase-dominant activity (unless there is bistability). Spatial organization can relieve this basic constraint: by having enzymes at different locations (with different balances of enzyme forms), it is possible for the bifunctional enzyme system to exhibit both extreme activities simultaneously, acting on substrates localized with them. Furthermore, if a fixed enzyme amount is now (re)distributed between multiple locations (where it may act on substrate localized there), the effect of enzyme dilution which is associated with this distribution in many instances is minimal, precisely because of ACR. Taken together, these points suggest an attractive advantage of distributing bifunctional enzymes between multiple locations to act on localized substrate there.

12.4.11 Note on Measurement

We comment, in the context of spatial organization, on the effect of spatially averaged measurements, for instance by lysing a cell. In the case of localization of enzymes in one location, with others non-localized, gradients are absent. This means that spatially averaged measurements do not introduce an essential qualitative distortion (though quantitative inferences about concentrations may be incorrect). However, as seen above, a purely kinetic model that does not account for the dilution effect can have important limitations. With localization in multiple locations, gradients may be present, and spatial averaging can introduce qualitative distortions, which has important implications for both inferences drawn therefrom and models built based on this. Another basic point is related to relative enzyme and substrate concentrations (expression levels) in cells, which implicitly ignores spatial organization. Depending on the localization pattern of enzymes and substrates (and associated sizes of localization regions), conclusions simply based on relative total protein levels could be misleading, as this may not reflect relative levels at locations of co-existence.

12.5 LOCALIZATION AND ACTIVITY

We explore new aspects of the bidirectional interplay between localization and chemistry: how chemistry is used to impose localization and how localization impacts chemistry. We previously focused on different patterns of localization of enzymes/substrates and their consequences for enzymatic modification. This was imposed by restricting the presence of relevant species to particular regions and having them non-diffusible. An important aspect of imposing patterns of localization in cells is for this to be done in a flexible and dynamic way. Thus, a basic point of interest is on how active localization (i.e. localization of diffusing species) is imposed, and how that impacts chemical modification spatially and temporally. This is relevant to both monofunctional and bifunctional enzymes. An important difference between these enzymes is that localizing a greater amount of enzyme can increase the activity of monofunctional enzymes, while for bifunctional enzymes the effect of total enzyme concentration may be weak. For bifunctional enzymes, an additional question which emerges is whether localization can be directly used to manipulate and switch enzyme activity. We study these aspects in the context of bifunctional enzymes, noting that some basic insights which emerge are also relevant to monofunctional enzymes (discussed below). We study how the dominant enzyme activity can be switched (from phosphatase to kinase or vice versa), an aspect of interest in natural and engineered contexts.

Different 'levers' are available to a cell for spatially manipulating activity of bifunctional enzymes: direct activity regulation (by a factor affecting the balance of enzyme forms), localization, possible feedback (where relevant) or a combination of factors (figure 8*a*). The impact of localization on enzymes is through the 'localization network' (figure 8) and associated localization parameters (binding/unbinding rate constants for each enzyme form). The localization parameters may be the same or different for the two enzyme forms, corresponding, respectively, to equal and biased localization. A bias in localization was introduced by altering the binding parameters of the enzyme forms (see Models and electronic supplementary material, §§2.3 and 3.3). It could also be incorporated by changing unbinding parameters of the two forms, and in any case analytical work explicitly reveals the effect of parameters in this localization network. Our dissection of these factors expands our analysis of spatial organization of bifunctional enzymes studied previously, and creates a bridge to multiple contexts in natural and engineered biology.

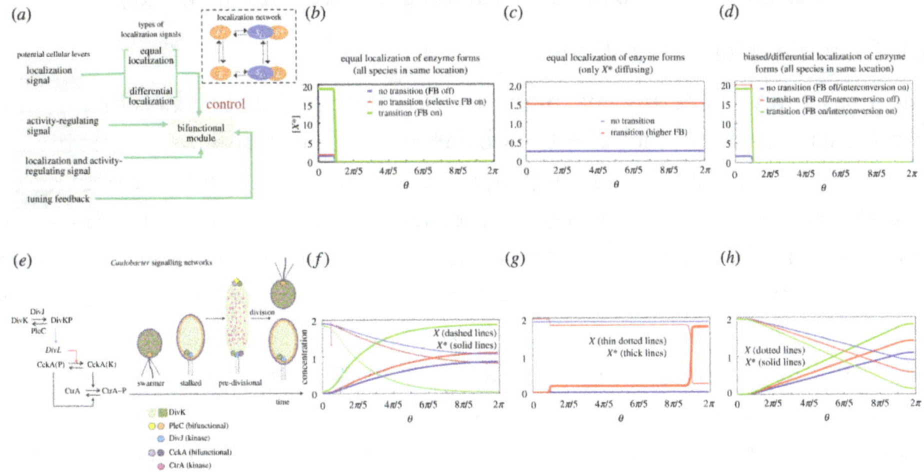

Fig. 8: Spatial manipulation of activity of bifunctional enzymes (CMC).

(*a*) A depiction of the key factors for spatially regulating the activity of the bifunctional enzyme. The 'localization network' is also depicted, along with interconversion of localized forms (figure 1*f*). (*b*) The manipulation of enzymatic activity (CMC) via equal localization with and without feedback: active localization does not by itself introduce a switching in activity (blue curve), but a feedback on both localized and free forms achieves a switch in activity indicated by an elevated X^* concentration (green curve). (*c*) This feedback is compromised if the substrate is no longer co-localized with enzymes (see text) and consequently a higher feedback strength is needed to achieve the switch in activity. (*d*) The case of biased/differential localization: a switch can be achieved from localization by itself without feedback (red curve), if there is no interconversion of localized forms; if interconversion of localized forms occurs, this does not happen, in which case a feedback on localized and free forms can achieve the desired switch in activity. (*e*) A schematic depiction of dynamic transitions in localization patterns in *Caulobacter* with associated bifunctional enzymes/substrates. In the legend, DivK and CtrA labelled represent the phosphorylated forms of these species. (*f*–*h*) Dynamic localization involving a transition from a single pole to a double pole localization with opposite activities motivated by what is observed in *Caulobacter* (here we depict how a phosphatase-dominant activity at one pole (θ = ·) transits to one involving a kinase-dominant activity at the opposite pole, though the principle is also applicable for the reverse case). (*f*) A scenario showing how opposite enzymatic activities can be seen in the two poles, through a combination of selective localization at one pole, and feedback. Panels (*g*,*h*) focus on activity regulation at the second pole and localization (no feedback). Panel (*g*) shows that even with an activity-regulating signal at the second pole, in the absence of localization in the second pole, the desired switch in activity cannot be seen (blue lines, substrates diffusing), unless the substrate is static (red lines). Panel (*h*) shows how weakening the localization strength at the first pole progressively (blue-red-green curves) can achieve the desired activity pattern (alternatively strengthening localization at the second pole works).

12.5.1. Direct Activity Regulation

The most basic way of altering enzymatic activity through the kinetics is by means of an activity-regulating signal which affects the balance of enzyme forms. In order to investigate to what extent an activity-regulating signal can switch the dominant activity of the enzyme spatially, we investigated a range of scenarios where the relative locations of three factors—the activating signal, the enzyme and the substrate—were varied (see electronic supplementary material). This corresponds to an expansion of spatial designs studied in the previous section by incorporating (i) the activating signal and (ii) localization of diffusible entities. Different possibilities for the location of these species were considered: (i) localized at either one location in the spatial domain (e.g. representing one pole of a cell), (ii) localized at two specific locations in the spatial domain, and (iii) present/spreading everywhere in the spatial domain. We examined the possibility that the relevant species was weakly diffusive/non-diffusive or was highly diffusive.

Analysing these various cases reveals the following. (i) If the enzyme is localized, then the presence of activating factor at that location is necessary. (ii) If the substrate is localized and the enzyme is weakly diffusible, then the localization of the activating factor at that location is sufficient. (iii) If the enzyme is not localized and highly diffusible, then the activating factor should be present along with the enzyme or at least in a sufficiently large fraction of the region where the enzyme is present (having activating factor present in the location of the substrate is not sufficient). However, this restricts the possibility of enzyme having different dominant activities at different locations.

12.5.2. Localization as A Regulator of Activity

We turn to the roles of localization as a tuner of activity, noting its potential role, which may extend to highly diffusible enzymes, and focus on the localization network. We start by examining the case of enzyme and substrate in the same location, but with an additional (active) localization mechanism for the enzymes. In all cases, parameters are chosen to result in a strong localization (of the relevant enzyme form), and this is done by having binding to the location, much stronger than unbinding, in particular having weak unbinding. The default state of the enzyme (balance of kinase/phosphatase) is assumed to be strongly towards the phosphatase. In this context, we also examine the potential role of (linear) feedback between substrate and enzyme as an additional ingredient where relevant. We first focus on bifunctional enzymes with covalent modification biochemistry, and then discuss TCSs. Unless otherwise mentioned, both localized and free forms of enzymes are assumed active.

12.5.3 Unbiased Localization

In the case of equal (unbiased) localization (figure 8*b*) (no feedback) localization by itself does not result in a switching of dominant activity (blue curve). This is true if the localized forms do not interconvert, and also if the localized forms interconvert, at the same rates as the free forms or even if the 'interconversion equilibrium constant' (ratio of forward and backward rate constants) is the same for localized and free forms of enzymes. This is shown analytically in electronic supplementary material, §3.3. Now suppose a feedback from the substrate X^* resulting in a greater balance of E^* (the kinase form) was incorporated. If this acts only on the free (non-localized) enzymes, again, this generally does not result in a transition (red curve) if localized forms can interconvert (and localization in strong): the essential insight is that most of the enzyme is localized, and this feedback acts only on the non-localized form. If feedback is applied to both localized and free enzymes, then a transition to a dominant kinase activity is indeed achieved (green curve), demonstrating an important requirement for feedback regulation.

This previous result assumed that both enzymes and substrate were present in the same location, and not diffusing. This result is also valid in the case where the substrate is localized, but enzyme is freely diffusible. Having the presence of a localization mechanism for a diffusible enzyme increases the local enzyme concentrations but does not affect the balance of enzyme forms. On the other hand, if enzymes were completely confined to a location/compartment (and non-diffusible) and substrate was diffusible, the dilution of the substrate in the location where the enzyme is present implies a reduced feedback. This then implies a stronger feedback constant is needed to achieve a transition (figure 8*c*).

12.5.4 Biased Localization

We now examine the effect of biased/differential localization (figure 8*d*). We see here that strong biased localization can by itself result in a transition of activity if no interconversion is allowed between localized forms (red curve, figure 8*d*). On the other hand, if interconversion between localized forms is possible (say at rates similar to the free forms), then the effect of biased localization can be reversed (blue curve). This is established analytically in electronic supplementary material, §3.3. In such a case, in the absence of activity-regulating signals, feedback from the substrate to both localized and free forms is needed (green curve). The diffusion of substrate (as opposed to co-localization with enzymes) can weaken the feedback, which would have to overcome dilution to achieve a switch in activity.

12.5.5 Localization of Bifunctional Enzymes with Two-Component Biochemistry

We revisit these results in the case of bifunctional enzymes in a TCS. The distinctive feature here, of the balance of enzyme forms being affected by the substrate, is crucial. A direct consequence of this is the fact that, as the substrate amount is increased, the balance of enzyme forms is not a constant and can shift more to the phosphatase. We now turn to this distinctive tendency of switching activity, from kinase to phosphatase, and how this is affected by localization. We restrict ourself to the case that enzymes are localized but substrates are not.

12.5.6 Equal Localization

We see in figure 9*a* that with equal localization the switching of activity from kinase to phosphatase occurs at a well-defined concentration of substrate, which incidentally is practically the same as that with no localization. To understand the interplay of different factors, we examined scenarios which involved the disabling of different factors (figure 9*b*–*d*). These factors are the activity of localized forms and the interconversion of localized forms. In the presence of localization, but with the localized forms inactive and non-interconverting, the level of substrate at which this switching was reduced to a small degree (figure 9*b*). This simply reflects the fact that here localization simply serves to sequester enzyme, reducing the available enzyme for conversion, and consequently the level of substrate required to effect switching is reduced. Now if the localized forms are enzymatically active, but do not interconvert, the level of substrate at which switching happens is dramatically reduced (figure 9*c*). This is due to the confluence of two factors. (i) A substantial fraction of enzyme is localized. (ii) The substrate modification biochemistry of a TCS results in a unidirectional conversion reaction between localized forms (which are no longer constrained by the basal interconversion balance). On the other hand, if the localized forms do interconvert but are not catalytically active, then the switching of activity happens at a substantially higher level of substrate (figure 9*d*). Here, a substantial amount of enzyme is localized and subject to interconversion while the effect of the substrate is present only via its action of free enzyme forms. Overall this reveals the subtle interplay of factors at work even in the case of equal localization.

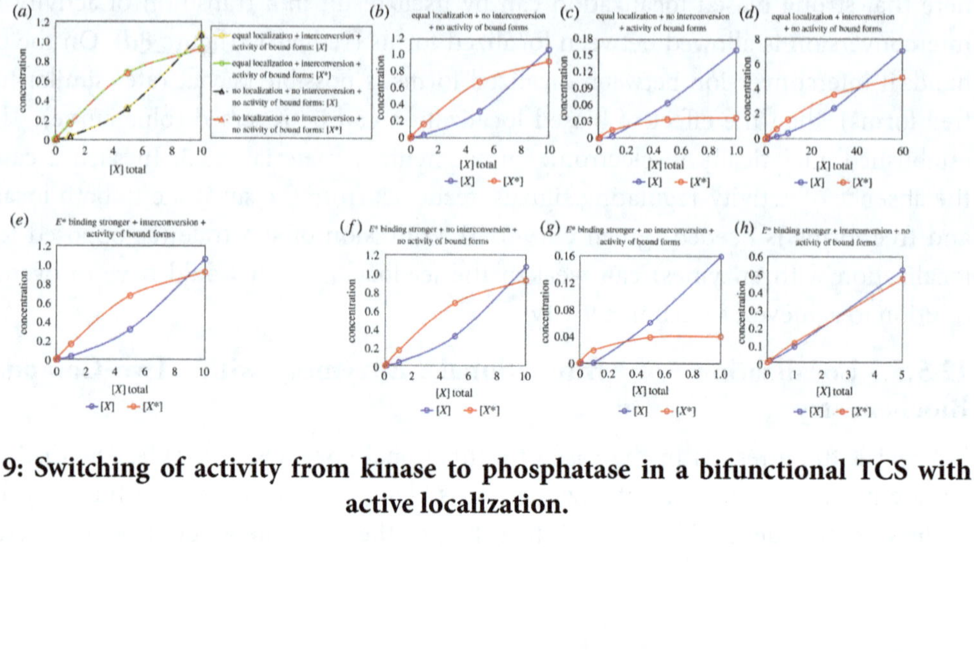

Fig. 9: Switching of activity from kinase to phosphatase in a bifunctional TCS with active localization.

The plots investigate how localization by itself can impact the switching from kinase to phosphatase activity in a bifunctional TCS (seen via the crossing of concentration curves for X and X* as the total substrate amount is varied). The top row (a–d) corresponds to equal localization of enzyme forms, while the bottom row (e–h) corresponds to biased localization (stronger for E*). For each row, we examine a number of scenarios corresponding to the absence of interconversion of localized forms, enzymatic activity of localized forms, or both (see text for details, and figure 8 for a depiction of the localization network of the enzymes).

12.5.7 Biased Localization

Figure 9e–h focuses on the case where the localization (binding constant) of the kinase form is clearly greater than that of the phosphatase. Many of the insights parallel the earlier case. A well-defined level of substrate determines the switching of activity from kinase to phosphatase (figure 9e). If there is localization with no activity or interconversion of the localized forms, there is a small reduction in the level of substrate needed to effect the switching (figure 9f). Notice that, in such a case, steady state implies a local equilibrium between each pair of reactions, and, as a consequence, the level of substrate at which switching occurs is practically the same as the corresponding case of unbiased localization. As before, if the localized forms are active but with no interconversion the level at which switching occurs is substantially reduced for the same reason (figure 9g). An interesting contrast (to unbiased localization) emerges when we examine the case of localized forms interconverting, while being inactive. In such a case, the level of substrate at which switching occurs is significantly reduced (figure 9h).

Analytically studying the effect of differential (strong) localization of enzymes in this case shows that the net effect is a reduction of free enzyme (the active enzyme), along with a shift towards phosphatase: this is because the free kinase form is more effectively localized (and sequestered from the substrate) than the free phosphatase form. Biased localization by itself, in such a case, can effect the switch from kinase to phosphatase. This behaviour does not depend on the TCS biochemistry in any essential way; in fact, if an analogous case of biased localization with inactive localized forms were considered in the CMC, a similar behaviour ensues and can be explained analytically (and the behaviour is not seen with equal localization).

12.5.8 Requirements for Localization to Tune Activity

A basic insight which emerges from the above studies is that localization (without feedback) can affect/switch activity if there is a difference between localized and non-localized forms (e.g. localized forms do not interconvert or are inactive). This depends on the interplay of (i) the 'localization network' (figure 8), (ii) the specific differences between localized and non-localized forms, and (iii) the biochemistry of substrate modification.

12.5.9 Switching Activity with Dynamic Spatial Reorganization

We examine a case of dynamic spatial regulation of bifunctional enzymes focusing on a question broadly motivated by observations in the cell cycle in Caulobacter, where (i) the establishment of a localized enzyme with a hypothesized opposite (kinase) activity to the original activity of the enzyme occurs (enzyme PleC) and (ii) the establishment of bifunctional enzymes in two opposite poles with opposing activities (for the enzyme CckA) occurs (figure 8e). The factors responsible for establishing the opposite/new activity at the second pole are unclear and being investigated. Our analysis allows for bridging the gap between the intrinsic kinetics of a bifunctional enzyme and its behaviour in such a scenario, to understand the ways in which dynamic spatial 'control' may be achieved by cells (also see Discussion); this is also relevant to the dynamic control of localization in synthetic biology. We computationally study this by examining a case where an enzyme with phosphatase activity at one pole can transit to localization at two poles with opposite activities (figure 8f–h) (this is also applicable to kinase activity at the first pole). A way of achieving the desired activity is by simply having a localized activity-regulating factor at the new pole; however, as seen in figure 8g (blue curve), if the localization of enzyme at this new pole is not strong enough, then such a factor is insufficient. In effect the strong localization of enzyme at the first pole and diffusion of enzyme swamps the effect of a localized activity-regulating factor at the second pole. Figure 8h shows that if the strength of enzyme localization in pole 1 is progressively reduced, then it is possible to use activity regulation in pole 2 (green curve). This consolidates the point that a strong disparity in localization strength of enzymes at the poles can limit the efficacy of an activity-regulating signal at a given pole. Alternatively, having a sufficiently strong localization with an activity-regulating signal at the new pole can have the desired effect. On the other hand, to employ localization by itself at the second pole to achieve a transition to an opposite activity, either (i) there should be stronger localization of the desired form at the new location with no interconversion, (ii) there should be stronger localization of the undesired form, with localized forms inactive, or (iii) if the newly created pole is required to have a phosphatase-dominant activity (i.e. the first pole is kinase dominant), then having localized forms active but not interconverting promotes this.

12.5.10 Relevance to Monofunctional Enzymes

The above insights on bifunctional enzymes are also relevant to monofunctional enzymes, especially when there is an enzyme activation step. In fact if we regard E and E^* as the inactive and active forms of an enzyme (say kinase), many of the essential insights carry through as they are primarily dependent on the balance of E and E^*, and the factors which affect it. In particular, (i) the different factors affecting E^*, (ii) the conditions under which localization can tune activity (both for a CMC and TCS), and (iii) the fact that localized activity regulation may not be sufficient, and localization of enzyme may be needed, are all relevant.

12.6 EVOLUTION AND SPATIAL ORGANIZATION

We now explore the interplay of biochemistry and spatial organization, through a new prism: that of evolution. Evolution, in this regard, can function on both the biochemistry as well as spatial organization, causing important alterations to each. At the outset (figure 10a), for our purposes we can identify four classes of scenarios depending on whether the original 'design' had a non-trivial spatial aspect or not, and whether a change introduced did or did not involve a non-trivial spatial aspect. Evolutionary changes are often studied as purely kinetic changes subsumed within kinetic descriptions, which is the one category which does not involve spatial aspects in any essential way. Here we focus on two other categories: (i) biochemical alterations built upon pre-existing spatial designs and (ii) spatial alterations built on a design which did not have a spatial aspect. We present focused insights pertaining to both these aspects.

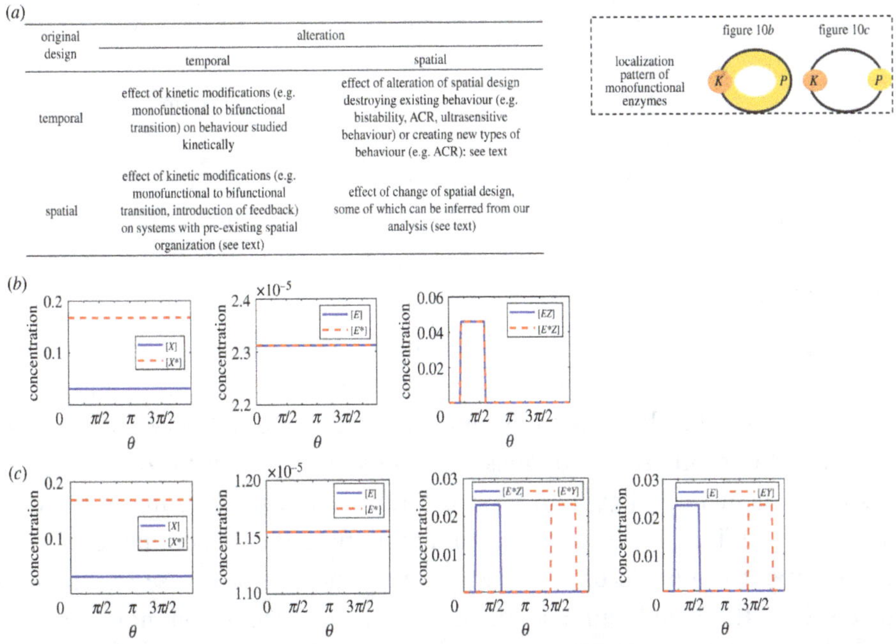

Fig. 10: The effect of pre-existing spatial organization on a transition from monofunctional to bifunctional enzymes.

(a) A broad classification table for alterations through evolution, depending on the presence or the absence of spatial aspects in the original 'design' and in the alterations imposed. The plot then investigates the impact of prior spatial organization on the evolution from monofunctional to bifunctional enzymes via gene fusion. Two scenarios for the monofunctional enzyme were considered (see schematic): (b) K (kinase) localized in one location and P (phosphatase) everywhere, or (c) K and P localized in two distinct locations. Both scenarios result in graded profiles of substrate (not shown: see green curves in figure 2b,c for a reference). If the resulting bifunctional enzyme can localize where either parent enzyme could, then the first scenario results in essentially complete localization of the bifunctional enzyme at one location, at a balance determined by the basal interconversion rates of the enzyme forms (and consequent absence of graded profiles of substrate) while in the second scenario both enzyme forms are localized at both locations (again at a balance determined by enzyme interconversion rates, implying that a gradient of substrate is also abolished)—also see electronic supplementary material, figure S8(A,B). EZ and E*Z denote localized forms of kinase and phosphatase forms of the enzyme, respectively, at one location (localizing species Z) and EY and E*Y denote the localized forms of these enzymes at another location (localizing species Y) here and in figure 11.

12.6.1 Biochemistry

We start with the biochemical alterations. We focus on one basic aspect of evolution which is relevant in the context of our study—that involving a transition from a monofunctional enzyme pair to a bifunctional enzyme. While, in TCSs, bifunctional enzymes have emerged from a common ancestor, other bifunctional enzymes have been experimentally demonstrated to have emerged via a process of gene fusion. This allows for a single transcript producing protein which is capable of both kinase and phosphatase activity. The evolutionary advantages of such a process have been discussed in. However no aspect of spatial organization has been considered. We now examine the effect of a transition from monofunctional to bifunctional enzymes, in the face of a pre-existing spatial organization.

To do this, we examine two scenarios for the monofunctional enzymes *in silico*. The first scenario is where one of the enzymes is localized: without loss of generality, we assume that the kinase is localized at one location of the cell, while the phosphatase is present everywhere. In the second scenario kinase and phosphatase are localized at two separate locations (see schematic in figure 10). We use this as a basis to examine the question: what effect does this have on evolvability from monofunctional to bifunctional enzymes? To make a comparison *in silico*, we assume equal amounts of kinase and phosphatase enzyme to start with, though our essential insights neither stem from nor require this.

The key aspect to be examined is the overall behaviour of the system, which in turn strongly depends on the localization pattern of the bifunctional enzymes created through this process. In all cases, we assume that both modified and unmodified substrates are diffusible and present everywhere in the domain: we keep this spatial 'design' of substrates fixed, and focus only on the enzymes. There are multiple possibilities regarding the localization pattern of the bifunctional enzymes. We examine these in turn.

The simplest and most obvious hypothesis regarding the localization of the bifunctional enzymes involves an 'OR' logic with regard to the localization of the parent enzymes: the bifunctional enzyme can localize at any particular location where either the parent kinase or phosphatase is localized. This, in turn, stems from a basic consideration that a binding motif which was present in the protein structure of either of the enzymes is now present in the protein structure of the bifunctional enzyme.

Computational analysis (figure 10*b*) reveals the basic fact that if there was only one localization present, then the bifunctional enzyme would localize exclusively (or almost exclusively) in that location. This simply follows from the fact that, with strong localization (comparable to that of the parent enzyme), almost all the enzyme ends up localized there. In terms of substrates, this creates a significant qualitative change from gradient to uniform substrate profiles (also see figures 2 and 4 for reference).

In the case of localization of the kinase and phosphatase at two locations (figure 10*c*), we find that the bifunctional enzyme ends up essentially equi-distributed at the two locations. In this case, the bifunctional enzyme system will exhibit the same balance of kinase and phosphatase at both poles (determined by the bifunctional enzyme interconversion rate; see electronic supplementary material, figure S8(B)), and the substrate will exhibit a uniform profile (figure 10*c*), a contrast from the monofunctional case where a gradient is obtained (as in figure 2). This then means that, to recapitulate the original behaviour, the activation signal (which converts the phosphatase form to kinase form) must have different levels at the two locations.

In the above scenario, the bifunctional enzyme localizes where either of the monofunctional enzymes could localize. This decouples localization from activity. We then examine conditions under which the bifunctional enzyme may approach scenarios analogous to those of a monofunctional enzyme pair (figure 11*a*,*b*). To do this, we incorporate a basic logic which couples binding at a location to its activity. Thus, in the first localization scenario, we assume (i) E^* alone can bind to the designated location (say pole of the cell) and (ii) no interconversion of enzyme forms occurs when this enzyme is bound (this is consistent with the requirement that only kinase is present in this location). With these assumptions computer simulations (figure 11*a*) reveal simply that (essentially) all enzyme ends up as kinase at this one location (blue curves of enzyme profiles). This is again a drastic change from the monofunctional enzyme results, and suggests that a simple mirroring of the localization logic of monofunctional enzymes does not guarantee the same behaviour in a bifunctional enzyme. We then allow for interconversion of localized forms, in which we incorporate a strong unbinding of the phosphatase. In such a case, we find that there is increased phosphatase accumulated near the designated location (but not bound). However if the diffusivity of enzymes is high, this results in the phosphatase form being present everywhere, while the kinase form is bound and primarily localized in one location (red curve of enzyme profiles). This is reminiscent of the localization pattern of the monofunctional enzymes, and results in a similar substrate behaviour.

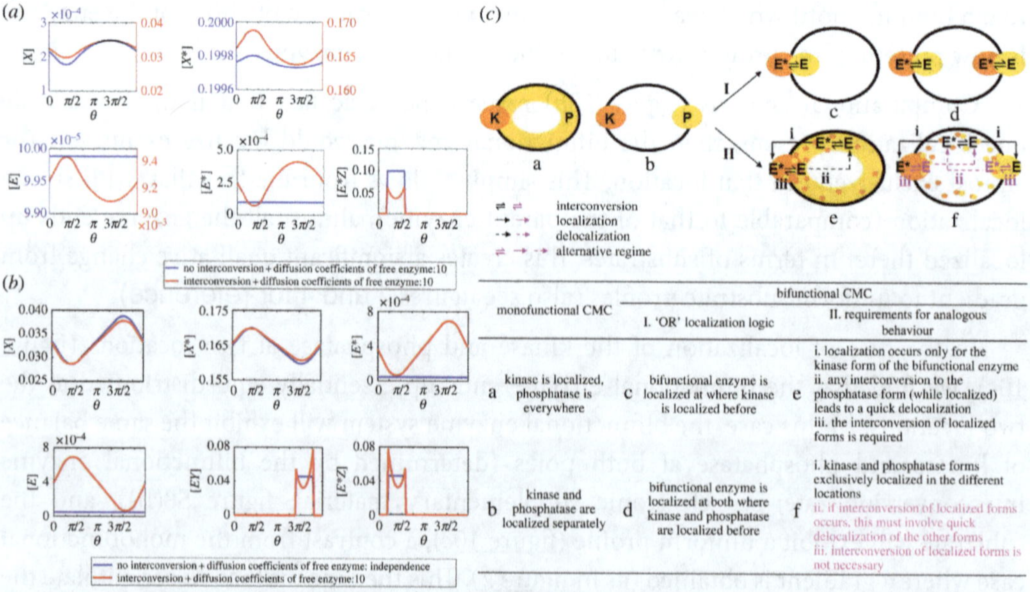

Fig. 11: Conditions where the bifunctional enzyme exhibits a comparable localization/activity to the monofunctional enzyme, building on the scenarios for the monofunctional enzymes in figure 10b,c.

(a,b) The species concentration profiles when only E* (kinase form) localizes (binds) in one location and when E (phosphatase form) and E* localize in two different locations, respectively. In (a), if there is no interconversion between localized E and E* (blue curve), the bifunctional enzyme forms end up essentially fully as E* at one pole; however, when the interconversion of localized forms with a strong unbinding of E is allowed (along with high diffusivity of enzymes), broadly similar results with monofunctional enzyme can be generated (red curve: note the difference in scales between blue and red curves, and that the total free phosphatase amount is comparable to the localized kinase amount). Furthermore, in (b), the bifunctional enzyme can also show a similar behaviour (E and E* localized in two different poles with graded substrate concentration profiles) to the monofunctional enzyme pair in the presence (red curve) or absence (blue curve) of the interconversion of localized forms. This however depends on relatively high diffusivity of free enzyme relative to substrate in the former case, while it is independent of diffusivity in the latter case. (c) A schematic summary of results in figures 10 and 11.

We revisit this when kinase and phosphatase are localized in two different locations (figure 11*b*). Here, employing this logic results in a situation where, even without any interconversion of localized forms, enzymes can localize at the two locations in a manner reminiscent of the monofunctional enzymes (blue curve, figure 11*b*). The relative concentration of kinase and phosphatase at the two poles is now affected by the basal interconversion rate constants between kinase and phosphatase forms (even if interconversion does not occur when localized). If interconversion of localized forms is allowed, similar localization patterns to those of the monofunctional enzymes can ensue, if there is immediate dissociation of the other form, and the enzymes are highly diffusible (red curve, figure 11*b*). This parallels the insight from the previous case.

These results (summarized in figure 11c) show how factors associated with the way in which molecules localize and the relationship between the localization of the bifunctional enzyme and that of the monofunctional enzyme can have a significant impact on the consequences of a transition in the biochemistry (also see electronic supplementary material, figures S8 and S9). This suggests that drastic changes in localization patterns can emerge, identifies conditions under which that happens and also identifies requirements under which there is no drastic qualitative change.

12.6.2 Evolution of Spatial Organization

We now investigate the effect of a transition in spatial organization (due to altered patterns of localization of species), for fixed biochemistry. This is relevant to transitions both in evolution and occurring during the lifetime of a cell. Our analysis of spatial designs facilitates a transparent analysis of this.

Monofunctional enzymes.F igure 2 has depicted the effect of different spatial designs on the overall substrate modification. A single change in localization pattern corresponds to a change of one index. As seen there, a change of substrate localization can be associated with dilution, while localizing substrates has the opposite effect. The identity of the substrate(s) therein is important as well. Transitions in enzyme locations can result in a spatial design where there is a purely transient response in modification or de-modification of proteins with a zero steady-state output. It remains to be seen if this has been exploited in cells. Alterations in enzyme localization may facilitate or prevent graded behaviour of substrates.

Bifunctional enzymes. An analogous set of plots are presented in figure 4 and electronic supplementary material, figure S2. Distinct aspects associated with bifunctional enzymes include the following. (i) Moving from a single location organization of enzymes to a double location organization (with no alteration in balance of enzyme forms) still continues to result in a spatially uniform profile (independent of diffusion). An alteration of balance of enzyme forms spatially is necessary for generating a robust gradient. (ii) In contrast to monofunctional enzymes, even with some substrate species non-diffusible/localized, zero concentrations are never encountered, as long as enzymes are not separated from substrates initially (discussed previously). (iii) Spatial alterations can actually facilitate ACR, even if it did not exist kinetically to start with. (iv) Spatial alterations can destroy different types of behaviour such as ACR, bistability and switch-like behaviour. Some of the changes introduced by a change of spatial organization may be compensated for kinetically.

A central theme which emerges from our studies of alterations in biochemistry and spatial organization is the possibility of fundamental qualitative alterations associated with altered enzyme localization patterns. If such a transition is deleterious to a cell (e.g. resulting in a zero steady-state concentration or abolishment of a gradient), then such transitions are not expected to occur. On the other hand, if some transitions are tolerated (say with attendant dilution effects) then (i) other subsequent kinetic alterations (e.g. increasing enzyme production) could compensate for this (and in some cases understood only in light of this), (ii) the 'memory' of the earlier spatial organization may be erased, and this may also facilitate new transitions, with further changes in spatial organization which may significantly diverge from the original spatial design. Finally, changes in spatial organization may also enable new desirable features.

12.7 SPATIAL ORGANIZATION: AT A GLANCE

Spatial organization is a central aspect of cells, while biochemical pathways are central to cellular information processing. This prompts many questions regarding their interplay which are relevant to systems and synthetic biology, and the emerging field bridging the two, with broader implications for chemical information processing. Advances in imaging and the advent of spatial proteomics and substantial recent experimental work in synthetic biology provide further impetus.

Our study employed models of basic biochemical building blocks of pathways and the most basic ways of imposing spatial organization: localization or delocalization of species. Since a basic way of imposing spatial organization in pathways is through single modification stages, this serves as a basis for understanding the effects of spatial organization on pathways and complex biomolecular networks. These contain additional ingredients, such as cascades, multisite substrate modification, feedback and so on. Even in these more complex cases, the effects of (de)localization based on chemical identity remain a core feature (for instance, a key effect of multisite substrate modification is determining localization).

Our study significantly expands on earlier work on CMC-based cascades in many essential ways: the consideration of a broad range of building blocks (CMCs and TCSs, monofunctional and bifunctional enzymes and augmentations involving feedback or dead-end complexes) and a range of information processing characteristics, the systematic consideration of spatial designs, the consideration of active localization and localization as a tool for tuning activity as well as the interplay of biochemistry and spatial organization from an evolutionary perspective.

By examining a range of spatial designs in basic modification cycles (localization scenarios for enzymes and substrates), a range of insights emerge. Static gradients are observed only in specific instances. A requirement for the steady-state modification extent not to be at either extreme (zero or complete modification) is that spatial organization must ensure that the modification cycle of events is not disrupted. Instances where such a disruption happens involves a particular form of modified substrate losing contact with the enzyme which modifies it (§3.1). An echo of this feature is seen in modification cascades too. Bifunctionality of enzymes both gains from and exhibits certain advantages, with regard to spatial organization (§3.2). On one hand, distinct extreme activities of the enzyme can be realized simultaneously, while, on the other hand, the bifunctionality ensures that, even with spatial organization, the modification cycle is never disrupted. We demonstrate conditions of spatial organization which are consistent with concentration robustness in modification cycles with bifunctional enzymes, and bistability and threshold behaviour. If a particular behaviour is seen experimentally either (i) spatial organization is absent and the behaviour can be understood purely in kinetic terms, (ii) specific forms of spatial organization may be present, with the kinetic regime/total protein concentrations bypassing constraints imposed by spatial organization, or (iii) the behaviour emerges from a non-trivial combination of spatial organization and kinetics, and is not attributable to kinetics alone (as shown in the case of robustness).

A basic way of imposing spatial organization is through chemistry itself. We demonstrate (§3.3) how the activity of a bifunctional enzyme may be manipulated by activity regulation, feedback (where applicable) and localization itself. We showed how activity regulation by itself may be insufficient to tune enzyme activity locally (especially where strong localization of the enzyme elsewhere may be present), and may need to act in concert with localization. On the other hand, localization by itself (without any activity-regulating factor) can be used to manipulate the activity of bifunctional enzymes when there is a difference between localized and non-localized forms (e.g. localized forms do not interconvert, or they are inactive). Differences in the nature of substrate modification (TCSs or CMCs) can be critical here. Many of these insights are also relevant to monofunctional enzymes involving an enzyme activation step. These insights are particularly relevant in instances where localization patterns change dynamically (seen in Caulobacter for instance), where reconciling local activity and overall behaviour has led to unresolved questions. From an evolutionary perspective, we examined changes in spatial organization for a fixed chemistry, and chemical changes for an existing spatial 'design' (§3.4). The former case highlights significant distortions in behaviour, but also new capabilities in information processing characteristics (e.g. ACR in §3.2). In the latter case, changes in chemistry (e.g. monofunctional to bifunctional enzymes, but more broadly relevant) may lead to significant unanticipated alterations in behaviour because this results in substantially altered localization patterns. Requiring a maintenance of qualitatively similar localization patterns in such cases may require (a confluence of) additional factors which are not guaranteed (and may be infeasible). This simply highlights important limitations in thinking of evolutionary changes purely in kinetic chemical terms.

12.8 UNDERSTANDING THE SPATIAL ORGANIZATION

4.2.1. Systems Biology

Understanding how cellular functions are realized through biomolecular networks is one of the central goals of systems biology. Understanding the many ways in which spatial organization may impact network behaviour is fundamental to this, and our analysis and extensions to more complex pathways is foundational to this, complementing data-driven approaches.

4.2.2. Space and Information Processing Biomolecular Networks

Our analysis is relevant to determining when experimentally observed behaviour may or may not be understood in purely kinetic terms. Spatial organization can cause modulatory changes (e.g. by dilution) or distort the sequence of modifications, in a way not encapsulated in the kinetics. In the latter case employing a purely kinetic description is fraught with risk, and reconciling data with the kinetic model is likely to introduce significant errors. Even in the former case, employing a (lumped) kinetic model may introduce quantitative and in certain cases qualitative distortions when the model is used extrapolatively, for instance to study dose–response curves. These aspects are crucial in understanding the robustness of biomolecular networks. Biomolecular network robustness is usually studied through kinetic parameter perturbations (often one at a time) to make inferences about system robustness. Spatial organization introduces new parameters which are associated with size and location: these parameters vary across the life cycle of the cell, and the interplay of these factors and kinetic factors determines the actual robustness of the biomolecular network.

4.2.3. Spatial Organization and Evolution

Spatial organization may either be a passive (present but not manipulated) or an active ingredient in evolution (§3.4). As a passive ingredient, it can exhibit a significant impact by providing constraints (and also capabilities) for the feasibility/advantages of kinetic changes (and for realizing particular behaviour). Furthermore, the order of transitions in chemistry and spatial organization are important: in some cases, alterations in spatial organization building on kinetic alterations could create an advantageous scenario without significantly compromising system behaviour (which can explain why it is observed), while reversing the order of events may be deleterious to the cell.

4.2.4. A Concrete Cellular Context

The cell-cycle network in Caulobacter involves two interacting modules with bifunctional enzymes: the DivK-DivJ-PleC and the CckA-CtrA modules, where PleC and CckA are bifunctional enzymes modulating the phosphorylation/dephosphorylation of DivK and CtrA (key cell-cycle drivers), respectively (figure 8). Interestingly, both modules are subject to temporal and spatial regulation, which is precisely choreographed. What relevance is our analysis here? There exist experimental studies in different cell-cycle phases and models which aim to consolidate known knowledge, explain aspects of observed behaviour and are predictive. The models make different assumptions about the behaviour of the bifunctional enzymes. Bistability and irreversible decision-making are at the core of a model at one stage of the cell cycle while other models treat distributed bifunctional enzymes as separate enzymes. Yet there are important questions associated with the kinetic assumptions and consistency of different models, the implications of bistability when extrapolated to different cell-cycle stages, the factors responsible for establishing a particular enzymatic activity pattern at the new pole, and understanding where bifunctionality is central.

While addressing these questions needs a dedicated study of its own, our approach can serve as a platform, with a distinct explorative framework. (i) Our analysis allows us to directly build from the study of a single module to two interacting modules (with augmentation to account for additional factors) with spatial and temporal regulation. This creates a library of scenarios for pathway behaviour in terms of assumptions about individual modules, and localization of respective components. Using this, it is possible to systematically tease out the effects of mutations which may alter spatial organization of different components in each of the modules, making a series of new testable predictions. Experiments on mutants arising from mislocalization have already been performed. (ii) Our analysis has implications for the possibility of bistability as a basis of a transition in one cell-cycle phase. We find that, once localization is lost, the bistability may be lost, but also that it may be retained (and propagated to other phases). The former case is guaranteed for sufficiently large growing cells, while the latter is possible with sufficient total substrate at the initial pole, or insufficient changes in cell size. This could be seen (or bypassed) in natural and mutant cells. (iii) Our study provides a number of insights into how opposite activity may be realized spatially and temporally at different locations (seen in *Caulobacter*), using basic modes of regulation, something which is not easily experimentally established.

4.2.5. Synthetic and Engineered Biology

Multiple synthetic biochemical circuits have been developed; for example, oscillators based on phosphorylation, and other biochemical logic modules. Considerable progress has been made recently in engineering compartmentalization through various means, including the use/creation of synthetic microcompartments and in general harnessing localization and compartmentalization, both statically and dynamically, and creating networks of compartments. Different components ranging from entire cell populations to gene regulatory machinery to enzymes have been localized. Compartmentalization has been applied to chemical pathways: this involves localization of selected components in compartments. Notable aspects of compartmentalization implemented in the bacterial cytoplasm include the reversible formation of compartments controlled by (de) phosphorylation, creation of membraneless compartments to facilitate cascade reactions and prevent substrate inhibition, synthetic scaffolds for manipulating spatial organization of enzymes and microcompartments for controlled encapsulation of enzymes. For bottom-up synthetic biology, the basic questions centre around the effect of tuneable dials, which affect transport within and between compartments and how different components may be localized. Our analysis demonstrates the effect of (de)localization of different combinations of components. This 'spatial design' approach can guide rational spatial engineering in biochemical pathways, including enzymatic cascades and multisite substrate modification and assessing multiple aspects of compartmentalization. Unique features of bifunctional enzymes in this regard include the distribution of bifunctional enzymes to multiple locations without significant loss of effect, their maintenance of local reversibility of reaction and the way robust output behaviour can be obtained. Each of these features could be exploited in synthetic biology. Another aspect is using localization as a way of tuning enzymatic activity temporally and spatially.

4.2.6. Chemical Information Processing

Subtle differences in chemistry (TCSs and CMCs) can determine whether spatial information is backpropagated 'spatial retroactivity') at steady state (§3.1).

The interplay of spatial organization and chemistry is bidirectional: spatial organization affects chemical behaviour, while chemistry can impact spatial organization. Both these aspects, and their confluence, can impact a pathway behaviour at different levels/scales, as exemplified by the following. (i) For bifunctional enzymes, details such as whether localization allows for enzymatic interconversion or enzymes being active, as well as any bias in localization of enzyme forms, can significantly impact the overall enzyme activity (§3.2). (ii) For a transition from monofunctional to bifunctional enzymes (by gene fusion), replicating the localization pattern of monofunctional enzymes is difficult if only one of the parent enzymes is localized: it can be achieved in bifunctional enzymes only if a similar correlation between enzyme activity (i.e whether the enzyme is in the conformation of a kinase or a phosphatase) and localization is maintained. This in turn imposes a significant constraint at a *molecular* level, requiring binding capabilities correlated with distinct enzyme conformational structures (§3.3).

A consolidated view of spatial regulation of enzymes/substrates is crucial for unravelling signalling networks in bacteria and eukaryotes, identifying design principles/ features, engineering pathways in synthetic biology and chemistry, with broader implications for chemical information processing. A systems analysis focusing on basic building blocks and modes of spatial organization creates a platform providing insights into all these areas, allowing for a seamless merging of the natural and the synthetic in this context.

13

STRUCTURE AND FUNCTION OF CELLS

13.1 INTRODUCTION: AN OVERVIEW

- The **cell** (from Latin *cella*, meaning "small room") is the basic structural, functional and biological unit of all known living organisms. Cells are the smallest unit of life that can replicate independently, and are often called the "building blocks of life" (http://en.wikipedia.org/wiki/Cell_biology). The cell is the structural integrity, functional. and hereditary smallest of living creatures in the form of a small space bounded by membranes and contains a concentrated liquid. In Becker. et al (2000:2) mentioned that the cell is the basic unit of biology.
- Cells consist of a protoplasm enclosed within a membrane, which contains many biomolecules such as proteins and nucleic acids. Organisms can be classified as unicellular (consisting of a single cell; including most bacteria) or multicellular (including plants and animals). While the number of cells in plants and animals varies from species to species, humans contain about 100 trillion (1014) cells. The cells come from preexisting cells and have a life of their own in addition to their joint role in the multicellular organism. Most living things are composed of single cells. or so-called unicellular organisms. such as bacteria and amoeba. Other living things. including plants. animals. and humans, are multicellular organisms composed of many specialized cell types with their respective functions Most plant and animal cells are visible only under the microscope, with dimensions between 1 and 100 micrometres. The human body is composed of more than 1013 cells. Nevertheless. the whole body of all organisms derived from a single cell division results. For example, the body of bacteria derived from the parent bacterial cell division, while the bodies of mice derived from cell division of the fertilized egg parent.
- The cell was discovered by Robert Hooke in 1665. The cell theory, first developed in 1839 by Matthias Jakob Schleiden and Theodor Schwann, states that all

organisms are composed of one or more cells, that all cells come from preexisting cells, that vital functions of an organism occur within cells, and that all cells contain the hereditary information necessary for regulating cell functions and for transmitting information to the next generation of cells. Cells emerged on Earth at least 3.5 billion years ago.

- Approximately 200 years later, Dutrochet. von Scheleiden. and Schwaan Hook confirms discovery. In 1824. R.J.H. Dutrochet cells expressed the principle which states that all animals and plants are composed of cells that stick together by the force of the adhesive. Then in 1838.
- M.J. Scheleiden published a book that includes an understanding of the genesis of plant tissue. Scheleiden finding suggests the presence of nucleoli and cell theory in plants. Meanwhile next year. T. Schwaan put forward the theory in animal cells. Cell theory states that living things are composed of cells. The discovery of the cell theory above Durjadin line with findings in 1835 that found that in the cell there is a viscous substance. which is now known as protoplasm. In 1839, Theodor Schwann, who after discussing with Schleiden realized that he had observed the nucleus of the animal cell as Schleiden studied in plants, suggesting that all animal parts are also made up of cells. According to him. the universal principle of formation of various body parts of all organisms is the cell formation. In the mid-19th century, in 1958. R. Virchow put forward the theory that corrects the theory of abiogenesis biogenesis. Biogenesis theory states that all living cells come from cells that already exist. The concept was popular with Omnis cellula cellulae. Later in the 20th century that many experts find various types of structures and formations contained within the cell. For example,in 1867, L. ST. George found the cell organelles are now called - Golgi complex. In 1869. F. Meischer find nuclein. and in 1887. van Beneden find centrioles.
- The cell is the smallest unit of life. All organisms alive today. derived from a stem cell existing in millions of years ago. These cells undergo a gradual evolution going to adjust to its environment. Based on these changes. it is now the cell can be grouped into two major groups. namely prokaryotic cells (prokaryotic) and eukaryotic cells (eukaryotic).

13.2 TYPES OF CELL

- Prokaryotic and eukaryotic term first used by Hans Ris in 1960.
- Eukaryotic cells are distinguished from the more primitive **prokaryotic cells** by the presence of:
    i. cytoplasmic membranous organelles,
    ii. a nuclear membrane (i.e. a true nucleus), and
    iii. chromosomal proteins. In this lab we will focus primarily on organelles, their functions within the cell and how they differ between plant and animal cells.

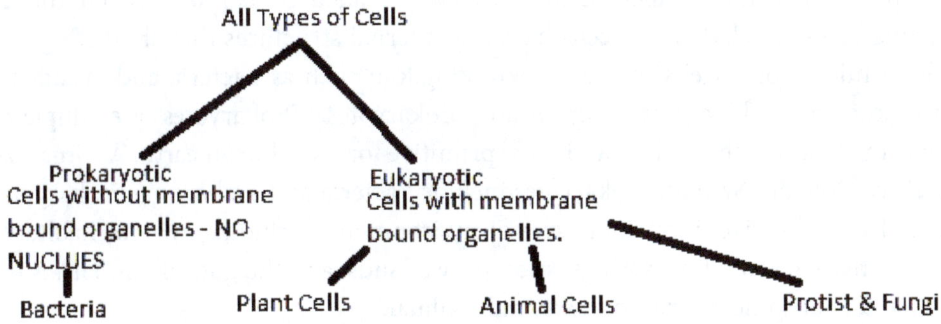

Fig. 1: All types of cells

Important discoveries about cells growing in line with advances in technology and the discovery of advanced tools. Until now it is known that the structure and cell activity is not as simple as previously thought. For more details schemes cell development and cell theory can be seen in the figure below.

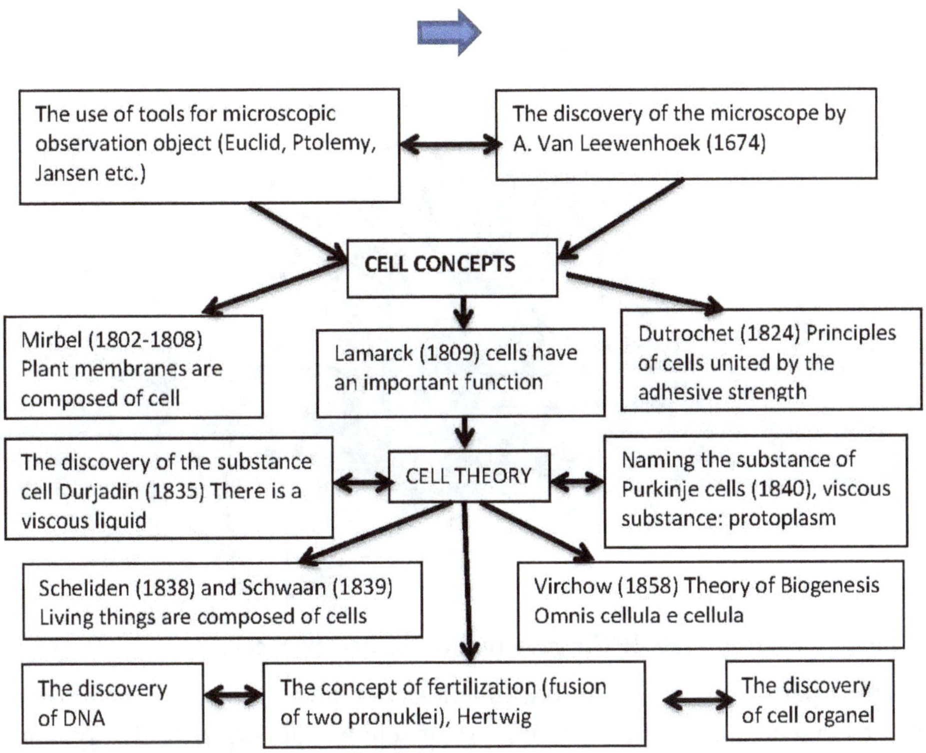

Fig. 2: Schemes cell development and cell theory

13.2.1 Prokaryotic Cells

Cells that lack a membrane-bound nucleus are called prokaryotes (from the Greek meaning before nuclei). These cells have few internal structures that are distinguishable under a microscope. Cells in the monera kingdom such as bacteria and cyanobacteria (also known as blue-green algae) are prokaryotes. Prokaryotes are single-celled organisms that are the earliest and most primitive forms of life on earth. As organized in the Three Domain System, prokaryotes include bacteria and archaeans. Prokaryotes are able to live and thrive in various types of environments including extreme habitats such as hydrothermal vents, hot springs, swamps, wetlands, and the guts of animals. Most are unicellular, but some prokaryotes are multicellular.

Prokaryotic cells are the simplest systems that exhibit all of the signs of life. They are the smallest types of cell, averaging 2-5 μm in length, which makes them just visible under the light microscope. Prokaryotic cells differ significantly from eukaryotic cells. They don't have a membrane-bound nucleus and instead of having chromosomal DNA, their genetic information is in a circular loop called a plasmid. Bacterial cells are very small, roughly the size of an animal mitochondrion (about 1-2μm in diameter and 10 μm long). Prokaryotic cells feature three major shapes: rod shaped, spherical, and spiral. Instead of going through elaborate replication processes like eukaryotes, bacterial cells divide by binary fission.

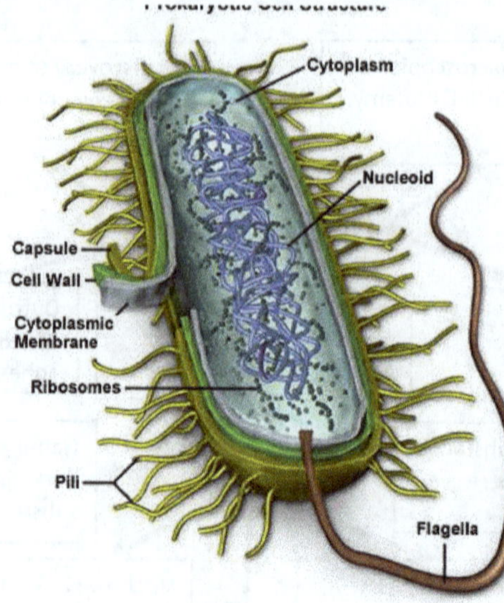

Fig. 3: The structure of the bacterial cell.

Bacteria perform many important functions on earth. They serve as decomposers, agents of fermentation, and play an important role in our own digestive system. Also, bacteria are involved in many nutrient cycles such as the nitrogen cycle, which restores nitrate into the soil for plants. Unlike eukaryotic cells that depend on oxygen for their metabolism, prokaryotic cells enjoy a diverse array of metabolic functions. For example, some bacteria use sulfur instead of oxygen in their metabolism.

Despite their small size, inside each cell there is the complete chemical and biochemical machinery necessary for growth, reproduction and the acquisition and utilization of energy. Prokaryotes have a large array of abilities. Some of them live in the absence of oxygen, some live in extreme conditions of heat or cold, others at the bottom of oceans where the only source of energy is hot hydrogen sulfide bubbling up from the core of the earth.

Prokaryotic cells are not as complex as eukaryotic cells. They have no true nucleus as the DNA is not contained within a membrane or separated from the rest of the cell, but is coiled up in a region of the cytoplasm called the nucleoid. Using bacteria as our sample prokaryote, the following structures can be found in bacterial cells (http://biology.about.com/od/cellanatomy/ss/prokaryotes.htm):

i. Capsule – A gelatinous capsule is present in some bacteria outside the cell membrane and cell wall. The capsule may be polysaccharide as in pneumococci, meningococci or polypeptide as Bacillus anthracis or hyaluronic acid as in streptococci. Capsules are not marked by normal staining protocols and can be detected by India ink or methyl blue; which allows for higher contrast between the cells for observation. Found in some bacterial cells, this additional outer covering protects the cell when it is engulfed by other organisms, assists in retaining moisture, and helps the cell adhere to surfaces and nutrients.
ii. Cell Wall - Outer covering of most cells that protects the bacterial cell and gives it shape. The cell wall acts to protect the cell mechanically and chemically from its environment, and is an additional layer of protection to the cell membrane. Different types of cell have cell walls made up of different materials; plant cell walls are primarily made up of pectin, fungi cell walls are made up of chitin and bacteria cell walls are made up of peptidoglycan.
iii. Cytoplasm - A gel-like substance composed mainly of water that also contains enzymes, salts, cell components, and various organic molecules. The cytoplasm in prokaryotic cells is a gel-like, yet fluid, substance in which all of the other cellular components are suspended. Jello for cells. It is very similar to the eukaryotic cytoplasm, except that it does not contain organelles. Recently, biologists have discovered that prokaryotic cells have a complex and functional cytoskeleton similar to that seen in eukaryotic cells.2 The cytoskeleton helps prokaryotic cells divide and helps the cell maintain its plump, round shape. As is the case in eukaryotic cells, the cytoskeleton is the framework along which particles in the cell, including proteins, ribosomes, and small rings of DNA called plasmids, move around.
iv. Cell Membrane or Plasma Membrane - Surrounds the cell's cytoplasm and regulates the flow of substances in and out of the cell. Prokaryotic cells can have multiple plasma membranes. Prokaryotes known as "gram-negative bacteria," for example, often have two plasma membranes with a space between them

known as the periplasm. Just inside the cell wall, the plasma membrane is a selective barrier which regulates the passage of materials to from the cell. It is through this membrane that a cell must exchange food molecules, gases and other vital ingredients. Composed of phospholipid and protein membranes form thin, flexible, self-sealing, highly selective barriers between the inside of the cell and the outside world. As in all cells, the plasma membrane in prokaryotic cells is responsible for controlling what gets into and out of the cell. A series of proteins stuck in the membrane (poor fellas) also aid prokaryotic cells in communicating with the surrounding environment. Among other things, this communication can include sending and receiving chemical signals from other bacteria and interacting with the cells of eukaryotic organisms during the process of infection. Infection is the kind of thing that you don't want prokaryotes doing to you. Keep in mind that the plasma membrane is universal to all cells, prokaryotic and eukaryotic. Because this cellular component is so important and so common, it is addressed in great detail in its own In Depth subsection.

v. The genetic information on the plasmids is transferrable between cells, allowing prokaryotes to share such abilities as antibiotic resistance. Humans have discovered that prokaryotic plasmids can be genetically engineered. Today, they are isolated, changed to carry other interesting information and then re-introduced into new cells. In this way unique and usefull little bacterial factories can be designed, created and put to work.

vi. Fimbriae (pili) - Hair-like structures on the surface of the cell that attach to other bacterial cells. Shorter pili called fimbriae help bacteria attach to surfaces. Fimbriae are responsible for attachment of bacteria to specific receptors of human cell (adherence). There are special types of pili called (sex pili) involved in conjunction.

vii. Flagella - Long, whip-like protrusion that aids in cellular locomotion. Flagella are organelles for cellular mobility. The bacterial flagellum stretches from cytoplasm through the cell membrane(s) and extrudes through the cell wall. They are long and thick thread-like appendages, protein in nature. Are most commonly found in bacteria cells but are found in animal cells as well. These are strands of protein that pass though the outer surface of the cell body either either singly or in tufts. Energy provided by the plasma membrane rotates the flagellum by means of a unique rotating 'joint' and this in turn moves the bacterium through its liquid world. Prokaryotic flagella are very different from similar looking structures used by eukaryotic cells.

viii. Ribosomes - Cell structures responsible for protein production. Prokaryotic ribosomes are smaller and have a slightly different shape and composition than those found in eukaryotic cells. Bacterial ribosomes, for instance, have about half of the amount of ribosomal RNA (rRNA) and one third fewer ribosomal proteins (53 vs.

ix. ~83) than eukaryotic ribosomes have.3 Despite these differences, the function of the prokaryotic ribosome is virtually identical to the eukaryotic version. Just like in eukaryotic cells, prokaryotic ribosomes build proteins by translating messages sent from DNA.

x. Plasmids - Gene carrying, circular DNA structures that are not involved in reproduction.

xi. Nucleiod Region - Area of the cytoplasm that contains the single bacterial DNA molecule. All prokaryotic cells contain large quantities of genetic material in the form of DNA and RNA. Because prokaryotic cells, by definition, do not have a nucleus, the single large circular strand of DNA containing most of the genes needed for cell growth, survival, and reproduction is found in the cytoplasm. The DNA tends to look like a mess of string in the middle of the cell:

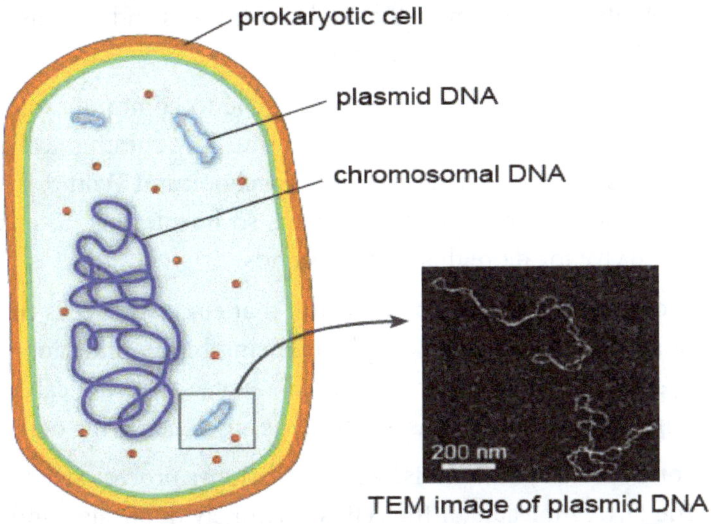

Fig. 4: Transmission electron micrograph image

Usually, the DNA is spread throughout the entire cell, where it is readily accessible to be transcribed into messenger RNA (mRNA) that is immediately translated by ribosomes into protein. Sometimes, when biologists prepare prokaryotic cells for viewing under a microscope, the DNA will condense in one part of the cell producing a darkened area called a nucleoid.

As in eukaryotic cells, the prokaryotic chromosome is intimately associated with special proteins involved in maintaining the chromosomal structure and regulating gene expression. In addition to a single large piece of chromosomal DNA, many prokaryotic cells also contain small pieces of DNA called plasmids. These circular rings of DNA are replicated independently of the chromosome and can be transferred from one prokaryotic cell to another through pili, which are small projections of the cell membrane that can form physical channels with the pili of adjacent cells.

The transfer of plasmids between one cell and another is often referred to as "bacterial sex." Sounds dirty. The genes for antibiotic resistance, or the gradual ineffectiveness of antibiotics in populations, are often carried on plasmids. If these plasmids get transferred from resistant cells to nonresistant cells, bacterial infection in populations can become much harder to control. For example, it was recently learned that the superbug MRSA, or multidrug-resistant Staphylococcus aureus, received some of its drug-resistance genes on plasmids.

Prokaryotic cells are often viewed as "simpler" or "less complex" than eukaryotic cells. In some ways, this is true: prokaryotic cells usually have fewer visible structures, and the structures they do have are smaller than those seen in eukaryotic cells. Don't be fooled, however, into thinking that just because prokaryotic cells seem "simple" that they are somehow inferior to or lower than eukaryotic cells and organisms. Making this assumption can get you into some serious trouble.

Biologists are now learning that bacteria are able to communicate and collaborate with one another on a level of complexity that rivals any communication system ever developed by humans. Prokaryotes showed you, Facebook and Twitter. In addition, some Archaean cells are able to thrive in environments so hostile that no eukaryotic cell or organism would survive for more than a few seconds.

Prokaryotic cells are also able to pull off stuff that eukaryotic cells could only dream of, in part because of their increased simplicity. Being bigger and more complex is not always better. These cells and organisms are just as adapted to their local conditions as any eukaryote, and in that sense, are just as "evolved" as any other living organism on Earth.

Reproduction in prokaryotic cells is by binary fission; a process of growth, enlargment and division. The DNA molecule of the cell is accurately duplicated and the two copies separated form each other by movement of the cell membrane to which they are attached. The cell then divides into two smaller but identical cells and each begins its own independent existence.

Fig. 5: Reproduction in prokaryotic cells is by binary fission

13.2.2 Eukaryotic Cells

Eukaryotes are organisms whose cells are organized into complex structures by internal membranes and a cytoskeleton. The most characteristic membrane bound structure is the nucleus. This feature gives them their name, (also spelled "eucaryote,") which comes from the Greek ευ, meaning good/true, and κάρυον, meaning nut, referring to the nucleus.

Animals, plants, fungi, and protists are eukaryotes. Eukaryotes is wrapped by a nuclear membrane so that it does not mix with the cytoplasm. The most striking difference from prokaryotic cells is the true nucleus that encloses most of the cell's DNA so that the DNA is stored in a different compartment of the cytoplasm.

To better understand the models shown below eukaryotic cells. in the form of three-dimensional and two- dimensional. Parts of the cell are described below in particular will be discussed further in this learning activity.

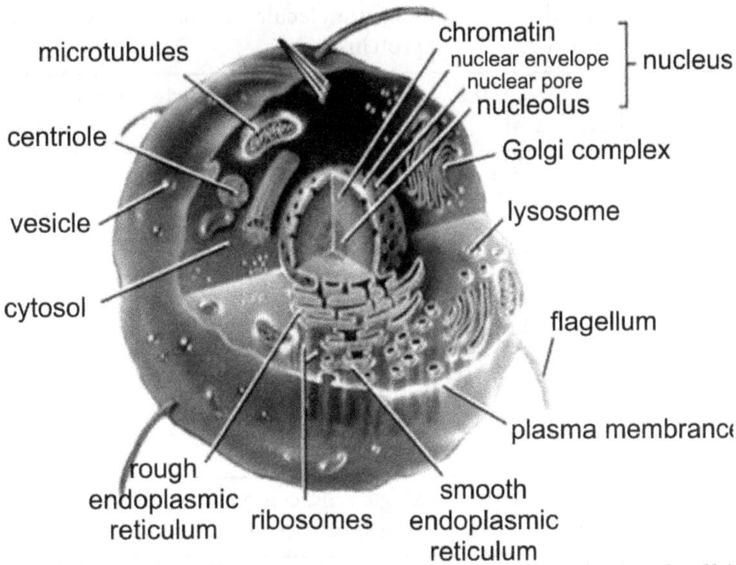

Fig. 6: Eukaryotic Cells models in three dimensions (animal cells).

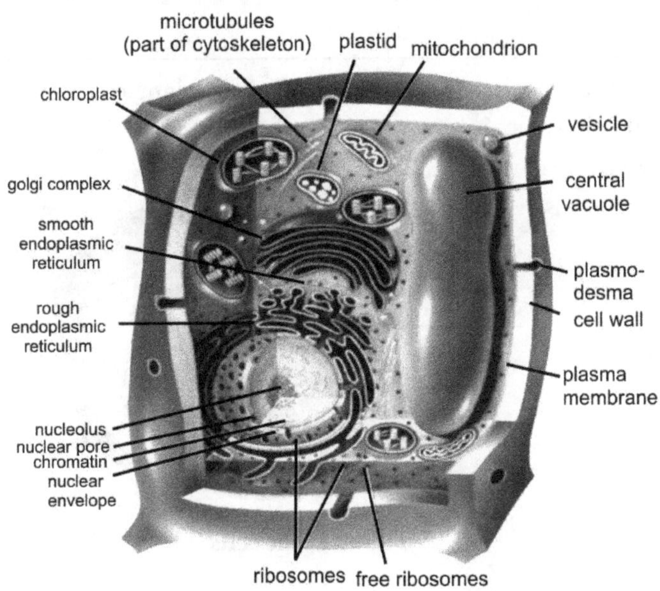

Fig. 7: Eukaryotic cell model in two dimensions.

Table 1. Comparison of features of prokaryotic and eukaryotic cells

	Prokaryotes	Eukaryotes
Typical organisms	Bacteria, archaea	Protists, fungi, plants, animals
Typical size	~ 1–5 Mm[9]	~ 10–100 Mm[9]
Type of nucleus	Nucleoid region; no true nucleus	True nucleus with double membrane
DNA	Circular (usually)	Linear molecules (chromosomes) with histone proteins
RNA/protein synthesis	Coupled in the cytoplasm	Rna synthesis in the nucleus protein synthesis in the cytoplasm
Ribosomes	50S and 30s	60S and 40s
Cytoplasmic structure	Very few structures	Highly structured by endomembranes and a cytoskeleton
Cell movement	Flagella made of flagellin	Flagella and cilia containing microtubules; lamellipodia and filopodia containing actin
Mitochondria	None	One to several thousand (though some lack mitochondria)
Chloroplasts	None	In algae and plants
Organization	Usually single cells	Single cells, colonies, higher multicellular organisms with specialized cells
Cell division	Binary fission (simple division)	Mitosis (fission or budding) meiosis

In addition we can also see the difference between plant and animal cells by observing the models listed below.

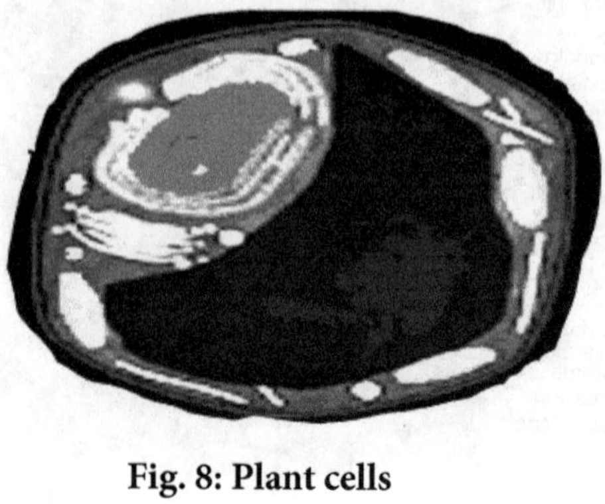

Fig. 8: Plant cells

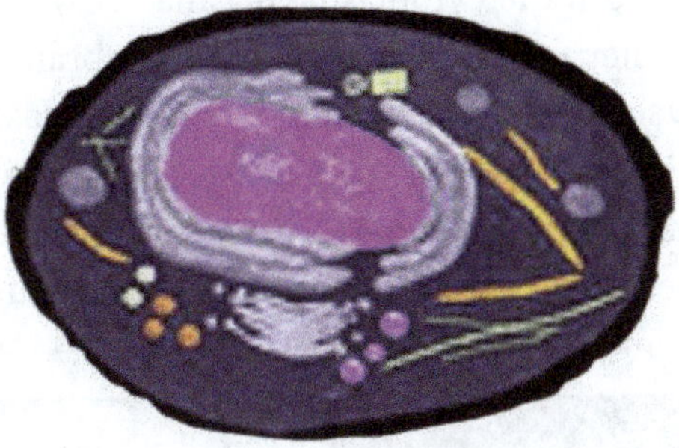

Fig. 9: Animal cells

A cell structure is illustrated in Figure 9. Plant cells and animal cells Figure 1.6. which describes the shape of various types of cells.

If we look at the plants and animals have a very big difference. where plants can not move with such active animals. This is because the shape of plant cells are rigid so it is not flexible. in contrast to animal cells that are flexible and can change shape. Aside from the shape. the differences plant cells and animal cells can also be differentiated from the following:

Table 2. Differences as plant and animal cells

Plant Cells	Animal Cells
Plant cells Animal cells larger than	Animal cells are smaller than plant cell
Do not have the lysosome	Not having plastids
Do not have the centrosome	Do not have cell walls
Have a cell wall and cell membrane	Having lysosomes
Generally have plastids	Having centrosome
Have a fixed shape	Have no fixed shape
Has a large vacuole size. Lot	Not having vakuala (although there also have vacuoles but small size)

There are two main parts of the cell. namely: core and its contents are often called nucleoplasm. and the remaining part is called the cytoplasm. Nucleus and cytoplasm were surrounded by a membrane. as well as smaller parts like mithokhondria and Golgi bodies. Broadly speaking, the structure and function of each cell component is as follows:

Nuclei

The cell nucleus consists of a nuclear membrane, nucleoplasm, nucleolus and chromosomes. The nuclear membrane is a double membrane that has four phospholipid layers and large pores through which materials pass. It also contains a viscous liquid known as the nucleoplasm. The nucleus is the most prominent organelle in the cell. This small organ is separated from the cytoplasm (plasma cells) by wrapping which consists of two membranes. the inner membrane and outer membrane. The nucleus contains the genetic material that is Deoxy Ribonucleic Acid (ADN) is encased in a nuclear membrane. All chromosomal DNA is stored in the nucleus. packed in chromatin fibers thanks to its alliance with the histone proteins that same mass. Fill nucleus communicates with the cytosol through the holes in the wrapper called pores nucleus. Nucleoli in the nucleus there is a place for ribosoma producing cells.

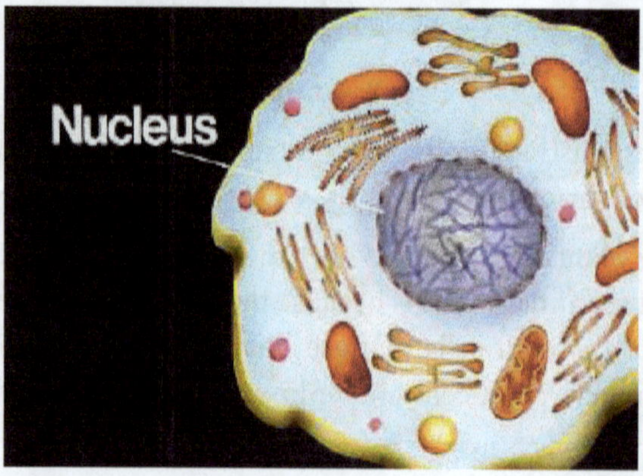

Fig. 10: The structure of the nucleus (micro.magnet.fsu.edu).

Plas5ma membrane

Cell membranes are found in animal cells whereas cell walls are found in plant cells. Cell walls and membranes have similar functions. Like a city perimeter, cell membranes surround the cell and have the ability to regulate entrance and exit of substances, thereby maintaining internal balance. These membranes also protect the inner cell from outside forces. Cell walls, as the city analogy implies, are much stronger than cell membranes and protect cells from lysing (exploding) in extremely hypotonic (diluted) solutionsMembrane is very thin and is selectively permeable to the size of 7.5- 10 nm. The plasma membrane is a lipid double layer (bilayer) is the molecular structure of two layers. Lipids are important are glycolipids and phospholipids and little chance of containing cholesterol. The structure of the plasma membrane of cells such support to be able to take advantage of changes in ion permeability control at the plasma membrane of cells for communication purposes. In addition, it also serves as a protective organelles within the cell.

Different from the plasma membrane of eukaryotic cells, the plasma membrane in eukaryotic cells can develop specialized capabilities or organelles. In eukaryotic cells, which do not have mitochondria. the plasma membrane is also in charge of implementing energy metabolism. The difference is what causes that in eukaryotic cells. the plasma membrane is not formed mesosom.

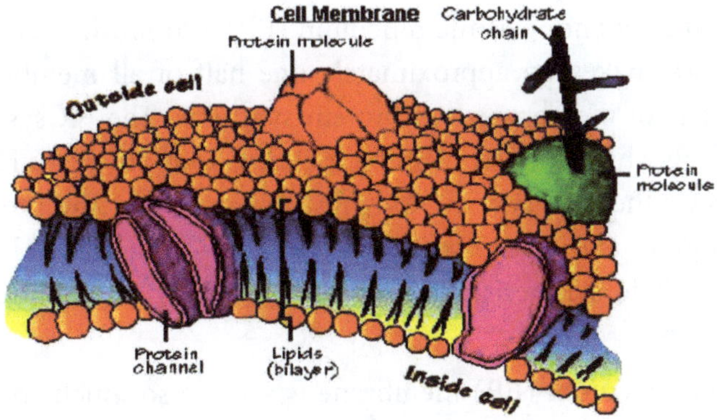

Fig. 11: Plasma Membrane

The cell organelles

Organelles are parts of the cell which are adapted and/or specialized for carrying out one or more vital functions, analogous to the organs of the human body (such as the heart, lung, and kidney, with each organ performing a different function). Both eukaryotic and prokaryotic cells have organelles, but prokaryotic organelles are generally simpler and are not membrane-bound.

There are several types of organelles in a cell. Some (such as the nucleus and golgi apparatus) are typically solitary, while others (such as mitochondria, chloroplasts, peroxisomes and lysosomes) can be numerous (hundreds to thousands). The cytosol is the gelatinous fluid that fills the cell and surrounds the organelles.

The number of organelles in the cytoplasm of eukaryotic cells are more complex than prokaryotic cells. The organelles eg mitochondria. endoplasmic reticulum. nucleus. ribosomes. microtubules. and others.

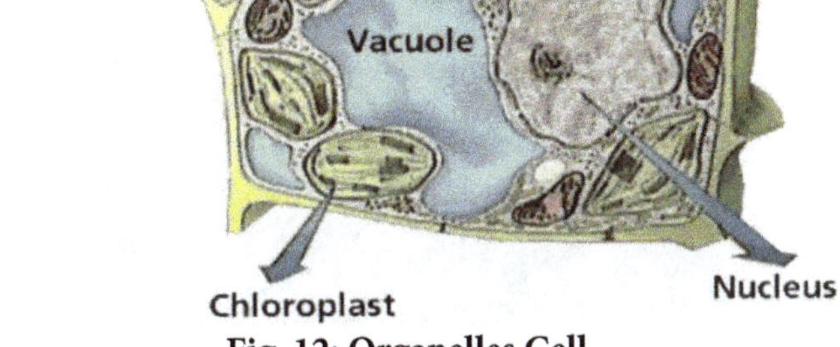

Fig. 12: Organelles Cell.

Endoplasmic Reticulum (Endoplasmic reticulum)

There are two types of endoplasmic reticulum (ER) – **Smooth ER** and **Rough ER**. This extensive network makes up approximately one half of all membranous tissue of the cell and is the site of membrane and protein synthesis. The ER system is much like a road system along which industry can be found. Goods are manufactured and shipped to needed areas via the road system. Rough ER is named for the presence of ribosomes along its membrane and is the source of proteins. Smooth ER lacks ribosomes and is responsible for lipid synthesis and processes a variety of metabolic processes such as drug detoxification.

Endoplasmic reticulum (ER) membrane is a maze so much so that it covers more than half the total membrane in eukaryotic cells. The word ` endoplasmic means within the cytoplasm ' and reticulum derived from the Latin word which means the network. RE consists of a network of tubules and membrane bubbles called sisternal or lumen. ER membrane separates the internal space. namely the space sisternal from the cytosol. And because the ER membrane continuous with the nuclear envelope. the room between the two membrane sheath was continuous with the RE sisternal room.

In general, the RE has the following functions;

- Performers synthetic metabolic activity. because it contains a variety of enzymes.
- denaturation and elongation of fatty acids.
- Providing a wide surface for enzymatic reactions.
- An ultra- structural skeleton that provides the mechanical strength of the cell. the cytoplasm koloidalnya matrix.
- As a place of exchange of molecules through a process of osmosis. diffusion and active transport to the ER membrane and eksosistosis.
- Establish a new core wrap on cell division.
- cell protection function for the ER membrane is able to eliminate the toxic effects of substances through the detoxification process.

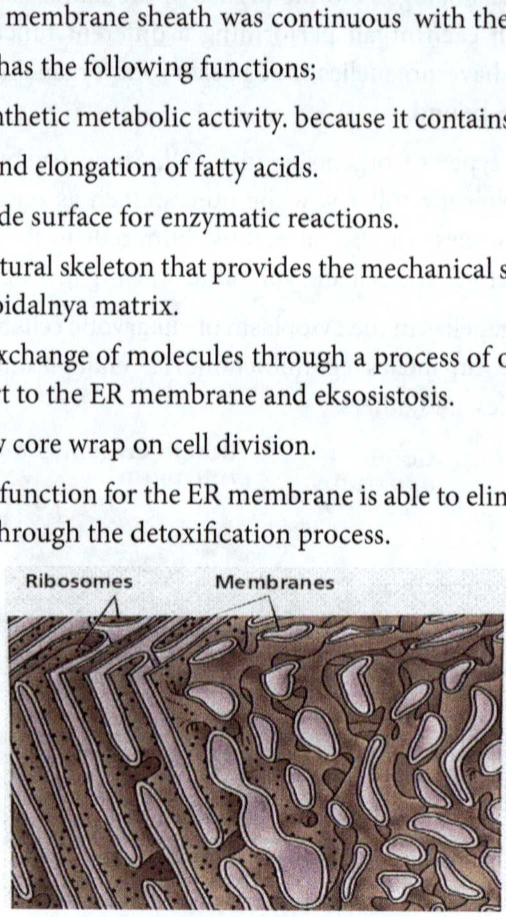

Fig. 13: Endoplasmic reticulum.

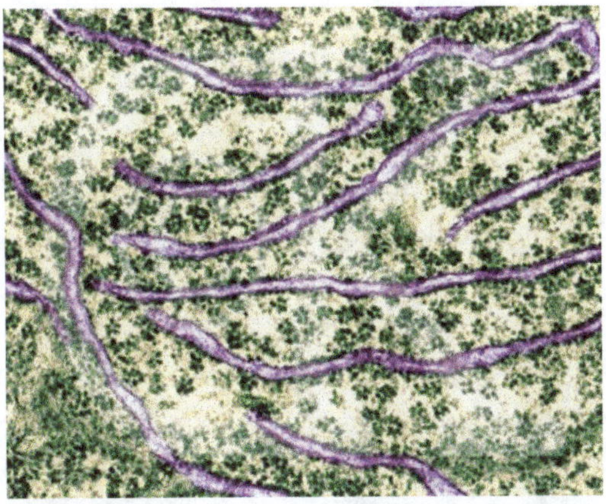

Fig. 14: Photos using Scanning Electron Microscopy of reticulum endoplasmic and the ribosomes. (TEM x 61.560).

xii. **Golgi apparatus (Golgi Apparatus/Golgi Complexes)**

Like a post office, the golgi apparatus is used for shipping those goods created by the ER and ribosomes to the rest of cell. These organelles were first discovered by Camilio Golgi. a scientist from Italy. Golgi apparatus is common on plant and animal cells. In animal cells are 10-20 Golgi apparatus. As with plants that have hundreds of Golgi bodies in each cell. Golgi apparatus has a length of about 1-3 lm and a height of about 0.5 lm. Golgi apparatus including cell vacuolar system and there are no ribosomes. On the polar structure of cells. single Golgi apparatus. large and occupies at the core and at the poles of the cell. for example in the glandular

cells eksokim prankeas. In liver cells in a cell. there are about 50 that form the Golgi complex varies between cells with one another. Golgi apparatus consists of a group of membrane bounded flattened bag called saccula. Near saccula contained secretory vesicles form spherical bubbles. The Golgi apparatus in plants called diktiosom. In the manufacture of polysaccharide diktiosom occurs in the form of cellulose that is used as the building blocks of the cell wall.

Based on morphological observations and in situ cytochemistry and biochemical studies indicate that the Golgi apparatus is involved in a large number of cell activities include assembly of proteins and lipids high carbohydrate or better known as glycosylation process. recovery of the cell membrane. and secretion.

In general, the function of the Golgi apparatus. among others:

- forming cell walls in plants
- produces lysosomes
- forming acrosome in spermatozoa containing enzymes to break down the cell wall of the egg.

- Places such as mucus synthesis of polysaccharides. cellulose. hemicellulose. and pectin (constituent of plant cell walls).
- Forming the plasma membrane.
- Forming bag to wrap secretion of substances to be issued a cell. such as proteins. glycoproteins. carbohydrates. and fats.

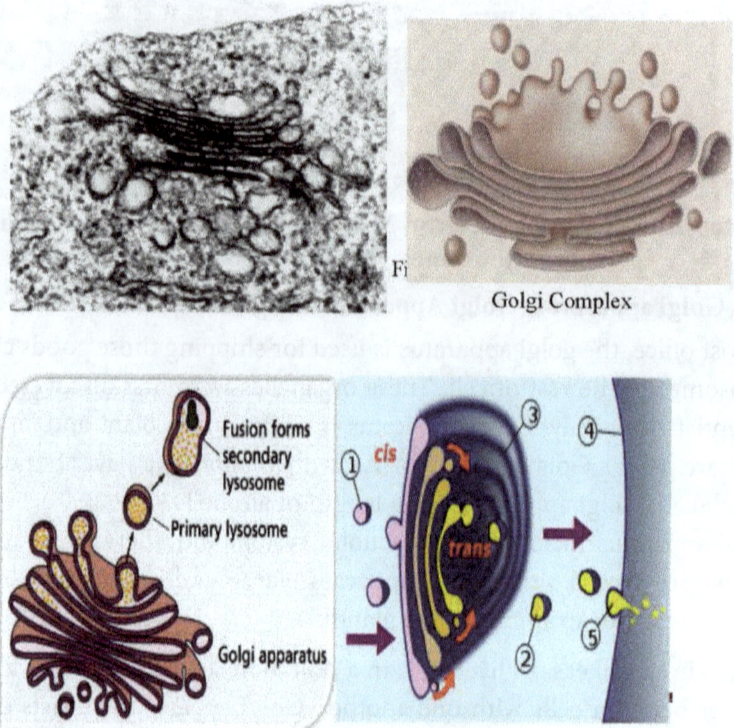

Fig. 15: Golgi apparatus.

Mitochondria

Mitochondria are found in both plant and animal cells and is the site of cellular respiration. Through this process that will be covered in the Photosynthesis and Respiration lab ATP is created which is used for energy by the cell. The size and shape of mitochondria. as well as the numbers in the cells. tissues and varies according to the physiological state of the cell by. By using visible light microscopy oval mitochondria. but mitochondria can also dumbbell -shaped. spherical. or racket. with a diameter of 0.5-1.0 and length up to 7 lm. Due to the very small size of new structures can be viewed using an electron microscope. Mitochondria contain small amounts of DNA. RNA and ribosomes. Mitochondrial DNA provide the password for the synthesis of certain specific proteins on the inner membrane. Most mitochondrial proteins are encoded by nuclear DNA and synthesized by ribosomes are present in the cytosol or in the endoplasmic reticulum. This shows that there is a connection / transfer of information from DNA to the nucleus of mitochondrial later emerged from DNA found in the mitochondria themselves.

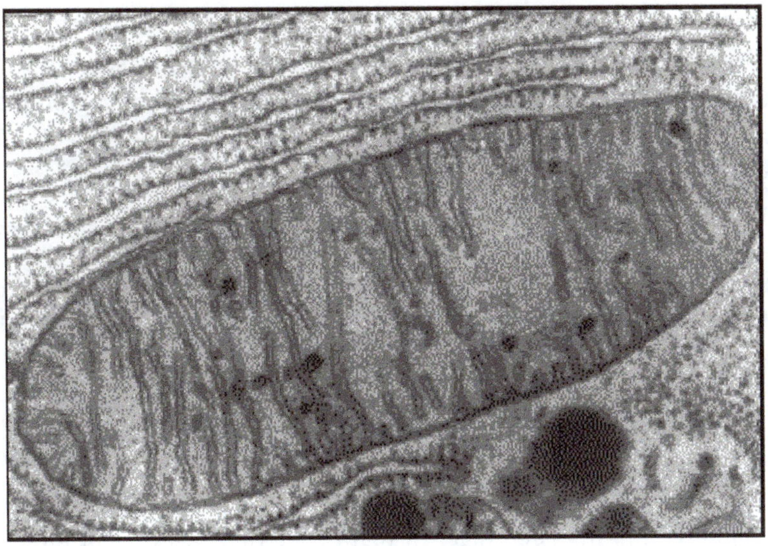

Fig. 16: Electron microscope picture of a mitochondria

Chloroplasts

Place photosynthesis is high organ subcellular structure called chloroplasts. The result of chemical changes in photosynthesis are CO_2 and H_2O into carbohydrates. Carbohydrates are stored and produced as a result of photosynthesis can be seen as grains of starch.

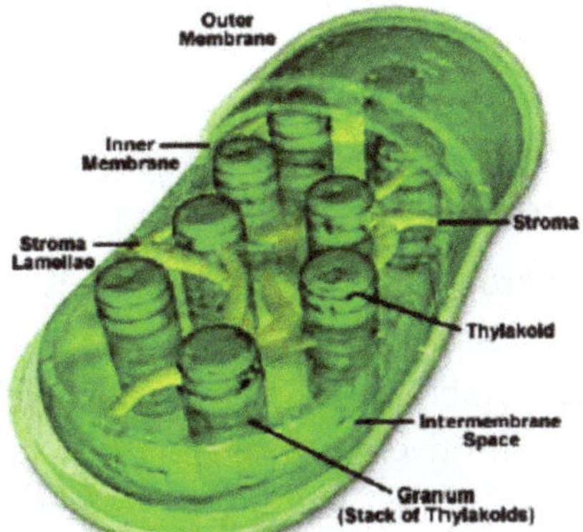

Fig. 17: Cloroplast

In the membrane

Eukaryotic cells typically have a much larger volume than prokaryotic cells. typically a thousand times or more. Material or any material of cells it contains many times more. For example that the human body cells contain DNA that is a thousand times more than that of bacteria.

Membranes in various organelles such as the mitochondria membrane. vacuole membrane (in plant cells). the Golgi apparatus and the other is the venue for important reactions. Due to the addition of the cell volume must be balanced with the addition of the cell surface area by maintaining the ratio of surface area to volume ratio. This explains why all eukaryotic cells have many characteristics of the basic form and complexity of the membrane in the form of:

- In the endoplasmic reticulum membrane that forms a maze -like compartments.
- Golgi membranes in the body that make up the pile of bags deflated play a role in the conversion of product molecules from the endoplasmic reticulum.
- Lysosomal membrane of cells that contains digestive enzymes.
- Peroxisome membrane wrapping where the formation and decomposition of H_2O_2 are reactive and dangerous during the oxidation of a variety of molecules by O_2.
- Vacuole membrane (tonoplas) in plant cells that form small bubbles and large cavity filled with fluid.

With the cell structure located next to the cell. can provide adequate surface area corresponding to the large volume. namely that between membrane-bound compartments within the cell and outside the cell environment occurs an exchange mechanisms (transport) is relentless. The mechanism of endocytosis and exocytosis is. which only occurs in eukaryotic cells.

In endocytosis. the parts forming the outer membrane curvature toward the later rounded and separated into bubbles of membrane-bound cytoplasm and contain substances that come from outside of cells and molecules that have been previously absorbed on the surface of cells.

Exocytosis is the reverse process of endocytosis. In this case. bubbles encased in a cell membrane. referring to the plasma membrane and release their contents into the outer environment. That way. the membranes around the compartment which is located deep within the cell to function effectively increase the cell surface area for the exchange of materials from outside.

Cytoskeleton

The cytoskeleton acts to organize and maintain the cell's shape; anchors organelles in place; helps during endocytosis, the uptake of external materials by a cell, and cytokinesis, the separation of daughter cells after cell division; and moves parts of the cell in processes of growth and mobility. The eukaryotic cytoskeleton is composed of microfilaments, intermediate filaments and microtubules. There are a great number of proteins associated with them, each controlling a cell's structure by directing, bundling, and aligning filaments. The prokaryotic cytoskeleton is less well-studied but is involved in the maintenance of cell shape, polarity and cytokinesis.

The **cyotoskeleton** represents the cell's skeleton. Like the bony skeletons that give us stability, the cytoskeleton gives our cells shape, strength, and the ability to move, but it does much more than that. The cytoskeleton is made up of three types of fibers that constantly shrink and grow to meet the needs of the cell: microtubules, microfilaments, and actin filaments. Each type of fiber looks, feels, and functions differently. **Microtubules** consists of a strong protein called tubulin and they are the 'heavy lifters' of the cytoskeleton. They do the tough physical labor of separating duplicate chromosomes when cells copy themselves and serve as sturdy railway tracks on which countless molecules and materials shuttle to and fro. They also hold the ER and Golgi neatly in stacks and form the main component of flagella and cilia.

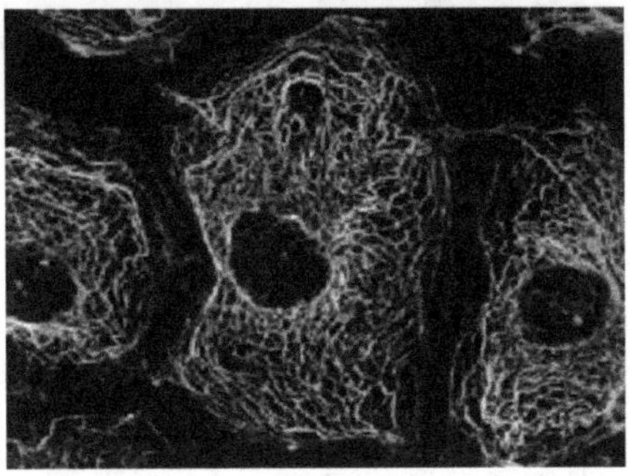

Fig. 18: Cytoskeleton (www.sinauer.com).

All eukaryotic cells are equipped with a cell skeleton (cytoskeleton) that functions give it shape. motility and ability to regulate the organelles and organelles move from one part of the cell to another. This is caused by the growing size of a cell. the more complicated and more specialized structures in it.Thus the greater the necessity to keep these structures remain as they are and adjust their movements.

Cell skeleton is composed of a network of protein filaments. Three of the most important of which are actin filaments (also called microfilaments). intermediate filaments. and Microtubules.

Microfilaments are long thin fibers with a diameter of 5-6 nm. Composed of a protein called actin. Many microfilaments form a collection or network at various places in the cell. That coupled with the presence of cell motion. When an animal cell divides into two. for example. a beam forming microfilaments and separates the two daughter cells.

In many cells. the cytoplasm moved and this phenomenon called cytoplasmic flow. Motion depends on the presence of microfilaments. Microfilaments is also a feature that is important in the cell move and change shape. This does not just apply to the free movement of independent cells as well as amoeba. but also in most animal cells during embryo formation.

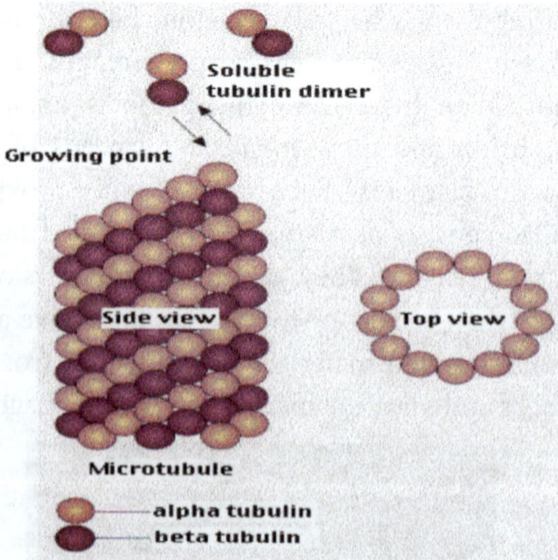

Fig. 19: Scheme of cytoskeleton.

Cytoplasmic intermediate filament is a long fiber with a diameter of about 10 nm. Called intermediate because its diameter is larger than the diameter of microfilaments (6 nm) and smaller than the diameter of microtubules (25 nm) and the filaments ' thick ' (15 nm) in skeletal muscle fibers. Intermediate filaments composed of fibrous protein molecules. Intermedia is a hollow filament yarn consisting of five protofilamen. parallel to one another and form a circle. Intermediate filaments found in many cells that often get mechanical stresses. such as cell epithelium. the axons of nerve cells or smooth muscle cells.

Cylindrical microtubule protein is found in most animal and plant cells. There are two kinds of - tubulin and - tubulin. Each with a molecular weight of about 55,000 daltons. Microtubules also play a very important role in cell division. Successful cell division requires proper distribution of chromosomes into each daughter cell. Each chromosome moves to the goal ended in a bundle of microtubules. Microtubules are also used in the formation of centrioles. basal bodies and flagella.

There are two groups of microtubules: 1) microtubules are stabilized microtubules can be preserved with a fixative solution of any sort. for example: OsO4. MnO4. or aldehydes at any temperature. 2) microtubules are labile microtubules can be preserved.

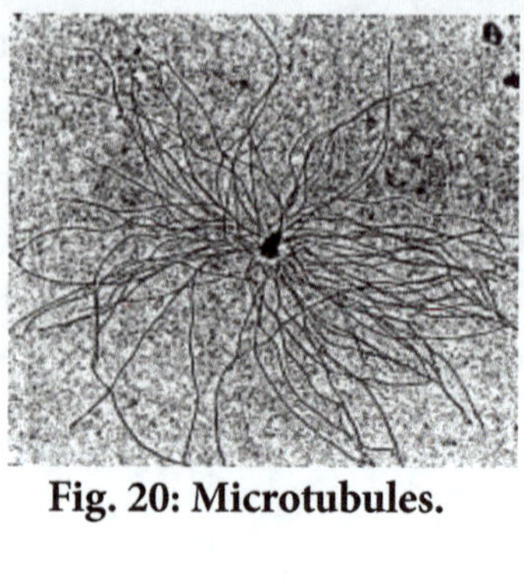

Fig. 20: Microtubules.

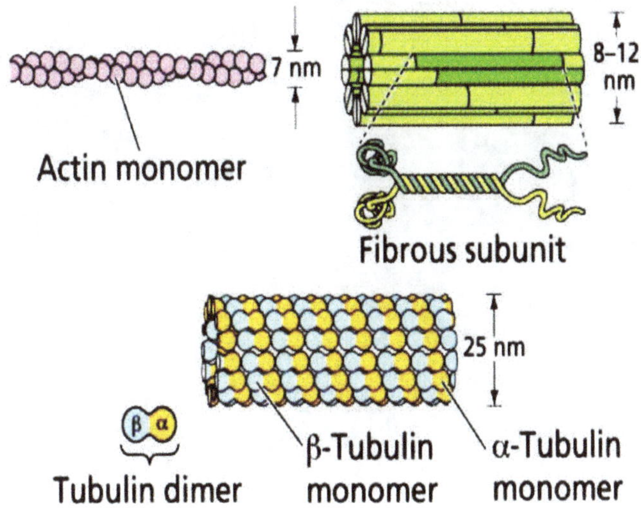

Fig. 21: Microtubules parts.

Table 3. Comparison between the properties of microtubules. microfilaments and intermediate filaments.

The nature / Signs	**Microtubule**	**Intermedia filaments**	**Microfilaments**
Structure	Hollow with wall consists of 13 protofilamen	Hollow with walls consisting of 4-5 rotofilamen	Incompressible (solid) consists of polymeric actin (Actin - F)
Midline (nm)	24	10	7
Unity monomer	α and β tubulin	5 kinds of proteins	Actin –G
ATP-ase activity	Located in dinein	----	---
Function	The ability of the movement in eukaryotes. Kromosoma movement. Intra- cell material movement Maintaining cell shape.	Combine unity contraction in the muscle cells	Berperan dalam kontraksi otot Plays a role in muscle contraction changes in cell shape of protoplasm sitokenesis.

14

MODELLING RADIATION FLUXES IN SIMPLE AND COMPLEX ENVIRONMENTS

14.1 INTRODUCTION: AN OVERVIEW

- Climate and air quality must be considered in urban and regional planning at regional level, as it is relevant to human health and well-being. This task falls within the field of human-biometeorology, which deals with the effects of weather conditions, climate and air quality on humans. Examples of inclusion of human-biometeorological results in urban planning processes are not rare any more; see for example; the Urban Climate Analysis of Freiburg.
- Linking human-biometeorology and town planning requires determining the human-biometeorological factors that are significant for the planning process. The humanbiometeorological evaluation of the atmospheric environment can be used, for example, to clarify whether or not planning instruments are available for maintaining and improving the human-biometeorological situation in the planning and, if they are, which are the appropriate ones to use.
- For the evaluation of the thermal component of urban climate and applied climatology, the methods have been applied and tested in many studies. The presented analysis can be undertaken in different time and space scales. In addition to the mobility of humans in urban areas, a diversity of microclimates in urban areas also exists. Short-wave radiation fluxes have the greatest variation of meteorological parameters in urban areas. The biometeorological influences of short- and long-wave radiation fluxes can be transformed into a synthetic parameter, the mean radiant temperature.
- Tmrt is defined as the uniform temperature of a hypothetical spherical surface surrounding the subject (emissivity ε=1) that would result in the same net

radiation energy exchange with the subject as the actual, complex radiative environment. The latter usually varies considerably under open space conditions.

- On the one hand, existing measurements and methods for the calculation of short- and long-wave radiation fluxes are always applied for horizontal surfaces, and on the other hand radiation fluxes are very complex to measure in urban areas. For human-biometeorological studies, the radiation environment of the human body has to be established as three-dimensional.
- The main purpose of this paper is to present a method for the calculation of Tmrt for thermal human-bioclimate studies in different time and space scales through the use of the RayMan model. The model is compatible with Microsoft Windows, and can analyse complex urban structures and other environments. The model requires only basic meteorological data (air temperature, air humidity and wind speed) for the calculation of radiation fluxes and common thermal indices for the thermal human-bioclimate. Applications and examples of Tmrt and thermal human-bioclimates for different structures and uses are demonstrated. Also shown in the present paper are validations of RayMan simulations with field measurements in the city of Freiburg, southwest Germany, in the summer of 2001.
- The most important meteorological parameter affecting the human energy balance during sunny weather conditions is the mean radiant temperature Tmrt. It considers the uniform temperature of a surrounding surface giving off blackbody radiation, which results in the same energy gain of a human body given the prevailing radiation fluxes. This energy gain usually varies considerably in open space conditions. In this paper, the model 'RayMan', used for the calculation of short- and long-wave radiation fluxes on the human body, is presented. The model, which takes complex urban structures into account, is suitable for several applications in urban areas such as urban planning and street design. The final output of the model is, however, the calculated Tmrt, which is required in the human energy balance model, and thus also for the assessment of the urban bioclimate, with the use of thermal indices such as predicted mean vote (PMV), physiologically equivalent temperature (PET) and standard effective temperature (SET*).
- The model has been developed based on the German VDI-Guidelines 3789, Part II (environmental meteorology, interactions between atmosphere and surfaces; calculation of short- and long-wave radiation) and VDI-3787 (environmental meteorology, methods for the human-biometeorological evaluation of climate and air quality for urban and regional planning. Part I: climate). The validation of the results of the RayMan model agrees with similar results obtained from experimental studies.

14.2 SCIENTIFIC BACKGROUND AND METHODS

14.2.1 Atmospheric Environment

- Cause and effect relations between the atmospheric environment And human health or human comfort can be analyzed by a human-biometeorological classification. The thermal component is an important factor and can be described and quantified by thermal indices. The thermal component comprises the meteorological parameters of air temperature, air humidity, wind velocity, and also short- and long-wave radiation, which thermo-physiologically affects human beings indoors and outdoors. This component is relevant to human health because of the close relationship between the thermoregulatory mechanism and the circulatory system.
- Human-biometeorological studies have been carried out for some time. In the past, thermal indices were frequently used to estimate the thermal environment. These indices, however, were based on single or composite meteorological parameters, such as wet-bulb or equivalent temperature. In the 1970s, several scientists began to use physiologically relevant indices that were derived from the human energy balance for the assessment of the thermal component). A common model for the human energy balance is MEMI (Munich energy balance for individuals), and the derived thermal assessment index PET (physiologically equivalent temperature). Other models are the predicted mean vote (PMV) and standard effective temperature. The three thermal indices PET, SET* and PMVare part of the RayMan model, as are energy fluxes and body parameters by MEMI. They all require mean radiant temperature Tmrt as input.
- The following meteorological parameters were taken into account in the thermal indices:
  1. air temperature
  2. vapour pressure
  3. wind velocity
  4. mean radiant temperature

Body parameters used in MEMI are:

i. human activity and body heat production
ii. heat transfer resistance of clothing.

The PMV, PET and SET* indices facilitate the thermophysiological acquisition of thermal conditions of indoor and surrounding outdoor air as point layers. Importance of radiation fluxes in human-biometeorological studies For the estimation of thermal indices, common meteorological data like air temperature, air humidity and wind speed are required. The mean radiant temperature Tmrt is the most important meteorological input parameter for obtaining the human energy balance during summer conditions.

Therefore, Tmrt has the strongest influence on thermo-physiological significant indices like PET or PMV.

Tmrt can be obtained through different measurement procedures and models. The procedure for determining Tmrt experimentally is very complex, time-intensive and expensive. This is due to the combination of pyranometer and pyrgeometer, which have to be orientated in six directions (4 cardinal directions, upwards, downwards) to measure the complete short- and long-wave radiation fluxes which are significant for a person in the 3D environment. Mean radiation flux densities of the human body can be calculated from the measured short- and long-wave radiation fluxes:

$$S_{str} = a_k \sum_{i-1}^{6} K_i F_i + a_1 \sum_{i-1}^{6} L_i F_i$$

where Ki is the short-wave (solar) and Li the long-wave (terrestrial) radiation fluxes, and ak and al are the absorption coefficients for short-wave and long-wave radiation. Fi is the angle factors of the solid surfaces. Following the calculation of Tmrt, it can be calculated through the use of the Stefan–Boltzmann radiation law (in °C), σ is the Stefan –Boltzmann constant (5.67*10–8 W/m–2K–1):

$$T_{mrt} = \sqrt[4]{(S_{str}/(a_l \sigma))} - 273.2$$

The measured radiation fluxes have to be multiplied with weighting factors for the human body. For a standing man, the incoming radiation fluxes have to be known in the four horizons (vertical position of the instruments) which are more important than the radiation fluxes in the vertical direction. The advantage of this method is that it is integral for measuring the short- and long-wave radiation fluxes. The disadvantage of this method is that it cannot be applied for long-term experimental investigations and studies. As a typical example, Fig. 1 shows the relationship between the measured Tmrt and the human biometeorological thermal index PET. The latter refers to measurements on both sides of a street and under a tree, which have been marked as the measurement sites for the human-biometeorological evaluations of urban structures. The measurements were carried out during summer weather on 17, 18, 19 and 25 July and 2 August 2001 in Freiburg, southwest Germany, within a radius of 100 m. The measurements were carried out daily from 5:00 CET to 22:00 CET.

For the estimation of long-term studies without directly measured radiation fluxes, Tmrt can be calculated through models like RayMan. In the literature, recommended methods for estimating radiation fluxes are based on parameters including air temperature, air humidity, degree of cloud cover, atmospheric turbidity, and time of the day and the year. The albedo of the surrounding surfaces and their solid angle proportions, however, must be specified.

Additionally, other factors like the geometrical properties of buildings, vegetation etc. have to be taken into consideration. For such models to be applied in simple situations, the following atmospheric parameters are required:

i. Direct solar radiation
ii. Diffuse solar radiation
iii. Reflected short-wave radiation
iv. Atmospheric radiation (long-wave)
v. Long-wave radiation from the solid surfaces

The following parameters describing the surroundings of the human body also have to be known:

i. Sky view factor
ii. View factor of the different solid surfaces
iii. Albedo of the different solid surfaces
iv. Emissivity of the different solid surfaces

All the above parameters have to be known in order to calculate mean radiant temperature. Many methods already exist for the calculation of short-wave radiation fluxes that apply simple and complex models with different time resolutions on the basis of different methods, and these are well-tested. Models using sunshine duration for the calculation of short-wave radiation fluxes are used, as well as simple parameterizations for turbidity. Methods for the calculation of long-wave radiation fluxes also exist.

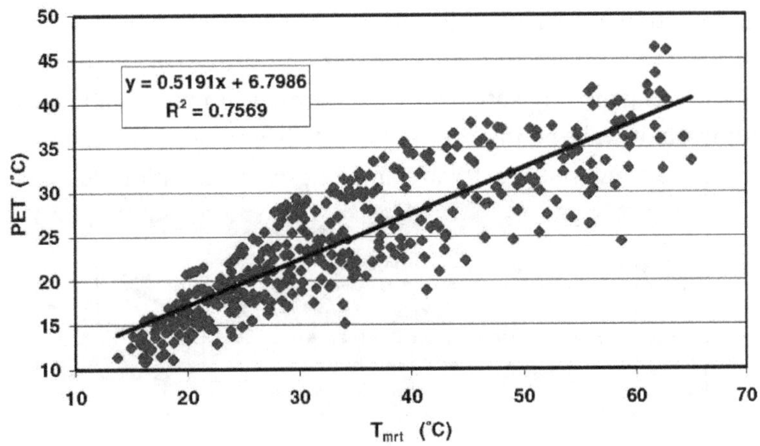

Fig. 1: Relationship between Tmrt and PET for 17, 18, 19 and 25 July and 2 August 2001 in Freiburg

The way in which radiation fluxes are modified in and around urban structures can be described by the sky view factor, which is also an important parameter for urban climate studies. There are many existing methods for its calculation. Also important on the local scale is the knowledge of shadowing. Finally, radiative properties such as emissivity and albedo for the human body also have to be known. To calculate the mean radiant temperature Tmrt, the relevant properties and dimensions of the radiating surfaces and of the visible section of the sky must be known. The posture of the human body (e.g. seated or standing) is also required. To calculate Tmrt, the entire surroundings of the human body are divided into n isothermal surfaces with the temperatures Ti (i=1 to n) and emissivities εi, to which the solid angle portions ("angle factors") Fi are to be allocated as weighting factors. Long-wave radiation (Ei=εi*σ*Ti 4) and diffuse short-wave radiation Di are emitted from each of the n surfaces of the surroundings. This results in a value for Tmrt as:

$$T_{mrt} = \left[\frac{1}{\sigma}\sum_{i-1}^{n}\left(E_i + a_k\frac{D_i}{\varepsilon_p}\right)F_i\right]^{0.25}$$

σ is the Stefan–Boltzmann constant (5.67*10–8 W/(m2K4)), and εp is the emission coefficient of the human body (standard value 0.97). Di comprises the diffuse solar radiation and the diffusely reflected global radiation. ak is the absorption coefficient of the irradiated body surface area of short-wave radiation (standard value 0.7). Tmrt is incremented to T*mrt, if there is also direct solar radiation:

$$T^*{}_{mrt} = \left[T^4_{mrt} + \frac{f_p a_k I^*}{(\varepsilon_p \sigma)}\right]^{0.25}$$

In this case, I* is the radiation intensity of the sun on a surface perpendicular to the incident radiation direction, and the surface projection factor fp is a function of the incident radiation direction and the body posture. For applications in human-biometeorology, it is generally sufficient to determine fp for a rotationally symmetrical person standing up or walking in order to have daily averages. The calculation of hourly, daily and monthly averages of short-wave and long-wave radiation fluxes is also possible.

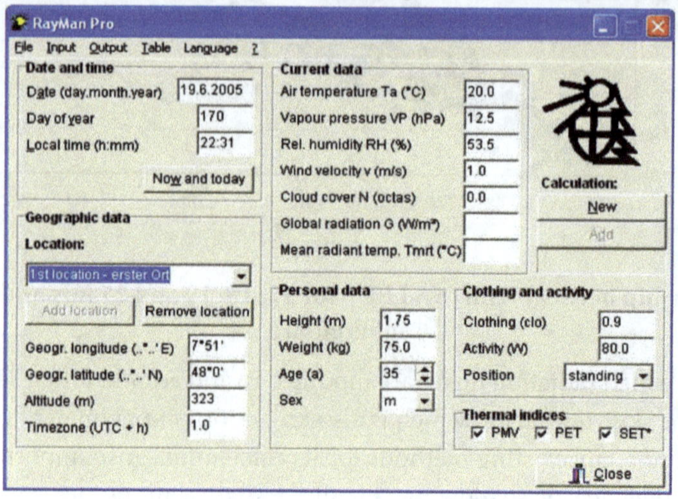

Fig. 2: Input window of RayMan and the relevant values for the calculation of mean radiant temperature and thermal indices

i. Calculation of thermal indices like PMV, PET and SET*.
ii. A full assessment of the thermal bioclimate for every type of landscape or urban structure is possible based on the output of the RayMan model.

14.2.2 Applications and Validations

RayMan has been tested for different purposes. In the following, some typical examples of pplications are given.

An aim of the developers is to check the results produced by RayMan against measurements in different environments.

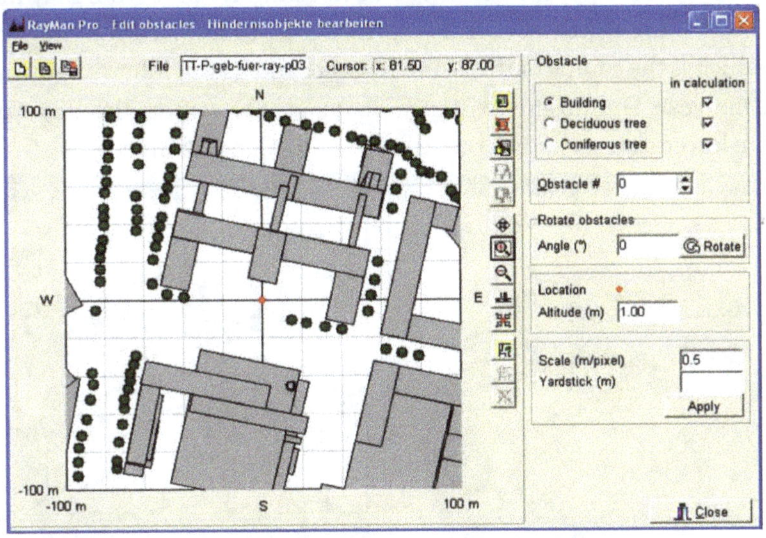

Fig. 3: Input window of urban structures (example: investigation area for Fig. 5 and 6)

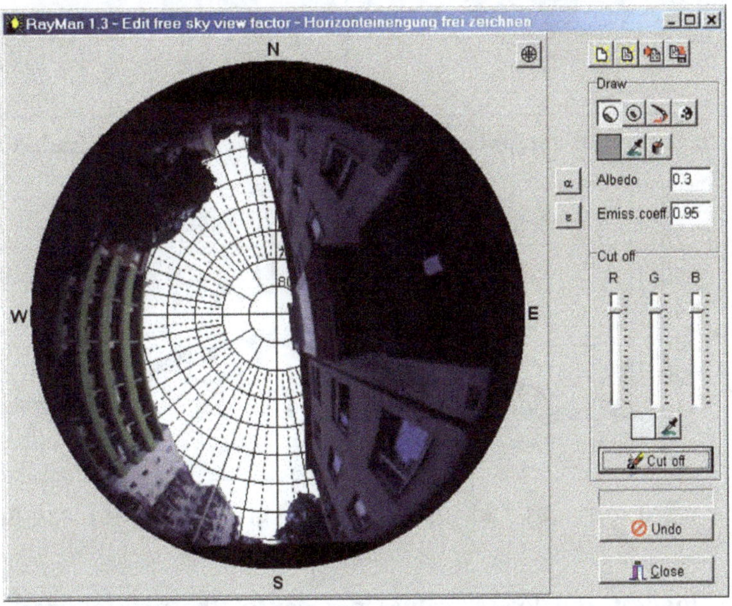

Fig. 4: Input window of free drawing and fish eye photographs

The thermal index PET, which can be calculated by RayMan, is suitable for the evaluation of the thermal environment not only in summer, but also throughout the year. As an example of such an application in a Mediterranean climate, Fig. 7 shows mean, highest and lowest PET values at 12:00 UTC each day at Heraklion (on the island of Crete) in Greece for the period 1980–1989. This kind of illustration provides good information on the variability of PET on each individual day of the year within the study period. The results of Fig. 7 show that different grades of cold stress (PET<18°C) occur mostly from October to April. Mean PET values over 30°C, indicating at least moderate heat stress, occur from June to September. On some hot summer days from May to September, PET at 12:00 UTC exceeded 40°C. This represents a pronounced thermal stress level in Heraklion, which is the capital of the island of Crete and has an intensive tourism industry throughout the year. Due to the existence of the etesian winds, the PET conditions in Heraklion are lower during summer than on the mainland of Greece.

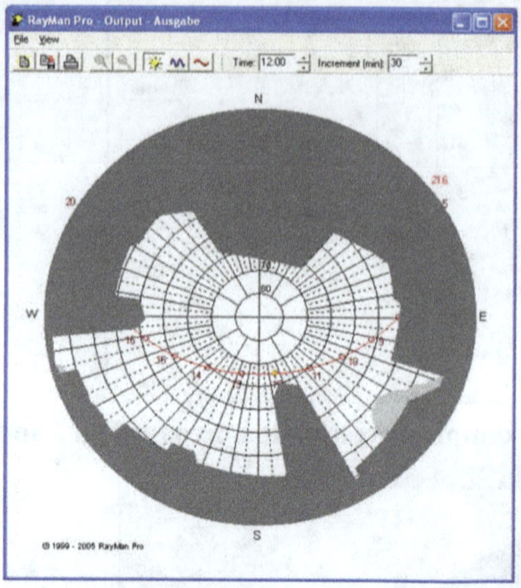

Fig. 5: Output of sun paths

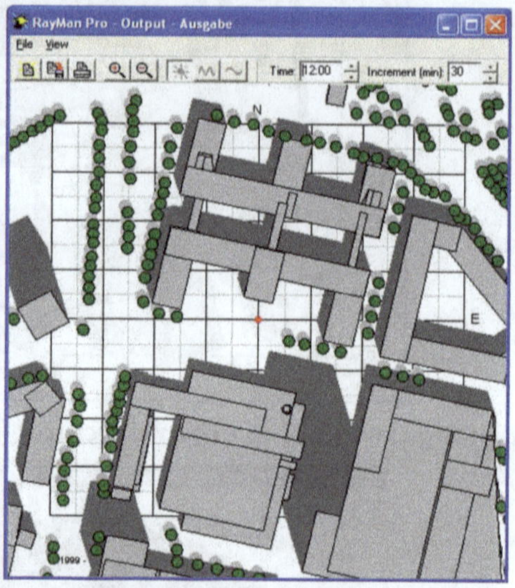

Fig. 6: Output of shaded areas

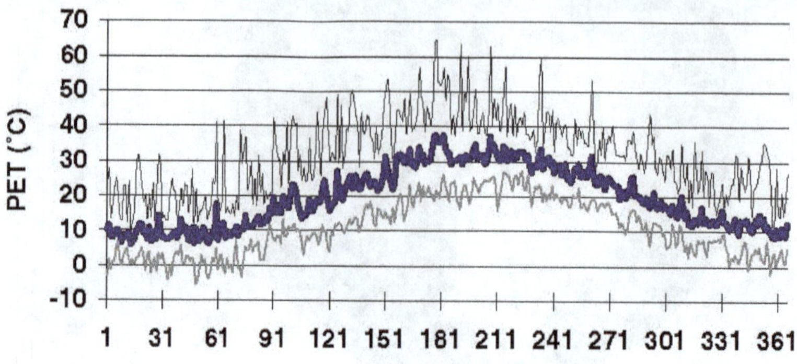

Fig. 7: Mean, highest and lowest values per day of the physiologically equivalent temperature (PET) at 12:00 UTC at Heraklion/Greece, for the years 1980 to 1989.

14.3 CONDITIONS IN URBAN AREAS

The thermal conditions in urban structures vary considerably, and the most influential thermal factor for human beings during summer results from the short- and longwave radiation fluxes. For the quantification of the influence of urban canyons on the radiation field of human beings, measurements have been carried out at several locations in differently oriented streets, and with a nearby reference measurement site in a green area under a tree. The measurements were carried out on August 2nd 2001 in Freiburg, southwest Germany. The structures represent a typical street and a small green area in Freiburg (Fig. 8).

The measurements have been carried out between 5:00 CET and 22:00 CET. In Fig. 8, fish-eye photographs of the five measurement sites are shown. During the period of measurements, the wind speed was lower than 3 m/s, and significant differences in the air temperature and air humidity between the different measurement sites were not expected. High variability was observed in the shortand long-wave radiation fluxes and the resulting human biometeorological parameter, mean radiant temperature. In Fig. 9, the diurnal courses of Tmrt at the five different measurement sites are presented. Additionally, values for Tmrt calculated from measured data from the urban climate station on the roof of a 50 m high building are displayed in Fig. 9. From Fig. 9 we obtain the ndication that there are differences between roof and ground level in the morningand evening. These differences in Tmrt occur because of the higher long-wave emission of the solid surfaces near the ground (see fish-eye photograph). During that time, the in Freiburg differences in Tmrt at ground level were very small. At the point when short-wave radiation fluxes start to be present, the differences in Tmrt between the four different measurement points in the two differently oriented urban canyons are beginning to increase, and they show higher discrepancies compared with the reference point under a tree.

As explained already, a complete assessment of the thermal environment can be described through a thermal index such as the physiologically equivalent temperature (PET). Figure 10 shows the conditions of PET for the measurement sites of Fig. 8. This figure demonstrates that there are differences between roof and street level during the morning and evening hours. We attribute this mainly to higher wind velocities at roof level.

For clear sky conditions and high short-wave radiation but no shadowing at the measurement points, we found that the thermal bioclimate in urban structures is very variable, and we found differences between the different measurement points that range from 10–15°C PET. In the afternoon (after 14:00 CET) the differences in PET between the sites in the urban canyon are not as high. This is mainly the result of the influence of air temperature, wind speed and, to a lesser extent, of the radiation fluxes.

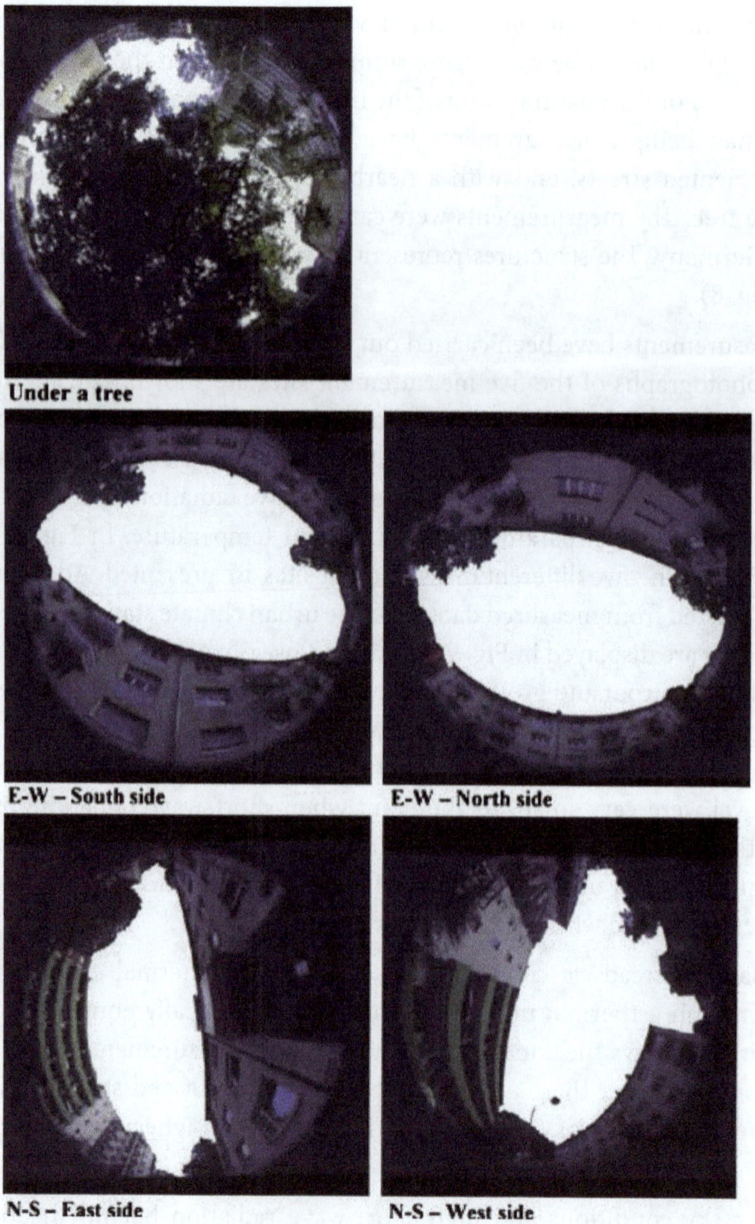

Fig. 8: Selected typical urban structures in Freiburg for human bio meteorological Analyses

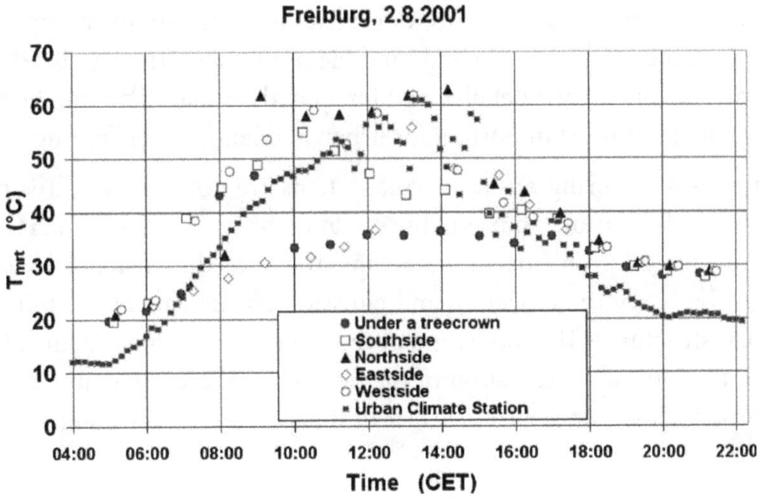

Fig. 9: Daily variation of Tmrt in five different urban structures for 2 August 2001.

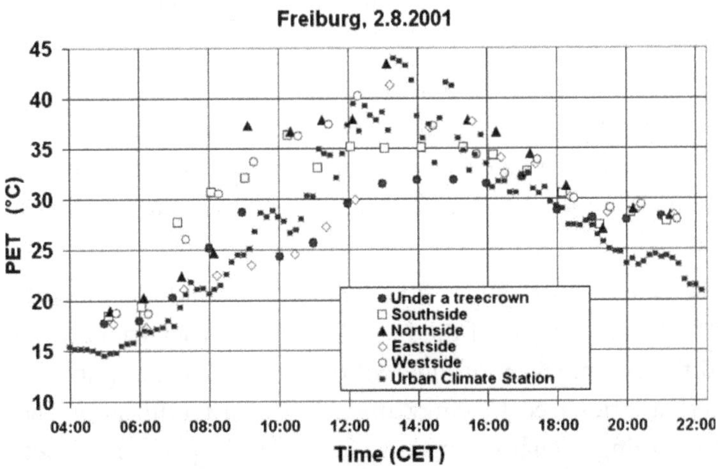

Fig. 10: Daily variation of PET in five different urban structures for 2 August 2001 in Freiburg Validation of RayMan with measurements.

To analyze the accuracy of the RayMan outputs, validating measurements have been carried out. Figure 11 shows the relationship between Tmrt modelled with RayMan and measurements that were carried out on 2 August 2001 in Freiburg, southwest Germany. Some scattering occurs, but the statistical relationship between Tmrt measured and modelled has an r2=0.7684 and is statistically significant at 95 % level. The scattering is an effect of clouds appearing, which caused some variation. Differences can also be due to the effect of the extremely complex structures at the measurement sites. Nevertheless, the simulation of Tmrt by RayMan shows reasonable agreement with the measured values.

As shown in the previous examples, the radiation fluxes are most important for the human energy balance of the human body and the resulting thermal index PET, especially during summer. Accordingly, a detailed and integrated estimation of the short- and long-wave radiation fluxes is most important for urban and landscape planning.

If continuous monitoring measurement systems are not available, it is necessary to estimate the three-dimensional fluxes of the radiation balance of the human body through modelling. The major problems in modelling the threedimensional radiation fluxes for the human body are the reflection and emission coefficients of the solid surfaces in complex urban structures. The modelling of the atmospheric components of short-wave radiation and the long-wave radiation of the free atmosphere is less important. It is not yet known how inclined or vertical surfaces influence the three-dimensional radiation budget of human beings.

Therefore, the model presented here needs to be supported with accompanying measurements. Furthermore, the quantification of clouds in urban areas and the turbidity estimation need to be enhanced. Improvements of atmospheric turbidity in other places in the world are also important.

One future step in the development of the model is permanent validation with measurements, especially on vertical surfaces. Validations in simple environments, such as plane areas, don't show large differences in the radiation fluxes. Further applications of the RayMan model and improvements for difficult climates and areas, e.g. tropical regions, are planned.

The presented model is developed for operational use and for case studies for different planning levels. The advantages of the model are in the user-friendly environment, the easy input and the combination of different options in fish-eye photographs, buildings and vegetation properties, free drawing with facilities for modification and the possibility of its use with varying scenarios, making for different planning cases. This includes the calculation of the energy balance of human beings and assessment of the thermal bioclimate with the use of thermal indices. Another advantage is the running time, which is very fast. RayMan is transparent, which means that all the input parameters are clearly known as well as the resulting detailed radiation and energy fluxes.

The main advantage of RayMan is in the requirements of meteorological data which have no high spatial variability, such as air temperature and air humidity. The absence of modification in wind parameters and cloud information, which have to be known, is a drawback, which can be solved easily by preparing the data in the input files or running simulations with different wind speeds. Furthermore, the calculations comprise only point results but no continuous surface data. Needed meteorological data can be produced by numerical models such as ENVI-MET and then imported in RayMan for specific studies. However, the implementation of spatial calculations would increase the run time tremendously and decrease the main advantage of RayMan.

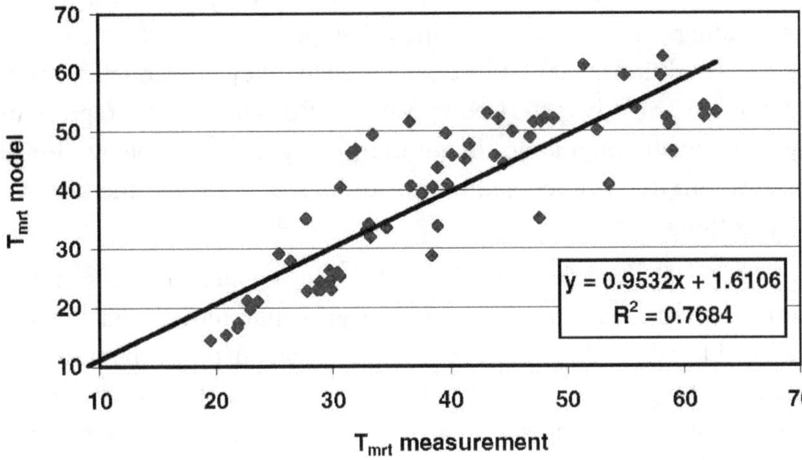

Fig. 11: Relationship of measured mean radiant temperature Tmrt,measurement and computed mean radiation temperature Tmrt,modeled by RayMan for 2 August 2001 in Freiburg

The development of the RayMan model has not been completed yet. It is planned to continue the development of RayMan, especially in the modification of atmospheric parameters and the detailed input information in the artificial and natural geometrical properties in urban areas.

14.4 HIGHLIGHTS: AT A GLANCE

- There exists an increasing demand to assess the thermal climate in a human-biometeorologically significant way. Therefore, assessment methods based on the human energy balance and resulting in thermal indices have been developed worldwide. To apply the thermal indices, suitable tools are necessary, which enable their calculation from the relevant meteorological and human variables. The RayMan model introduced in this study represents one of these tools. Due to its clear structure, the RayMan model can be applied not only by experts in human-biometeorology, but also by people with less experience in this field of science. By using thermal indices calculated with the RayMan model, the stationary perception of the thermal surroundings by people can be transformed into a numerical value.
- Some extensions of the RayMan model are planned in the future. They include a more accurate consideration of the radiation fluxes within three-dimensional urban structures, e.g. street canyons or courtyards, with the consequence that the site-dependent correlations between simulated Tmrt
- By use of the RayMan model and Tmrt obtained from direct measurements of all relevant radiation fluxes will have higher coefficients of determination.

- To consider adaptation and acclimatisation of people with their thermal surroundings, it is necessary to interview people about their perception of the thermal conditions. The results can be used to set up graded scales for the thermal indices. They are characterised by varying threshold values representing (i) the different adaptation and acclimatisation processes of people worldwide and (ii) varying behaviour of people in terms of their activity and the heat resistance of their clothing.
- As a stationary model, RayMan cannot be applied to assess non-stationary thermal conditions of single persons, e.g. if they move from the sunny to the shaded sidewalk within a street canyon. For that purpose other models have to be used (e.g. Höppe 1993, 1999). The assessment of onstationary thermal conditions, however, represents a special case, which is necessary for specific objectives; but for general application purposes like in urban planning, it is sufficient to operate with the stationary RayMan model.
- The RayMan model allows the integration of new thermal indices developed especially for application in complex environments. One of these indices is UTCI (universal thermal climate index), which is under development by an Expert Team at the WMO, and a Commission of the International Society for Biometeorology.

15

CONCEPTS OF RADIATION

15.1 RADIATION: AN INTRODUCTION

Radiation is energy that comes from a source and travels through space and may be able to penetrate various materials. Light, radio, and microwaves are types of radiation that are called nonionizing. The kind of radiation discussed in this document is called ionizing radiation because it can produce charged particles (ions) in matter.

Radiation is energy. It can come from unstable atoms that undergo radioactive decay, or it can be produced by machines. Radiation travels from its source in the form of energy waves or energized particles. There are different forms of radiation and they have different properties and effects.

- Ionizing radiation is produced by unstable atoms. Unstable atoms differ from stable atoms because unstable atoms have an excess of energy or mass or both. Radiation can also be produced by high-voltage devices (e.g., x-ray machines).
- Unstable atoms are said to be ***radioactive***. In order to reach stability, these atoms give off, or emit, the excess energy or mass. These emissions are called ***radiation***. The kinds of radiation are electromagnetic (like light) and particulate (i.e., mass given off with the energy of motion). Gamma radiation and x rays are examples of electromagnetic radiation. Gamma radiation originates in the nucleus while x rays come from the electronic part of the atom. Beta and alpha radiation are examples of particulate radiation.
- Interestingly, there is a "***background***" of natural radiation everywhere in our environment. It comes from space (i.e., cosmic rays) and from naturally occurring radioactive materials contained in the earth and in living things.

15.2 RADIATION EXPOSURE FROM VARIOUS SOURCES

Source	Exposure
External Background Radiation	0.60 mSv y^{-1}, U.S. Average
Natural K-40 and Other Radioactivity in Body	0.4 mSv y^{-1}
Air Travel Round Trip (NY-LA)	0.05 mSv
Chest X-Ray Effective Dose	0.10 mSv per view
Radon in the Home	2.00 mSv y^{-1} (variable)
Man-Made (medical x rays, etc.)	0.60 mSv y^{-1} (average)

15.3 TYPES OF RADIATION

The radiation one typically encounters is one of four types: alpha radiation, beta radiation, gamma radiation, and x radiation. Neutron radiation is also encountered in nuclear power plants and high-altitude flight and is emitted from some industrial radioactive sources.

15.3.1 Alpha Radiation

Alpha radiation is a heavy, very short-range particle and is actually an ejected helium nucleus. Some characteristics of alpha radiation are:

- Most alpha radiation is not able to penetrate human skin.
- Alpha-emitting materials can be harmful to humans if the materials are inhaled, swallowed, or absorbed through open wounds.
- A variety of instruments has been designed to measure alpha radiation. Special training in the use of these instruments is essential for making accurate measurements.
- A thin-window Geiger-Mueller (GM) probe can detect the presence of alpha radiation.
- Instruments cannot detect alpha radiation through even a thin layer of water, dust, paper, or other material, because alpha radiation is not penetrating.
- Alpha radiation travels only a short distance (a few inches) in air, but is not an external hazard.
- Alpha radiation is not able to penetrate clothing.
- Examples of some alpha emitters: radium, radon, uranium, thorium.

15.3.2 Beta Radiation

Beta radiation is a light, short-range particle and is actually an ejected electron. Some characteristics of beta radiation are:

- Beta radiation may travel several feet in air and is moderately penetrating.
- Beta radiation can penetrate human skin to the "germinal layer," where new skin

cells are produced. If high levels of beta-emitting contaminants are allowed to remain on the skin for a prolonged period of time, they may cause skin injury.

- Beta-emitting contaminants may be harmful if deposited internally.
- Most beta emitters can be detected with a survey instrument and a thin-window G-M probe (e.g., "pancake" type). Some beta emitters, however, produce very low-energy, poorly penetrating radiation that may be difficult or impossible to detect. Examples of these difficult-to-detect beta emitters are hydrogen-3 (tritium), carbon-14, and sulfur-35.
- Clothing provides some protection against beta radiation.
- Examples of some pure beta emitters: strontium-90, carbon-14, tritium, and sulfur-35.
- **Gamma and X Radiation** Gamma radiation and x rays are highly penetrating electromagnetic radiation. Some characteristics of these radiations are:
- Gamma radiation or x rays are able to travel many feet in air and many inches in human tissue. They readily penetrate most materials and are sometimes called "penetrating" radiation.
- X rays are like gamma rays. X rays, too, are penetrating radiation. Sealed radioactive sources and machines that emit gamma radiation and x rays, respectively, constitute mainly an external hazard to humans.
- Gamma radiation and x rays are electromagnetic radiation like visible light, radiowaves, and ultraviolet light. These electromagnetic radiations differ only in the amount of energy they have. Gamma rays and x rays are the most energetic of these.
- Dense materials are needed for shielding from gamma radiation. Clothing provides little shielding from penetrating radiation, but will prevent contamination of the skin by gamma-emitting materials.
- Gamma radiation is easily detected by survey meters with a sodium iodide detector probe.
- Gamma radiation and/or characteristic x rays frequently accompany the emission of alpha and beta radiation during radioactive decay. Examples of some gamma emitters: iodine-131, cesium-137, cobalt-60, radium-226, and technetium-99m.

15.4 MEASUREMENT OF RADIATION

The International System of Units (SI) for radiation measurement is now the official system of measurement and uses the "gray" (Gy) and "sievert" (Sv) for absorbed dose and equivalent dose, respectively.

In the United States, radiation absorbed dose, equivalent, and exposure used to be measured and stated in traditional units called rad, rem, or roentgen (R), respectively. For practical purposes with gamma and x rays, these units of measure for exposure or dose are considered equal. Exposure can be from an external source irradiating the whole body, an extremity, or other organ or tissue resulting in an external radiation dose. Alternately, internally deposited radioactive material may cause an internal radiation dose to the whole body or other organ or tissue.

Smaller fractions of these measured quantities often have a prefix, e.g., milli (m) means 1/1,000. For example, 1 Sv = 1,000 mSv. Micro (μ) means 1/1,000,000. So, 1 Sv = 1,000,000 μSv.

Conversions are as follows:

- 1 Gy = 100 rad
- 1 mGy = 100 mrad
- 1 Sv = 100 rem
- 1 mSv = 100 mrem

With radiation counting systems, radioactive transformation events can be measured in units of "disintegrations per minute" (dpm) and, because instruments are not 100 percent efficient, "counts per minute" (cpm). Background radiation levels are typically less than 0.10 μSv per hour, but due to differences in detector size and efficiency, the cpm reading on fixed monitors and various handheld survey meters will vary considerably.

15.4.1 How Much Radioactive Material Is Present?

The size or weight of a quantity of material does not indicate how much radioactivity is present. A large quantity of material can contain a very small amount of radioactivity, or a very small amount of material can have a lot of radioactivity.

For example, uranium-238, with a 4.5-billion-year half-life, has only 5.5 MBq of activity per pound, while cobalt-60, with a 5.3-year half-life, has nearly 19,000 TBq of activity per pound. This "specific activity," or curies per unit mass, of a radioisotope depends on the unique radioactive half-life and dictates the time it takes for half the radioactive atoms to decay.

The SI system uses the unit of becquerel (Bq) as its unit of radioactivity. The older, traditional unit previously used in the United States is the curie (Ci).

Common multiples of the becquerel are the megabecquerel (1 MBq = 1,000,000 Bq) and the gigabecquerel (1 GBq = 1,000,000,000 Bq). One curie is 37 billion Bq. Since the Bq represents such a small amount, one is likely to see a prefix noting a large multiplier used with the Bq as follows:

- 37 GBq = 37 billion Bq = 1 curie
- 1 MBq = 1 million Bq = ~ 27 microcuries
- 1 GBq = 1 billion Bq = ~ 27 millicuries
- 1 TBq = 1 trillion Bq = ~ 27 curies

15.4.2 How Can You Detect Radiation?

Radiation cannot be detected by human senses. A variety of handheld and laboratory instruments is available for detecting and measuring radiation. The most common handheld or portable instruments are:

- **Geiger Counter, with Geiger-Mueller (G-M) Tube or Probe.** A G-M tube is a gas-filled device that, when a high voltage is applied, creates an electrical pulse when radiation interacts with the wall or gas in the tube. These pulses are converted to a reading on the instrument meter. If the instrument has a speaker, the pulses also give an audible click. Common readout units are roentgens per hour (R/hr), milliroentgens per hour (mR/hr), rem per hour (rem/hr), millirem per hour (mrem/hr), and counts per minute (cpm). G-M probes (e.g., "pancake" type) are most often used with handheld radiation survey instruments for contamination measurements. However, energy-compensated G-M tubes may be employed for exposure measurements. Further, often the meters used with a G-M probe will also accommodate other radiation-detection probes. For example, a zinc sulfide (ZnS) scintillator probe, which is sensitive to just alpha radiation, is often used for field measurements where alpha-emitting radioactive materials need to be measured.
- MicroR Meter, with Sodium Iodide Detector. A solid crystal of sodium iodide creates a pulse of light when radiation interacts with it. This pulse of light is converted to an electrical signal by a photomultiplier tube (PMT), which gives a reading on the instrument meter. The pulse of light is proportional to the amount of light and the energy deposited in the crystal. These instruments most often have upper and lower energy discriminator circuits and, when used correctly as single-channel analyzers, can provide information on the gamma energy and identify the radioactive material. If the instrument has a speaker, the pulses also give an audible click, a useful feature when looking for a lost source. Common readout units are microroentgens per hour (μR/hr) and/or counts per minute (cpm). Sodium iodide detectors can be used with handheld instruments or large stationary radiation monitors. Special plastic or other inert crystal "scintillator" materials are also used in place of sodium iodide.
- Portable Multichannel Analyzer. A sodium iodide crystal and PMT described above, coupled with a small multichannel analyzer (MCA) electronics package, are becoming much more affordable and common. When gamma-ray data libraries

and automatic gamma-ray energy identification procedures are employed, these handheld instruments can automatically identify and display the type of radioactive materials present. When dealing with unknown sources of radiation, this is a very useful feature.

- Ionization (Ion) Chamber. This is an air-filled chamber with an electrically conductive inner wall and central anode and a relatively low applied voltage. When primary ion pairs are formed in the air volume, from x-ray or gamma radiation interactions in the chamber wall, the central anode collects the electrons and a small current is generated. This in turn is measured by an electrometer circuit and displayed digitally or on an analog meter. These instruments must be calibrated properly to a traceable radiation source and are designed to provide an accurate measure of absorbed dose to air which, through appropriate conversion factors, can be related to dose to tissue. In that most ion chambers are "open air," they must be corrected for change in temperature and pressure. Common readout units are milliroentgens and roentgen per hour (mR/hr or R/hr). (Note: For practical purposes, consider the roentgen, rad, and the rem to be equal with gamma or x rays. So, 1 mR/hr is equivalent to 1 mrem/hr.)
- Neutron REM Meter, with Proportional Counter. A boron trifluoride or helium-3 proportional counter tube is a gas-filled device that, when a high voltage is applied, creates an electrical pulse when a neutron radiation interacts with the gas in the tube. The absorption of a neutron in the nucleus of boron-10 or helium-3 causes the prompt emission of a helium-4 nucleus or proton respectively. These charged particles can then cause ionization in the gas, which is collected as an electrical pulse, similar to the G-M tube. These neutron-measuring proportional counters require large amounts of hydrogenous material around them to slow the neutron to thermal energies. Other surrounding filters allow an appropriate number of neutrons to be detected and thus provide a flat-energy response with respect to dose equivalent. The design and characteristics of these devices are such that the amount of secondary charge collected is proportional to the degree of primary ions produced by the radiation. Thus, through the use of electronic discriminator circuits, the different types of radiation can be measured separately. For example, gamma radiation up to rather high levels is easily rejected in neutron counters.
- Radon Detectors. A number of different techniques are used for radon measurements in home or occupational settings (e.g., uranium mines). These range from collection of radon decay products on an air filter and counting, exposing a charcoal canister for several days and performing gamma spectroscopy for absorbed decay products, exposure of an electret ion chamber and read-out, and long-term exposure of CR39 plastic with subsequent chemical etching and alpha track counting. All these approaches have different advantages and disadvantages which should be evaluated prior to use.

The most common laboratory instruments are:

- **Liquid Scintillation Counters.** A liquid scintillation counter (LSC) is a traditional laboratory instrument with two opposing PMTs that view a vial that contains a sample and liquid scintillator fluid, or cocktail. When the sample emits a radiation (often a low-energy beta) the cocktail itself, being the detector, causes a pulse of light. If both PMTs detect the light in coincidence, the count is tallied. With the use of shielding, cooling of PMTs, energy discrimination, and this coincidence counting approach, very low background counts can be achieved, and thus low minimum detectable activities (MDA). Most modern LSC units have multiple sample capability and automatic data acquisition, reduction, and storage.
- **Proportional Counter.** A common laboratory instrument is the standard proportional counter with sample counting tray and chamber and argon/methane flow through counting gas. Most units employ a very thin (microgram/cm2) window, while some are windowless. Shielding and identical guard chambers are used to reduce background and, in conjunction with electronic discrimination, these instruments can distinguish between alpha and beta radiation and achieve low MDAs. Similar to the LSC units noted above, these proportional counters have multiple sample capability and automatic data acquisition, reduction, and storage. Such counters are often used to count smear/wipe or air filter samples. Additionally, large-area gas flow proportional counters with thin (milligram/cm2) mylar windows are used for counting the whole body and extremities of workers for external contamination when exiting a radiological control area.
- **Multichannel Analyzer System.** A laboratory MCA with a sodium iodide crystal and PMT (described above), a solid-state germanium detector, or a silicon-type detector can provide a powerful and useful capability for counting liquid or solid matrix samples or other prepared extracted radioactive samples. Most systems are used for gamma counting, while some silicon detectors are used for alpha radiation. These MCA systems can also be utilized with well-shielded detectors to count internally deposited radioactive material in organs or tissue for bioassay measurements. In all cases, the MCA provides the capability to bin and tally counts by energy and thus identify the emitter. Again, most systems have automatic data acquisition, reduction, and storage capability.

15.4.3 How Can You Keep Radiation Exposure Low and Measure It?

Although some radiation exposure is natural in our environment, it is desirable to keep radiation exposure as low as reasonably achievable (ALARA) in an occupational setting. This is accomplished by the techniques of time, distance, and shielding.

Time: The shorter the time in a radiation field, the less the radiation exposure you will receive. Work quickly and efficiently. Plan your work before entering the radiation field.

Distance: The farther a person is from a source of radiation, the lower the radiation dose. Levels decrease by a factor of the square of the distance. Do not touch radioactive materials. Use remote handling devices, etc., to move materials to avoid physical contact.

Shielding: Placing a radioactive source behind a massive object provides a barrier that can reduce radiation exposure.

Administrative and Engineering Controls: The use of administrative and engineering controls is essential for keeping radiation exposure ALARA.

Monitoring occupational radiation exposure is a fundamental aspect of radiation protection. This can be done by measuring radiation fields with handheld instruments described above and, if exposure conditions are predictable and relatively low (i.e., less than 10% of the regulatory limit), expected exposures can be calculated and documented. Alternately, regular radiation field survey measurements can be performed, and personnel dosimeters are issued to workers.

- Film Badge: A film badge is one of the earliest devices used to measure worker exposure to gamma radiation from radium and x rays. Initially packets of dental X-ray film were worn and developed periodically to view the degree of darkening. Later special metal filters were used in a X-ray film holder, with an open window to provide unattenuated film area for high-energy beta measurement. With appropriate calibration of exposure versus optical density, these devices provide an accurate measure of worker external exposure and a permanent record.
- Thermoluminescent Dosimeter (TLD) Badge: The TLD badge is a personnel monitoring device with special chemical compounds (e.g., lithium fluoride) in powder of solid form that retain deposited energy from radiation exposure. These TL materials emit light when subsequently heated in a reader. The light is detected by a PMT, and through calibration the electrical current provides a proportional measure of radiation exposure. However once read out, the signal from these devices is erased for the most part. Thus, quality control on measurements must be to the strictest standards. The National Institute of Standards and Technology (NIST) has developed a national voluntary laboratory accreditation program (NAVLAP) for all external dosimetry (e.g., film, TLD) processors. Cross-checks, reviewed procedures, on-site inspections, etc., all provide assurance that dosimeter results are of the highest quality.
- Optically Stimulated Luminescence (OSL) Badge: The OSL dosimeter/reader technology is relatively new and uses a laser to stimulate an aluminum oxide material that was in the badge for personnel radiation monitoring of occupationally exposed workers. With optically stimulated luminescence, a tiny crystal traps and stores energy from exposure to ionizing radiation fields. The amount of exposure can be determined by shining a green light on the crystal and measuring the

intensity of the blue light emitted. OSL systems allow instantaneous readings that can be repeated, as opposed to TLDs which take 20 or 30 seconds for a one-time-only reading.

- Pocket Ionization (Ion) Chamber: This is a sealed cylindrical air-filled chamber, sometimes called a direct reading dosimeter (DRD) or quartz fiber dosimeter (QFD), with a charged quartz fiber that can be directly viewed through a built-in microscope. This filament can be seen against a scale from typically 0 to 200 milliroentgen or 0 to 5 R. Ionizing gamma radiation passing through the chamber causes a discharge of the device and a deflection of the fiber upscale. When properly manufactured, maintained, and calibrated, these devices provide a fairly accurate direct measure of external exposure. In the late 1980s a thin-walled type was introduced that was more sensitive to diagnostic-energy x rays. The advantage of these DRDs or QFDs is instantaneous indication of radiation exposure. However, they are fragile devices and subject to leakage. Frequent calibration and leakage checks are recommended, as well as the use of two dosimeters side by side. Readings that do not reasonably match should be suspect.
- Electronic Dosimeters: Electronic dosimeters have been available since the early 1980s. These devices use energy-compensated Geiger-Mueller tubes or solid-state detectors with supporting electronics in a package typically the size of a deck of playing cards. Features vary with respect to size, ruggedness, user control, display of accumulated dose and/or dose rate, alarm set point, battery life, computer interface, etc.

15.5 RADIOACTIVE CONTAMINATION

If radioactive material is not in a sealed source container, it might be spread onto other objects. Contamination occurs when material that contains radioactive atoms is deposited on materials, skin, clothing, or any place where it is not desired. It is important to remember that radiation does not spread or get "on" or "in" people; rather, it is radioactive contamination that can be spread. A person contaminated with radioactive material will receive radiation exposure until the source of radiation (the radioactive material) is removed.

- A person is externally contaminated if radioactive material is on the skin or clothing.
- A person is internally contaminated if radioactive material is breathed in, swallowed, or absorbed through wounds.
- The environment is contaminated if radioactive material is spread about or is unconfined.

15.5.1 How Can You Work Safely Around Radiation or Contamination?

You can work safely around radiation and/or contamination by following a few simple precautions:

- Use time, distance, shielding, and containment to reduce exposure.
- Wear dosimeters (e.g., film or TLD badges) if issued.
- Avoid contact with the contamination.
- Wear protective clothing that, if contaminated, can be removed.
- Wash with nonabrasive soap and water any part of the body that may have come in contact with the contamination.
- Assume that all materials, equipment, and personnel that came in contact with the contamination are contaminated. Radiological monitoring is recommended before leaving the scene.

15.5.2 Is It Safe to Be Around Sources of Radiation?

A single high-level radiation exposure (i.e., greater than 100 mSv) delivered to the whole body over a very short period of time may have potential health risks. From follow-up of the atomic bomb survivors, we know acutely delivered very high radiation doses can increase the occurrence of certain kinds of disease (e.g., cancer) and possibly negative genetic effects. To protect the public and radiation workers (and environment) from the potential effects of chronic low-level exposure (i.e., less than 100 mSv), the current radiation safety practice is to prudently assume similar adverse effects are possible with low-level protracted exposure to radiation. Thus, the risks associated with low-level medical, occupational, and environmental radiation exposure are conservatively calculated to be proportional to those observed with high-level exposure. These calculated risks are compared to other known occupational and environmental hazards, and appropriate safety standards and policies have been established by international and national radiation protection organizations (e.g., International Commission on Radiological Protection and National Council on Radiation Protection and Measurements) to control and limit potential harmful radiation effects.

Both public and occupational regulatory dose limits are set by federal agencies (i.e., Environmental Protection Agency, Nuclear Regulatory Commission, and Department of Energy) and state agencies (e.g., agreement states) to limit cancer risk. Other radiation dose limits are applied to limit other potential biological effects with workers' skin and lens of the eye.

Annual Radiation Dose Limits	Agency
Radiation Worker - 50 mSv	(NRC, "occupationally" exposed)
General Public - 1 mSv	(NRC, member of the public)
General Public - 0.25 mSv	(NRC, D&D all pathways)
General Public - 0.10 mSv	(EPA, air pathway)
General Public - 0.04 mSv	(EPA, drinking-water pathway)

16

TEMPERATURE AND HEAT

16.1 INTRODUCTION TO TEMPERATURE AND HEAT

Heat and temperature are important concepts for each of us, every day. How we dress in the morning depends on whether the day is hot or cold, and most of what we do requires energy that ultimately comes from the Sun. The study of heat and temperature is part of an area of physics known as thermodynamics. The laws of thermodynamics govern the flow of energy throughout the universe. They are studied in all areas of science and engineering, from chemistry to biology to environmental science.

Heat is the flow of energy from one object to another. This flow of energy is caused by a difference in temperature. The transfer of heat can change temperature, as can work, another kind of energy transfer that is central to thermodynamics.

Fig. 1: These snowshoers on Mount Hood in Oregon are enjoying the heat flow and light caused by high temperature.

All three mechanisms of heat transfer are relevant to this picture. The heat flowing out of the fire also turns the solid snow to liquid water and vapor. (credit: "Mt. Hood

Territory"/Flickr). In this chapter, we explore heat and temperature . It is not always easy to distinguish these terms. Heat is the flow of energy from one object to another. This flow of energy is caused by a difference in temperature . The transfer of heat can change temperature , as can work, another kind of energy transfer that is central to thermodynamics. We return to these basic ideas several times throughout the next four chapters, and you will see that they affect everything from the behavior of atoms and molecules to cooking to our weather on Earth to the life cycles of stars.

16.2 TEMPERATURE AND THERMAL EQUILIBRIUM

Heat is familiar to all of us. We can feel heat entering our bodies from the summer Sun or from hot coffee or tea after a winter stroll. We can also feel heat leaving our bodies as we feel the chill of night or the cooling effect of sweat after exercise.

What is heat ? How do we define it and how is it related to temperature ? What are the effects of heat and how does it flow from place to place? We will find that, in spite of the richness of the phenomena, a small set of underlying physical principles unites these subjects and ties them to other fields. We start by examining temperature and how to define and measure it.

16.2.1 Temperature

The concept of temperature has evolved from the common concepts of hot and cold. The scientific definition of temperature explains more than our senses of hot and cold. As you may have already learned, many physical quantities are defined solely in terms of how they are observed or measured, that is, they are efined operationally.

Temperature is operationally defined as the quantity of what we measure with a thermometer. As we will see in detail in a later chapter on the kinetic theory of gases, temperature is proportional to the average kinetic energy of translation, a fact that provides a more physical definition. Differences in temperature maintain the transfer of heat , or heat transfer, throughout the universe.

Heat transfer is the movement of energy from one place or material to another as a result of a difference in Temperature . (You will learn more about heat transfer later in this chapter.)

16.2.2 Thermal Equilibrium

An important concept related to temperature is thermal equilibrium . Two objects are in thermal equilibrium if they are in close contact that allows either to gain energy from the other, but nevertheless, no net energy is transferred between them. Even when not in contact, they are in thermal equilibrium if, when they are placed in contact, no net energy is transferred between them. If two objects remain in contact for a long time, they typically come to equilibrium . In other words, two objects in thermal equilibrium do not exchange energy.

Experimentally, if object A is in equilibrium with object B, and object B is in equilibrium with object C, then (as you may have already guessed) object A is in equilibrium with object C. That statement of transitivity is called the zeroth law of thermodynamics . (The number "zeroth" was suggested by British physicist Ralph Fowler in the 1930s. The first, second, and third laws of thermodynamics were already named and numbered then. The zeroth law had seldom been stated, but it needs to be discussed before the others, so Fowler gave it a smaller number.) Consider the case where A is a thermometer. The zeroth law tells us that if A reads a certain temperature when in equilibrium with B, and it is then placed in contact with C, it will not exchange energy with C; therefore, its temperature reading will remain the same. In other words, if two objects are in thermal equilibrium , they have the same temperature .

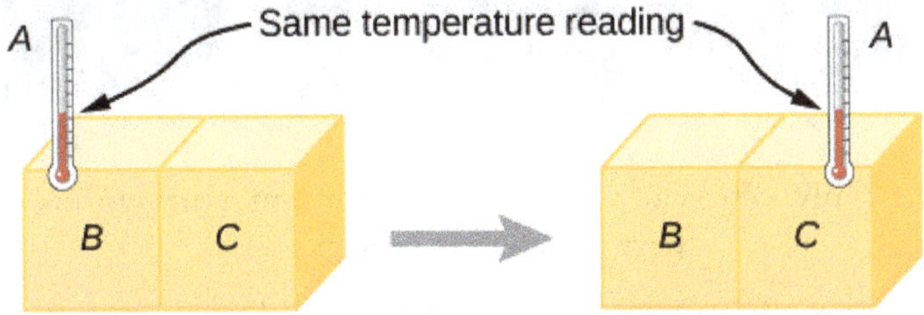

Figure 2: If thermometer A is in thermal equilibrium with object B, and B is in thermal equilibrium with C, then A is in thermal equilibrium with C. Therefore, the reading on A stays the same when A is moved over to make contact with C.

A thermometer measures its own temperature. It is through the concepts of thermal equilibrium and the zeroth law of thermodynamics that we can say that a thermometer measures the temperature of something else, and to make sense of the statement that two objects are at the same temperature .

In the rest of this chapter, we will often refer to "systems" instead of "objects." As in the chapter on linear momentum and collisions, a system consists of one or more objects—but in thermodynamics, we require a system to be macroscopic, that is, to consist of a huge number (such as 10231023) of molecules. Then we can say that a system is in thermal equilibrium with itself if all parts of it are at the same temperature . (We will return to the definition of a thermodynamic system in the chapter on the first law of thermodynamics) .

16.3 THERMOMETERS AND TEMPERATURE SCALES

Any physical property that depends consistently and reproducibly on temperature can be used as the basis of a thermometer. For example, volume increases with temperature for most substances. This property is the basis for the common alcohol thermometer and the original mercury thermometers. Other properties used to measure temperature include electrical resistance , color, and the emission of infrared radiation (Figure 3).

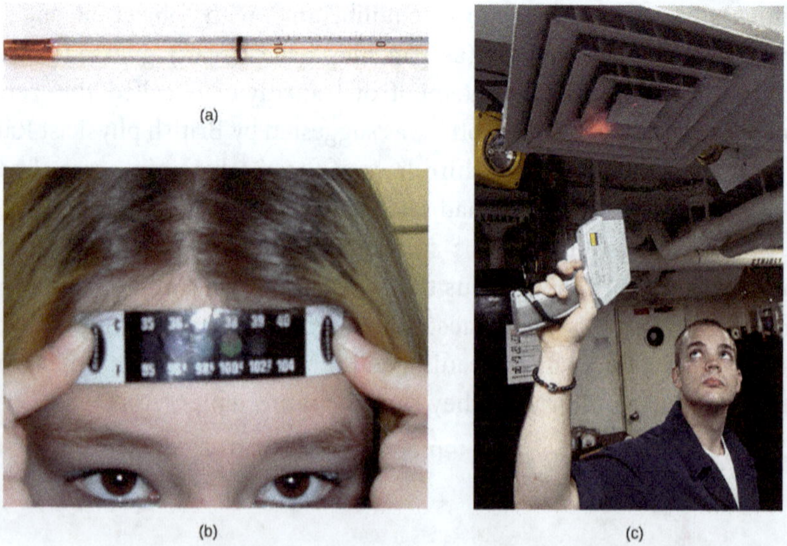

Fig. 3: Because many physical properties depend on temperature, the variety of thermometers is remarkable.

(a) In this common type of thermometer, the alcohol, containing a red dye, expands more rapidly than the glass encasing it. When the thermometer's temperature increases, the liquid from the bulb is forced into the narrow tube, producing a large change in the length of the column for a small change in temperature. (b) Each of the six squares on this plastic (liquid crystal) thermometer contains a film of a different heat-sensitive liquid crystal material. Below 95oF , all six squares are black. When the plastic thermometer is exposed to a temperature of 95oF , the first liquid crystal square changes color. When the temperature reaches above 96.8oF , the second liquid crystal square also changes color, and so forth. (c) A firefighter uses a pyrometer to check the temperature of an aircraft carrier's ventilation system. The pyrometer measures infrared radiation (whose emission varies with temperature) from the vent and quickly produces a temperature readout. Infrared thermometers are also frequently used to measure body temperature by gently placing them in the ear canal. Such thermometers are more accurate than the alcohol thermometers placed under the tongue or in the armpit.

Thermometers measure temperature according to well-defined scales of measurement. The three most common temperature scales are Fahrenheit, Celsius, and Kelvin. Temperature scales are created by identifying two reproducible temperatures. The freezing and boiling temperatures of water at standard atmospheric pressure are commonly used.

On the Celsius scale, the freezing point of water is 0°C and the boiling point is 100°C. The unit of temperature on this scale is the degree Celsius (°C). The Fahrenheit scale (still the most frequently used for common purposes in the United States) has the freezing point of water at 32oF and the boiling point a. 212oF Its unit is the degree Fahrenheit (oF). You can see that 100 Celsius degrees span the same range as 180 Fahrenheit degrees.

Thus, a temperature difference of one degree on the Celsius scale is 1.8 times as large as a difference of one degree on the Fahrenheit scale, or

$$\Delta T_F = \frac{9}{5}\Delta T_C.$$

The definition of temperature in terms of molecular motion suggests that there should be a lowest possible temperature, where the average kinetic energy of molecules is zero (or the minimum allowed by quantum mechanics). Experiments confirm the existence of such a temperature, called absolute zero. An absolute temperature scale is one whose zero point is absolute zero. Such scales are convenient in science because several physical quantities, such as the volume of an ideal gas, are directly related to absolute temperature.

The Kelvin scale is the absolute temperature scale that is commonly used in science. The SI temperature unit is the kelvin, which is abbreviated K (not accompanied by a degree sign). Thus 0 K is absolute zero. The freezing and boiling points of water are 273.15 K and 373.15 K, respectively. Therefore, temperature differences are the same in units of kelvins and degrees Celsius, or $\Delta TC=\Delta TK$.

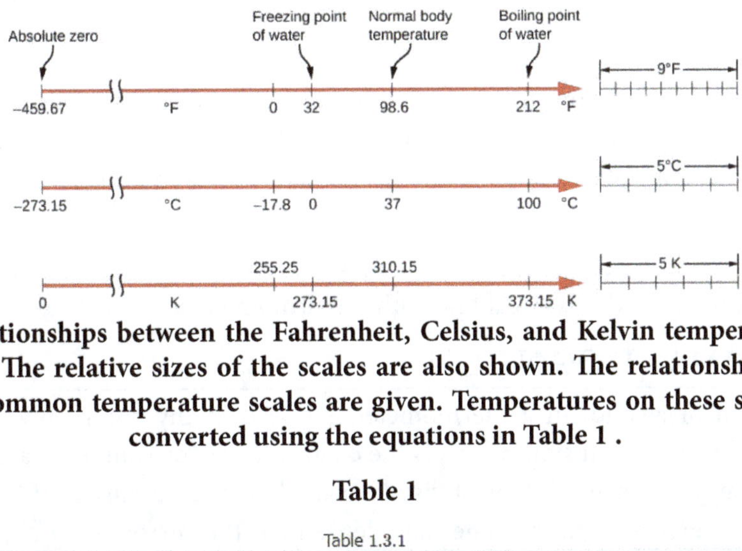

Fig. 4: Relationships between the Fahrenheit, Celsius, and Kelvin temperature scales are shown. The relative sizes of the scales are also shown. The relationships between the three common temperature scales are given. Temperatures on these scales can be converted using the equations in Table 1 .

Table 1

Table 1.3.1

To convert from...	Use this equation...
Celsius to Fahrenheit	$T_F = \frac{9}{5}T_C + 32$
Fahrenheit to Celsius	$T_C = \frac{5}{9}(T_F - 32)$
Celsius to Kelvin	$T_K = T_C + 273.15$
Kelvin to Celsius	$T_C = T_K - 273.15$
Fahrenheit to Kelvin	$T_K = \frac{5}{9}(T_F - 32) + 273.15$
Kelvin to Fahrenheit	$T_F = \frac{9}{5}(T_K - 273.15) + 32$

The Kelvin scale is part of the SI system of units, so its actual definition is more complicated than the one given above. First, it is not defined in terms of the freezing and boiling points of water, but in terms of the triple point. The triple point is the unique combination of temperature and pressure at which ice, liquid water, and water vapor can coexist stably. As will be discussed in the section on phase changes, the coexistence is achieved by lowering the pressure and consequently the boiling point to reach the freezing point. The triple-point temperature is defined as 273.16 K. This definition has the advantage that although the freezing temperature and boiling temperature of water depend on pressure, there is only one triple-point temperature.

Second, even with two points on the scale defined, different thermometers give somewhat different results for other temperatures. Therefore, a standard thermometer is required. Metrologists (experts in the science of measurement) have chosen the constant-volume gas thermometer for this purpose. A vessel of constant volume filled with gas is subjected to temperature changes, and the measured temperature is proportional to the change in pressure. Using "TP" to represent the triple point,

$$T = \frac{p}{p_{TP}} T_{TP}.$$

The results depend somewhat on the choice of gas, but the less dense the gas in the bulb, the better the results for different gases agree. If the results are extrapolated to zero density, the results agree quite well, with zero pressure corresponding to a temperature of absolute zero.

Constant-volume gas thermometers are big and come to equilibrium slowly, so they are used mostly as standards to calibrate other thermometers.

16.4 THERMAL EXPANSION

The expansion of alcohol in a thermometer is one of many commonly encountered examples of thermal expansion, which is the change in size or volume of a given system as its temperature changes. The most visible example is the expansion of hot air. When air is heated, it expands and becomes less dense than the surrounding air, which then exerts an (upward) force on the hot air and makes steam and smoke rise, hot air balloons float, and so forth. The same behavior happens in all liquids and gases, driving natural heat transfer upward in homes, °Ceans, and weather systems, as we will discuss in an upcoming section. Solids also undergo thermal expansion. Railroad tracks and bridges, for example, have expansion joints to allow them to freely expand and contract with temperature changes, as shown in Figure 5 .

(a) (b)

Fig. 5: (a) Thermal expansion joints like these in the (b) Auckland Harbour Bridge in New Zealand allow bridges to change length without buckling.

What is the underlying cause of thermal expansion? As previously mentioned, an increase in temperature means an increase in the kinetic energy of individual atoms. In a solid, unlike in a gas, the molecules are held in place by forces from neighboring molecules; as we saw in Oscillations, the forces can be modeled as in harmonic springs described by the Lennard-Jones potential. Energy in Simple Harmonic Motion shows that such potentials are asymmetrical in that the potential energy increases more steeply when the molecules get closer to each other than when they get farther away. Thus, at a given kinetic energy, the distance moved is greater when neighbors move away from each other than when they move toward each other. The result is that increased kinetic energy (increased temperature) increases the average distance between molecules—the substance expands.

For most substances under ordinary conditions, it is an excellent approximation that there is no preferred direction (that is, the solid is "isotropic"), and an increase in temperature increases the solid's size by a certain fraction in each dimension. Therefore, if the solid is free to expand or contract, its proportions stay the same; only its overall size changes.

Table 2: Thermal Expansion Coefficients

Material	Coefficient of Linear Expansion $\alpha(1/°C)$	Coefficient of Volume Expansion $\beta(1/°C)$
Solids		
Aluminum	25×10^{-6}	75×10^{-6}
Brass	19×10^{-6}	56×10^{-6}
Copper	17×10^{-6}	51×10^{-6}
Gold	14×10^{-6}	42×10^{-6}
Iron or steel	12×10^{-6}	35×10^{-6}

Material	Coefficient of Linear Expansion α(1/°C)	Coefficient of Volume Expansion β(1/°C)
Invar (nickel-iron alloy)	0.9×10^{-6}	2.7×10^{-6}
Lead	29×10^{-6}	87×10^{-6}
Silver	18×10^{-6}	54×10^{-6}
Glass (ordinary)	9×10^{-6}	27×10^{-6}
Glass (Pyrex®)	3×10^{-6}	9×10^{-6}
Quartz	0.4×10^{-6}	1×10^{-6}
Concrete, brick	-12×10^{-6}	-36×10^{-6}
Marble (average)	2.5×10^{-6}	7.5×10^{-6}
Liquids		
Ether		1650×10^{-6}
Ethyl alcohol		1100×10^{-6}
Gasoline		950×10^{-6}
Glycerin		500×10^{-6}
Mercury		180×10^{-6}
Water		210×10^{-6}
Gases		
Air and most other gases at atmospheric pressure		3400×10^{-6}

16.4.1 Thermal Expansion in Two and Three Dimensions

Unconstrained objects expand in all dimensions, as illustrated in Figure 3. That is, their areas and volumes, as well as their lengths, increase with temperature. Because the proportions stay the same, holes and container volumes also get larger with temperature. If you cut a hole in a metal plate, the remaining material will expand exactly as it would if the piece you removed were still in place. The piece would get bigger, so the hole must get bigger too.

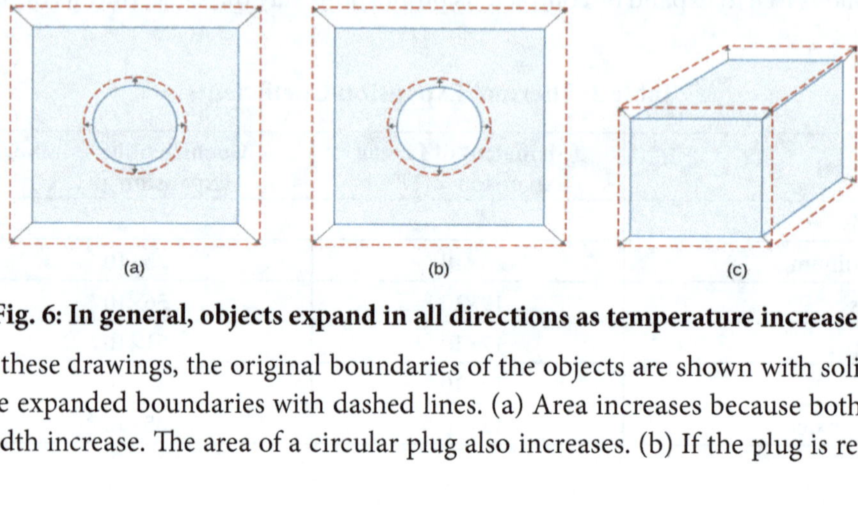

Fig. 6: In general, objects expand in all directions as temperature increases.

In these drawings, the original boundaries of the objects are shown with solid lines, and the expanded boundaries with dashed lines. (a) Area increases because both length and width increase. The area of a circular plug also increases. (b) If the plug is removed, the hole it leaves becomes larger with increasing temperature, just as if the expanding plug were still in place. (c) Volume also increases, because all three dimensions increase.

Volume expansion is defined for liquids, but linear and area expansion are not, as a liquid's changes in linear dimensions and area depend on the shape of its container. Thus, **Table 2** shows liquids' values of β but not α .

In general, objects expand with increasing temperature. Water is the most important exception to this rule. Water does expand with increasing temperature (its density decreases) at temperatures greater than $4°C(40°F)$. However, it is densest at $+4°C$ and expands with decreasing temperature between $+4°C$ and $0°C$ (40oFto32oF), as shown in **Figure 7.** A striking effect of this phenomenon is the freezing of water in a pond. When water near the surface cools down to $4°C$, it is denser than the remaining water and thus sinks to the bottom. This "turnover" leaves a layer of warmer water near the surface, which is then cooled. However, if the temperature in the surface layer drops below $4°C$, that water is less dense than the water below, and thus stays near the top. As a result, the pond surface can freeze over. The layer of ice insulates the liquid water below it from low air temperatures. Fish and other aquatic life can survive in $4°C$ water beneath ice, due to this unusual characteristic of water.

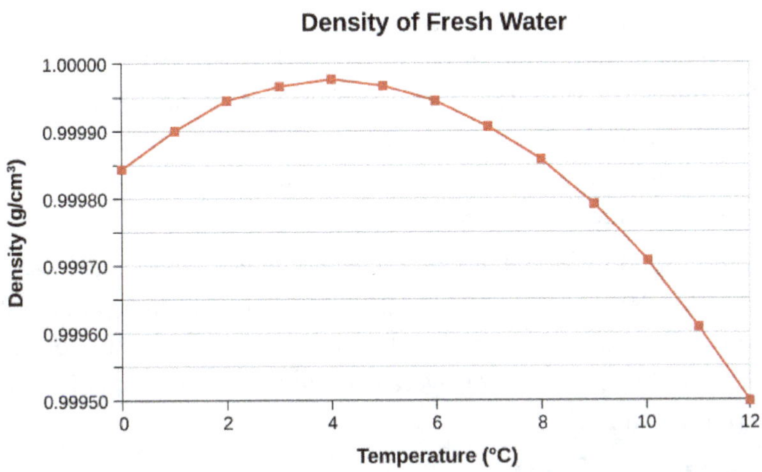

Fig. 7: This curve shows the density of water as a function of temperature.

Note that the thermal expansion at low temperatures is very small. The maximum density at $4°C$ is only 0.0075% greater than the density at $2°C$, and 0.012% greater than that at $0°C$. The decrease of density below $4°C$ occurs because the liquid water approachs the solid crystal form of ice, which contains more empty space than the liquid.

16.4.2 Thermal Stress

If you change the temperature of an object while preventing it from expanding or contracting, the object is subjected to stress that is compressive if the object would expand in the absence of constraint and tensile if it would contract. This stress resulting from temperature changes is known as thermal stress. It can be quite large and can cause damage.

To avoid this stress, engineers may design components so they can expand and contract freely. For instance, in highways, gaps are deliberately left between blocks to prevent thermal stress from developing. When no gaps can be left, engineers must consider thermal stress in their designs. Thus, the reinforcing rods in concrete are made of steel because steel's coefficient of linear expansion is nearly equal to that of concrete.

To calculate the thermal stress in a rod whose ends are both fixed rigidly, we can think of the stress as developing in two steps. First, let the ends be free to expand (or contract) and find the expansion (or contraction). Second, find the stress necessary to compress (or extend) the rod to its original length by the methods you studied in Static Equilibrium and Elasticity on static equilibrium and elasticity. In other words, the ΔL of the thermal expansion equals the ΔL of the elastic distortion (except that the signs are opposite).

16.5 HEAT TRANSFLEARNING OBJECTIVES

16.5.1 Solve Problems Involving Heat Transfer

We have seen in previous chapters that energy is one of the fundamental concepts of physics. Heat is a type of energy transfer that is caused by a temperature difference, and it can change the temperature of an object. As we learned earlier in this chapter, heat transfer is the movement of energy from one place or material to another as a result of a difference in temperature. Heat transfer is fundamental to such everyday activities as home heating and cooking, as well as many industrial proCesses. It also forms a basis for the topics in the remainder of this chapter.

We also introduce the concept of internal energy, which can be increased or decreased by heat transfer. We discuss another way to change the internal energy of a system, namely doing work on it. Thus, we are beginning the study of the relationship of heat and work, which is the basis of engines and refrigerators and the central topic (and origin of the name) of thermodynamics.

16.5.2 Internal Energy and Heat

A thermal system has internal energy (also called thermal energy), which is the sum of the mechanical energies of its molecules. A system's internal energy is proportional to its temperature. As we saw earlier in this chapter, if two objects at different temperatures are brought into contact with each other, energy is transferred from the hotter to the colder object until the bodies reach thermal equilibrium (that is, they are at the same temperature). No work is done by either object because no force acts through a distance. These observations reveal that heat is energy transferred spontaneously due to a temperature difference. Figure 8 shows an example of heat transfer.er, Specific Heat, and Calorimetry.

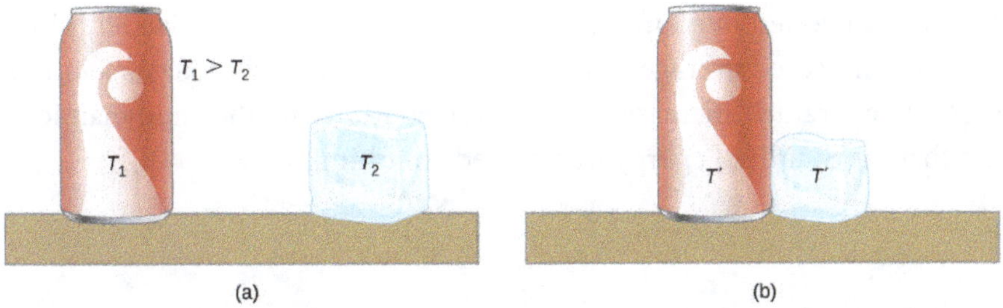

Fig. 8: (a) Here, the soft drink has a higher temperature than the ice, so they are not in thermal equilibrium. (b) When the soft drink and ice are allowed to interact, heat is transferred from the drink to the ice due to the difference in temperatures until they reach the same temperature, T', achieving equilibrium. In fact, since the soft drink and ice are both in contact with the surrounding air and the bench, the ultimate equilibrium temperature will be the same as that of the surroundings.

The meaning of "heat" in physics is different from its ordinary meaning. For example, in conversation, we may say "the heat was unbearable," but in physics, we would say that the temperature was high. Heat is a form of energy flow, whereas temperature is not. Incidentally, humans are sensitive to heat flow rather than to temperature.

Since heat is a form of energy, its SI unit is the joule (J). Another common unit of energy often used for heat is the calorie (cal), defined as the energy needed to change the temperature of 1.00 g of water by $1.00°C$ —specifically, between $14.5°C$ and $15.5°C$ since there is a slight temperature dependence. Also commonly used is the kil°Calorie (kcal), which is the energy needed to change the temperature of 1.00 kg of water by $1.00°C$. Since mass is most often specified in kilograms, the kil°Calorie is convenient. Confusingly, food calories (sometimes called "big calories," abbreviated Cal) are actually kil°Calories, a fact not easily determined from package labeling.

16.6 MECHANICAL EQUIVALENT OF HEAT

It is also possible to change the temperature of a substance by doing work, which transfers energy into or out of a system. This realization helped establish that heat is a form of energy. James Prescott Joule (1818–1889) performed many experiments to establish the mechanical equivalent of heat—the work needed to produce the same effects as heat transfer. In the units used for these two quantities, the value for this equivalence is

$$1.000\text{kcal} = 4186\text{J}.$$

We consider this equation to represent the conversion between two units of energy. (Other numbers that you may see refer to calories defined for temperature ranges other than $14.5°C$ to $15.5°C$.)

As shows one of Joule's most famous experimental setups for demonstrating that work and heat can produce the same effects and measuring the mechanical equivalent of heat. It helped establish the principle of conservation of energy. Gravitational potential energy (U) was converted into kinetic energy (K), and then randomized by viscosity and turbulence into increased average kinetic energy of atoms and molecules in the system, producing a temperature increase. Joule's contributions to thermodynamics were so significant that the SI unit of energy was named after him.

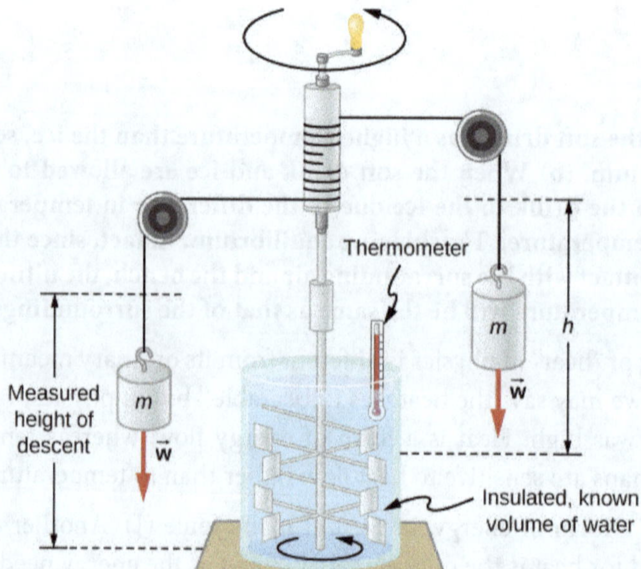

Fig. 9: Joule's experiment established the equivalence of heat and work. As the masses descended, they caused the paddles to do work, W=mgh , on the water. The result was a temperature increase, ΔT , measured by the thermometer. Joule found that ΔT was proportional to W and thus determined the mechanical equivalent of heat.

Increasing internal energy by heat transfer gives the same result as increasing it by doing work. Therefore, although a system has a well-defined internal energy, we cannot say that it has a certain "heat content" or "work content." A well-defined quantity that depends only on the current state of the system, rather than on the history of that system, is known as a state variable. Temperature and internal energy are state variables. To sum up this paragraph, heat and work are not state variables.

Incidentally, increasing the internal energy of a system does not necessarily increase its temperature. As we'll see in the next section, the temperature does not change when a substance changes from one phase to another. An example is the melting of ice, which can be accomplished by adding heat or by doing frictional work, as when an ice cube is rubbed against a rough surface.

16.7 TEMPERATURE CHANGE AND HEAT CAPACITY

We have noted that heat transfer often causes temperature change. Experiments show that with no phase change and no work done on or by the system, the transferred heat is typically directly proportional to the change in temperature and to the mass of the system, to a good approximation. (Below we show how to handle situations where the approximation is not valid.) The constant of proportionality depends on the substance and its phase, which may be gas, liquid, or solid. We omit discussion of the fourth phase, plasma, because although it is the most common phase in the universe, it is rare and short-lived on Earth.

We can understand the experimental facts by noting that the transferred heat is the change in the internal energy, which is the total energy of the molecules. Under typical conditions, the total kinetic energy of the molecules Ktotal is a constant fraction of the internal energy (for reasons and with exceptions that we'll see in the next chapter). The average kinetic energy of a molecule Kave is proportional to the absolute temperature. Therefore, the change in internal energy of a system is typically proportional to the change in temperature and to the number of molecules, N. Mathematically, $\Delta U \propto \Delta Ktotal = NKave \propto N \Delta T$. The dependence on the substance results in large part from the different masses of atoms and molecules. We are considering its heat capacity in terms of its mass, but as we will see in the next chapter, in some cases, heat capacities per molecule are similar for different substances. The dependence on substance and phase also results from differences in the potential energy associated with interactions between atoms and molecules.

Table 3: Specific Heats of Various Substances

Substance	Specific Heat (c) J/kg°C	Specific Heat (c) kcal/kg°C|2|
Solids		
Aluminum	900	0.215
Asbestos	800	0.19
Concrete, granite (average)	840	0.20
Copper	387	0.0924
Glass	840	0.20
Gold	129	0.0308
Human body (average at 37°C)	3500	0.83
Ice (average −50°C to 0°C)	2090	0.50
Iron, steel	452	0.108
Lead	128	0.0305
Silver	235	0.0562
Wood	1700	0.40
Liquids		
Benzene	1740	0.415
Ethanol	2450	0.586
Glycerin	2410	0.576

Substance	Specific Heat (c) J/kg°C	Specific Heat (c) kcal/kg°C[2]
Mercury	139	0.0333
Water (15.0°C)	4186	1.000
Gases[3]		
Air (dry)	721 (1015)	0.172 (0.242)
Ammonia	1670 (2190)	0.399 (0.523)
Carbon dioxide	638 (833)	0.152 (0.199)
Nitrogen	739 (1040)	0.177 (0.248)
Oxygen 1.5.1	651 (913)	0.156 (0.218)
Steam (100°C)	1520 (2020)	0.363 (0.482)

In general, specific heat also depends on temperature. Thus, a precise definition of c for a substance must be given in terms of an infinitesimal change in temperature. To do this, we note that c=1mΔQΔT and replace Δ with d:

$$c = \frac{1}{m}\frac{dQ}{dT}.$$

Except for gases, the temperature and volume dependence of the specific heat of most substances is weak at normal temperatures. Therefore, we will generally take specific heats to be constant at the values given in the table.

In a common kind of problem, objects at different temperatures are placed in contact with each other but isolated from everything else, and they are allowed to come into equilibrium. A container that prevents heat transfer in or out is called a calorimeter, and the use of a calorimeter to make measurements (typically of heat or specific heat capacity) is called calorimetry.

We will use the term "calorimetry problem" to refer to any problem in which the objects concerned are thermally isolated from their surroundings. An important idea in solving calorimetry problems is that during a heat transfer between objects isolated from their surroundings, the heat gained by the colder object must equal the heat lost by the hotter object, due to conservation of energy:

$$Q_{cold} + Q_{hot} = 0.$$

We express this idea by writing that the sum of the heats equals zero because the heat gained is usually considered positive; the heat lost, negative.

16.8 PHASE CHANGES

- Phase transitions play an important theoretical and practical role in the study of heat flow. In melting (or "fusion"), a solid turns into a liquid; the opposite process is freezing. In evaporation, a liquid turns into a gas; the opposite process is condensation.

- A substance melts or freezes at a temperature called its melting point, and boils (evaporates rapidly) or condenses at its boiling point. These temperatures depend on pressure. High pressure favors the denser form, so typically, high pressure raises the melting point and boiling point, and low pressure lowers them. For example, the boiling point of water is 100°C at 1.00 atm. At higher pressure, the boiling point is higher, and at lower pressure, it is lower.

16.9 PHASE DIAGRAMS

The phase of a given substance depends on the pressure and temperature. Thus, plots of pressure versus temperature showing the phase in each region provide considerable insight into thermal properties of substances. Such a pT graph is called a phase diagram.

Figure 1 shows the phase diagram for water. Using the graph, if you know the pressure and temperature, you can determine the phase of water. The solid curves—boundaries between phases—indicate phase transitions, that is, temperatures and pressures at which the phases coexist. For example, the boiling point of water is 100°C at 1.00 atm. As the pressure increases, the boiling temperature rises gradually to 374°C at a pressure of 218 atm. A pressure cooker (or even a covered pot) cooks food faster than an open pot, because the water can exist as a liquid at temperatures greater than 100°C without all boiling away. (As we'll see in the next section, liquid water conducts heat better than steam or hot air.) The boiling point curve ends at a certain point called the critical point—that is, a critical temperature, above which the liquid and gas phases cannot be distinguished; the substance is called a supercritical fluid. At sufficiently high pressure above the critical point, the gas has the density of a liquid but does not condense. Carbon dioxide, for example, is supercritical at all temperatures above 31°C. Critical pressure is the pressure of the critical point.

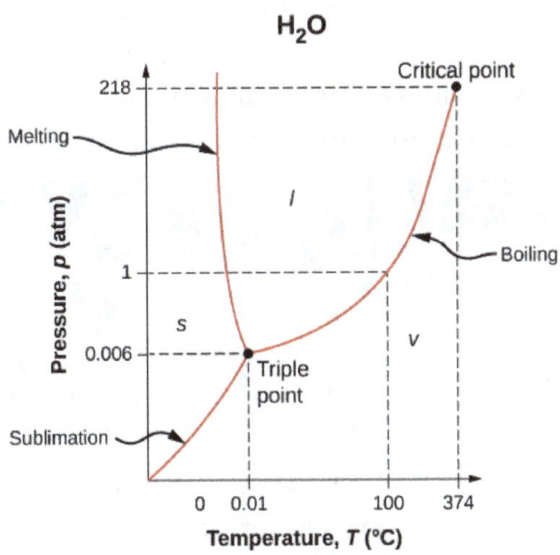

Fig. 10:The phase diagram (pT graph) for water shows solid (s), liquid (l), and vapor (v) phases. At temperatures and pressure above those of the critical point, there is no distinction between liquid and vapor.

Note that the axes are nonlinear and the graph is not to scale. This graph is simplified—it omits several exotic phases of ice at higher pressures. The phase diagram of water is unusual because the melting-point curve has a negative slope, showing that you can melt ice by increasing the pressure.

Similarly, the curve between the solid and liquid regions in Figure 10 gives the melting temperature at various pressures. For example, the melting point is 0°C at 1.00 atm, as expected. Water has the unusual property that ice is less dense than liquid water at the melting point, so at a fixed temperature, you can change the phase from solid (ice) to liquid (water) by increasing the pressure. That is, the melting temperature of ice falls with increased pressure, as the phase diagram shows. For example, when a car is driven over snow, the increased pressure from the tires melts the snowflakes; afterwards, the water refreezes and forms an ice layer.

As you learned in the earlier section on thermometers and temperature scales, the triple point is the combination of temperature and pressure at which ice, liquid water, and water vapor can coexist stably—that is, all three phases exist in equilibrium. For water, the triple point Occurs at $273.16K(0.01°C)$ and 611.2 Pa; that is a more accurate calibration temperature than the melting point of water at 1.00 atm, or $273.15K(0.0°C)$.

At pressures below that of the triple point, there is no liquid phase; the substance can exist as either gas or solid. For water, there is no liquid phase at pressures below 0.00600 atm. The phase change from solid to gas is called sublimation. You may have noticed that snow can disappear into thin air without a trace of liquid water, or that ice cubes can disappear in a freezer. Both are examples of sublimation. The reverse also happens: Frost can form on very cold windows without going through the liquid stage. Carbon dioxide has no liquid phase at atmospheric pressure. Solid CO2 is known as dry ice because instead of melting, it sublimes. Its sublimation temperature at atmospheric pressure is −78°C. Certain air fresheners use the sublimation of a solid to spread a perfume around a room. Some solids, such as osmium tetroxide, are so toxic that they must be kept in sealed containers to prevent human exposure to their sublimation-produced vapors.

(a) (b)

Fig. 11: Direct transitions between solid and vapor are common, sometimes useful, and even beautiful.

(a) Dry ice sublimes directly to carbon dioxide gas. The visible "smoke" consists of water droplets that condensed in the air cooled by the dry ice. (b) Frost forms patterns on a very cold window, an example of a solid formed directly from a vapor.

16.10 EQUILIBRIUM

At the melting temperature, the solid and liquid phases are in equilibrium. If heat is added, some of the solid will melt, and if heat is removed, some of the liquid will freeze. The situation is somewhat more complex for liquid-gas equilibrium. Generally, liquid and gas are in equilibrium at any temperature. We call the gas phase a vapor when it exists at a temperature below the boiling temperature, as it does for water at 20.0°C . Liquid in a closed container at a fixed temperature evaporates until the pressure of the gas reaches a certain value, called the vapor pressure, which depends on the gas and the temperature. At this equilibrium, if heat is added, some of the liquid will evaporate, and if heat is removed, some of the gas will condense; molecules either join the liquid or form suspended droplets. If there is not enough liquid for the gas to reach the vapor pressure in the container, all the liquid eventually evaporates.

If the vapor pressure of the liquid is greater than the total ambient pressure, including that of any air (or other gas), the liquid evaporates rapidly; in other words, it boils. Thus, the boiling point of a liquid at a given pressure is the temperature at which its vapor pressure equals the ambient pressure. Liquid and gas phases are in equilibrium at the boiling temperature (Figure 12). If a substance is in a closed container at the boiling point, then the liquid is boiling and the gas is condensing at the same rate without net change in their amounts.

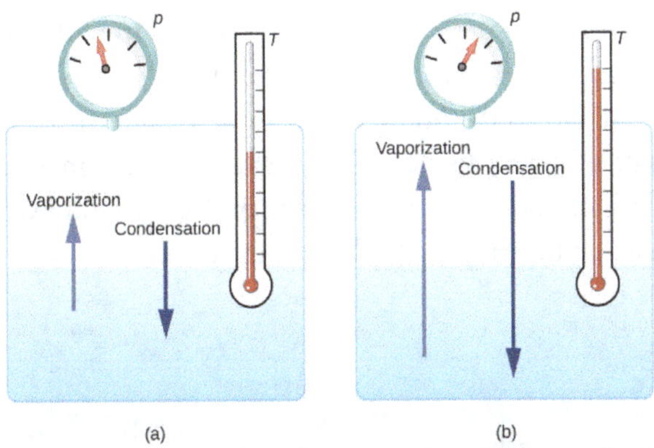

Fig. 12: Equilibrium between liquid and gas at two different boiling points inside a closed container.

(a) The rates of boiling and condensation are equal at this combination of temperature and pressure, so the liquid and gas phases are in equilibrium. (b) At a higher temperature, the boiling rate is faster, that is, the rate at which molecules leave the liquid and enter the gas is faster. This increases the number of molecules in the gas, which increases the gas pressure, which in turn increases the rate at which gas molecules condense and enter the liquid. The pressure stops increasing when it reaches the point where the boiling rate and the condensation rate are equal. The gas and liquid are in equilibrium again at this higher temperature and pressure.

For water, 100°C is the boiling point at 1.00 atm, so water and steam should exist in equilibrium under these conditions. Why does an open pot of water at 100°C boil completely away? The gas surrounding an open pot is not pure water: it is mixed with air. If pure water and steam are in a closed container at 100°C and 1.00 atm, they will coexist—but with air over the pot, there are fewer water molecules to condense, and water boils away. Another way to see this is that at the boiling point, the vapor pressure equals the ambient pressure. However, part of the ambient pressure is due to air, so the pressure of the steam is less than the vapor pressure at that temperature, and evaporation continues. Incidentally, the equilibrium vapor pressure of solids is not zero, a fact that accounts for sublimation.

16.11 PHASE CHANGE AND LATENT HEAT

So far, we have discussed heat transfers that cause temperature change. However, in a phase transition, heat transfer does not cause any temperature change.

For an example of phase changes, consider the addition of heat to a sample of ice at −20°C (Figure 13) and atmospheric pressure. The temperature of the ice rises linearly, absorbing heat at a constant rate of 2090J/kg·°C until it reaches 0°C . Once at this temperature, the ice begins to melt and continues until it has all melted, absorbing 333 kJ/kg of heat. The temperature remains constant at 0°C during this phase change. Once all the ice has melted, the temperature of the liquid water rises, absorbing heat at a new constant rate of 4186J/kg·°C . At 100°C the water begins to boil. The temperature again remains constant during this phase change while the water absorbs 2256 kJ/kg of heat and turns into steam. When all the liquid has become steam, the temperature rises again, absorbing heat at a rate of 2020J/kg·°C . If we started with steam and cooled it to make it condense into liquid water and freeze into ice, the pr°Cess would exactly reverse, with the temperature again constant during each phase transition.

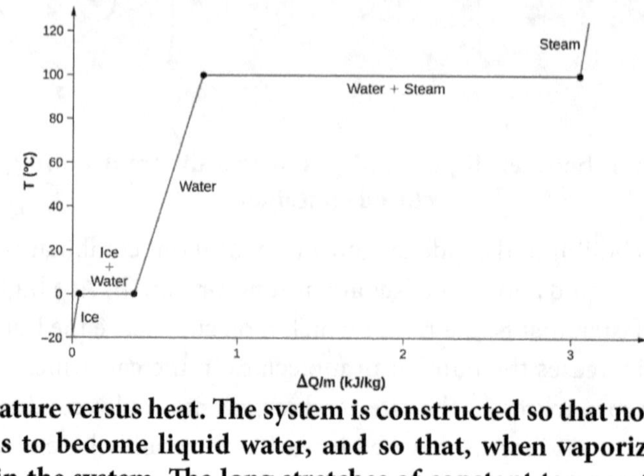

Fig. 13: Temperature versus heat. The system is constructed so that no vapor evaporates while ice warms to become liquid water, and so that, when vaporization occurs, the vapor remains in the system. The long stretches of constant temperatures at 0°C and 100°C reflect the large amounts of heat needed to cause melting and vaporization, respectively.

Where does the heat added during melting or boiling go, considering that the temperature does not change until the transition is complete? Energy is required to melt a solid, because the attractive forces between the molecules in the solid must be broken apart, so that in the liquid, the molecules can move around at comparable kinetic energies; thus, there is no rise in temperature. Energy is needed to vaporize a liquid for similar reasons. Conversely, work is done by attractive forces when molecules are brought together during freezing and condensation. That energy must be transferred out of the system, usually in the form of heat, to allow the molecules to stay together. Thus, condensation occurs in ass°Ciation with cold objects—the glass in Figure 14 , for example.

Fig. 14: Condensation forms on this glass of iced tea because the temperature of the nearby air is reduced. The air cannot hold as much water as it did at room temperature, so water condenses. Energy is released when the water condenses, speeding the melting of the ice in the glass. (credit: Jenny Downing)

The energy released when a liquid freezes is used by orange growers when the temperature approaches 0°C . Growers spray water on the trees so that the water freezes and heat is released to the growing oranges. This prevents the temperature inside the orange from dropping below freezing, which would damage the fruit (Figure 15).

Fig. 15: The ice on these trees released large amounts of energy when it froze, helping to prevent the temperature of the trees from dropping below 0°C . Water is intentionally sprayed on orchards to help prevent hard frosts. (credit: Hermann Hammer)

The energy involved in a phase change depends on the number of bonds or force pairs and their strength. The number of bonds is proportional to the number of molecules and thus to the mass of the sample. The energy per unit mass required to change a substance from the solid phase to the liquid phase, or released when the substance changes from liquid to solid, is known as the heat of fusion. The energy per unit mass required to change a substance from the liquid phase to the vapor phase is known as the heat of vaporization. The strength of the forces depends on the type of molecules. The heat Q absorbed or released in a phase change in a sample of mass m is given by

$$Q=mL_f\text{(melting/freezing)}$$

$$Q=mL_v\text{(vaporization/condensation)}$$

where the latent heat of fusion L_f and latent heat of vaporization L_v are material constants that are determined experimentally. (Latent heats are also called latent heat coefficients and heats of transformation.) These constants are "latent," or hidden, because in phase changes, energy enters or leaves a system without causing a temperature change in the system, so in effect, the energy is hidden.

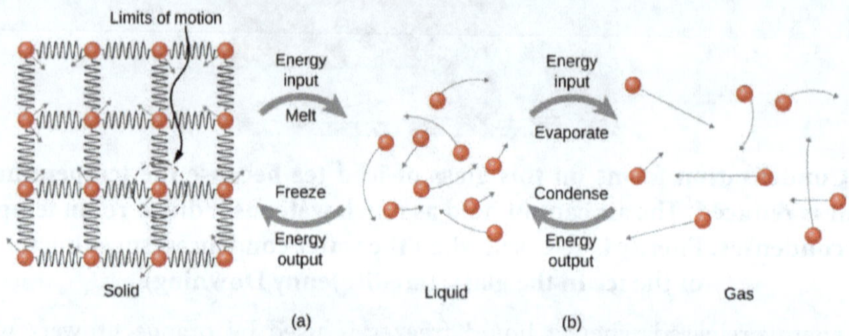

Fig. 16: (a) Energy is required to partially overcome the attractive forces (modeled as springs) between molecules in a solid to form a liquid. That same energy must be removed from the liquid for freezing to take place. (b) Molecules become separated by large distances when going from liquid to vapor, requiring significant energy to completely overcome molecular attraction. The same energy must be removed from the vapor for condensation to take place.

Table 1 lists representative values of L_f and L_v in kJ/kg, together with melting and boiling points. Note that in general, $L_v > L_f$. The table shows that the amounts of energy involved in phase changes can easily be comparable to or greater than those involved in temperature changes, as Figure 7 and the accompanying discussion also showed.

Values quoted at the normal melting and boiling temperatures at standard atmospheric pressure (1atm). Helium has no solid phase at atmospheric pressure. The melting point given is at a pressure of 2.5 MPa. [3]At 37.0°C (body temperature), the heat of vaporization L_v for water is 2430 kJ/kg or 580 kcal/kg. [4]At 37.0°C (body temperature), the heat of vaporization, L_v for water is 2430 kJ/kg or 580 kcal/kg.

Table 4 : Heats of Fusion and Vaporization

Lf			Lv			
Substance	Melting Point (°C)	kJ/kg	kcal/kg	Boiling Point (°C)	kJ/kg	kcal/kg
Helium[2]	−272.2(0.95K)	5.23	1.25	−268.9(4.2K)	20.9	4.99
Hydrogen	−259.3(13.9K)	58.6	14.0	−252.9(20.2K)	452	108
Nitrogen	−210.0(63.2K)	25.5	6.09	−195.8(77.4K)	201	48.0
Oxygen	−218.8(54.4K)	13.8	3.30	−183.0(90.2K)	213	50.9
Ethanol	-114	104	24.9	78.3	854	204
Ammonia	-75	332	79.3	-33.4	1370	327
Mercury	-38.9	11.8	2.82	357	272	65.0
Water	0.00	334	79.8	100.0	2256[3]	539[4]
Sulfur	119	38.1	9.10	444.6	326	77.9
Lead	327	24.5	5.85	1750	871	208
Antimony	631	165	39.4	1440	561	134
Aluminum	660	380	90	2450	11400	2720
Silver	961	88.3	21.1	2193	2336	558
Gold	1063	64.5	15.4	2660	1578	377
Copper	1083	134	32.0	2595	5069	1211
Uranium	1133	84	20	3900	1900	454
Tungsten	3410	184	44	5900	4810	1150

Phase changes can have a strong stabilizing effect on temperatures that are not near the melting and boiling points, since evaporation and condensation occur even at temperatures below the boiling point. For example, air temperatures in humid climates rarely go above approximately 38.0°C because most heat transfer goes into evaporating water into the air. Similarly, temperatures in humid weather rarely fall below the dew point—the temperature where condensation occurs given the concentration of water vapor in the air—because so much heat is released when water vapor condenses.

More energy is required to evaporate water below the boiling point than at the boiling point, because the kinetic energy of water molecules at temperatures below 100°C is less than that at 100°C, so less energy is available from random thermal motions. For example, at body temperature, evaporation of sweat from the skin requires a heat input of 2428 kJ/kg, which is about 10% higher than the latent heat of vaporization at 100°C. This heat comes from the skin, and this evaporative cooling effect of sweating helps reduce the body temperature in hot weather. However, high humidity inhibits evaporation, so that body temperature might rise, while unevaporated sweat might be left on your brow.

Like solid-liquid and liquid-vapor transitions, direct solid-vapor transitions or sublimations involve heat. The energy transferred is given by the equation $Q=mL_s$, where Ls is the heat of sublimation, analogous to Lf and Lv . The heat of sublimation at a given temperature is equal to the heat of fusion plus the heat of vaporization at that temperature.

We can now calculate any number of effects related to temperature and phase change. In each case, it is necessary to identify which temperature and phase changes are taking place. Keep in mind that heat transfer and work can cause both temperature and phase changes.

Snow is formed from ice crystals and thus is the solid phase of water. Because enormous heat is necessary for phase changes, it takes a certain amount of time for this heat to be transferred from the air, even if the air is above 0°C

16.12 MECHANISMS OF HEAT TRANSFER

Just as interesting as the effects of heat transfer on a system are the methods by which it occurs. Whenever there is a temperature difference, heat transfer occurs. It may occur rapidly, as through a cooking pan, or slowly, as through the walls of a picnic ice chest. So many processes involve heat transfer that it is hard to imagine a situation where no heat transfer occurs. Yet every heat transfer takes place by only three methods:

Conduction is heat transfer through stationary matter by physical contact. (The matter is stationary on a macroscopic scale—we know that thermal motion of the atoms and molecules occurs at any temperature above absolute zero.) Heat transferred from the burner of a stove through the bottom of a pan to food in the pan is transferred by conduction.

Convection is the heat transfer by the macroscopic movement of a fluid. This type of transfer takes place in a forced-air furnace and in weather systems, for example.

Heat transfer by radiation occurs when microwaves, infrared radiation, visible light, or another form of electromagnetic radiation is emitted or absorbed. An obvious example is the warming of Earth by the Sun. A less obvious example is thermal radiation from the human body.

In the illustration at the beginning of this chapter, the fire warms the snowshoes' faces largely by radiation. Convection carries some heat to them, but most of the air flow from the fire is upward (creating the familiar shape of flames), carrying heat to the food being cooked and into the sky. The snowshoes wear clothes designed with low conductivity to prevent heat flow out of their bodies.

In this section, we examine these methods in some detail. Each method has unique and interesting characteristics, but all three have two things in common: They transfer heat solely because of a temperature difference, and the greater the temperature difference, the faster the heat transfer (Figure 17).

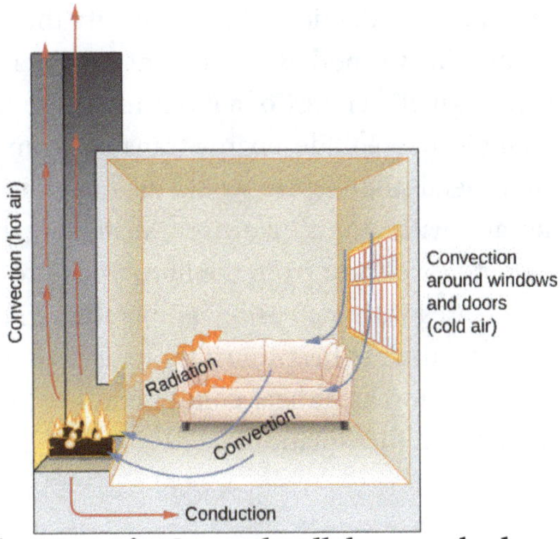

Fig. 17: In a fireplace, heat transfer Occurs by all three methods: conduction, convection, and radiation. Radiation is responsible for most of the heat transferred into the room. Heat transfer also occurs through conduction into the room, but much slower. Heat transfer by convection also occurs through cold air entering the room around windows and hot air leaving the room by rising up the chimney.

Conduction

As you walk barefoot across the living room carpet in a cold house and then step onto the kitchen tile floor, your feet feel colder on the tile. This result is intriguing, since the carpet and tile floor are both at the same temperature. The different sensation is explained by the different rates of heat transfer: The heat loss is faster for skin in contact with the tiles than with the carpet, so the sensation of cold is more intense.

Some materials conduct thermal energy faster than others. Figure 18 shows a material that conducts heat slowly—it is a good thermal insulator, or poor heat conductor—used to reduce heat flow into and out of a house.

Fig. 18: Insulation is used to limit the conduction of heat from the inside to the outside (in winter) and from the outside to the inside (in summer). (credit: Giles Douglas)

A molecular picture of heat conduction will help justify the equation that describes it. Figure 3 shows molecules in two bodies at different temperatures, Th and Tc for "hot" and "cold." The average kinetic energy of a molecule in the hot body is higher than in the colder body. If two molecules collide, energy transfers from the high-energy to the low-energy molecule. In a metal, the picture would also include free valence electrons colliding with each other and with atoms, likewise transferring energy. The cumulative effect of all collisions is a net flux of heat from the hotter body to the colder body. Thus, the rate of heat transfer increases with increasing temperature difference $\Delta T = Th - Tc$. If the temperatures are the same, the net heat transfer rate is zero. Because the number of collisions increases with increasing area, heat conduction is proportional to the cross-sectional area—a second factor in the equation.

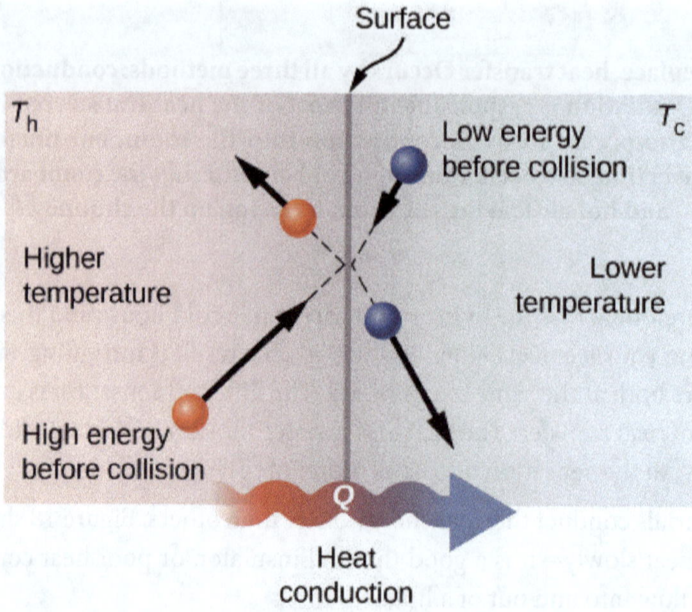

Fig. 19: Molecules in two bodies at different temperatures have different average kinetic energies. Collisions occurring at the contact surface tend to transfer energy from high-temperature regions to low-temperature regions. In this illustration, a molecule in the lower-temperature region (right side) has low energy before collision, but its energy increases after colliding with a high-energy molecule at the contact surface. In contrast, a molecule in the higher-temperature region (left side) has high energy before collision, but its energy decreases after colliding with a low-energy molecule at the contact surface.

A third quantity that affects the conduction rate is the thickness of the material through which heat transfers. Figure 20 shows a slab of material with a higher temperature on the left than on the right. Heat transfers from the left to the right by a series of molecular collisions. The greater the distance between hot and cold, the more time the material takes to transfer the same amount of heat.

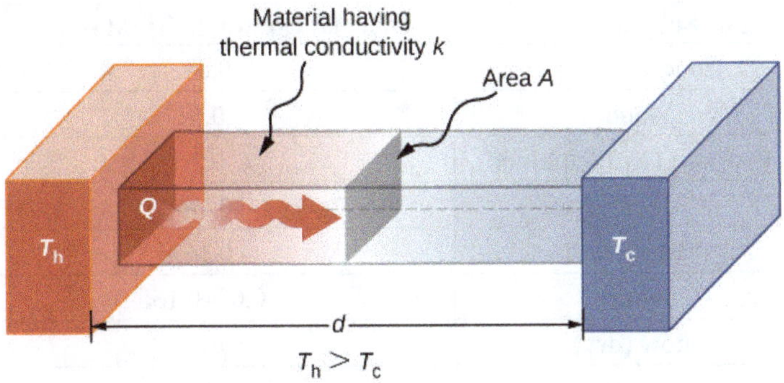

Fig. 20: Heat conduction occurs through any material, represented here by a rectangular bar, whether window glass or walrus blubber.

All four of these quantities appear in a simple equation deduced from and confirmed by experiments. The rate of conductive heat transfer through a slab of material, such as the one in Figure 20, is given by:

$$P = \frac{dQ}{dT} = \frac{kA(T_h - T_c)}{d}$$

where P is the power or rate of heat transfer in watts or in kil°Calories per second, A and d are its surface area and thickness, as shown in Figure 4, Th−Tc is the temperature difference across the slab, and k is the thermal conductivity of the material. Table 1 gives representative values of thermal conductivity.

More generally, we can write

$$P = -kA\frac{dT}{dx},$$

where x is the coordinate in the direction of heat flow. The power and area are constant, dT/dx is constant, and the temperature decreases linearly from Th to Tc.

Table 5 : Thermal Conductivities of Common Substances Values are given for temperatures near 0°C.

Substance	Thermal Conductivity k(W/m°C)
Diamond	2000
Silver	420
Copper	390
Gold	318
Aluminum	220
Steel iron	80
Steel (stainless)	14
Ice	2.2
Glass (average)	0.84

Substance	Thermal Conductivity k(W/m°C)
Concrete brick	0.84
Water	0.6
Fatty tissue (without blood)	0.2
Asbestos	0.16
Plasterboard	0.16
Wood	0.08–0.16
Snow (dry)	0.10
Cork	0.042
Glass wool	0.042
Wool	0.04
Down feathers	0.025
Air	0.023
Polystyrene foam	0.010

In developing insulation, the smaller the conductivity k and the larger the thickness d, the better. Thus, the ratio d/k, called the R factor, is large for a good insulator. The rate of conductive heat transfer is inversely proportional to R. R factors are most commonly quoted for household insulation, refrigerators, and the like. Unfortunately, in the United States, R is still in non-metric units of ft2·oF·h/Btu , although the unit usually goes unstated [1 British thermal unit (Btu) is the amount of energy needed to change the temperature of 1.0 lb of water by 1.0oF which is 1055.1 J]. A couple of representative values are an R factor of 11 for 3.5-inch-thick fiberglass batts (pieces) of insulation and an R factor of 19 for 6.5-inch-thick fiberglass batts (Figure 21). In the US, walls are usually insulated with 3.5-inch batts, whereas ceilings are usually insulated with 6.5-inch batts. In cold climates, thicker batts may be used.

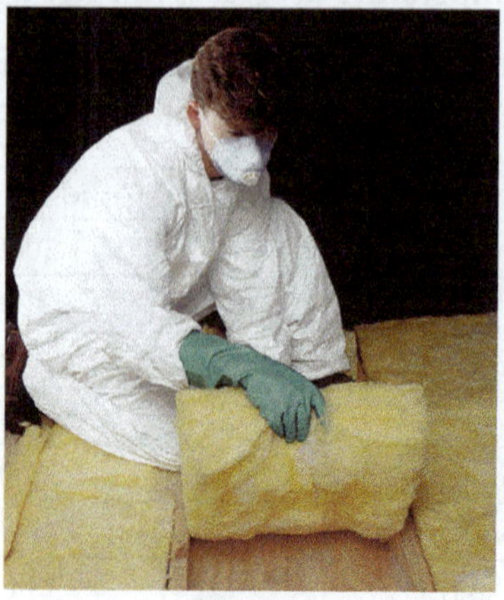

Fig. 21: The fiberglass batt is used for insulation of walls and ceilings to prevent heat transfer between the inside of the building and the outside environment.

Note that in Table 1 , most of the best thermal conductors—silver, copper, gold, and aluminum—are also the best electrical conductors, because they contain many free electrons that can transport thermal energy. (Diamond, an electrical insulator, conducts heat by atomic vibrations.) Cooking utensils are typically made from good conductors, but the handles of those used on the stove are made from good insulators (bad conductors).

Conduction is caused by the random motion of atoms and molecules. As such, it is an ineffective mechanism for heat transport over macroscopic distances and short times. For example, the temperature on Earth would be unbearably cold during the night and extremely hot during the day if heat transport in the atmosphere were only through conduction. Also, car engines would overheat unless there was a more efficient way to remove excess heat from the pistons. The next module discusses the important heat-transfer mechanism in such situations.

Convection

In convection, thermal energy is carried by the large-scale flow of matter. It can be divided into two types. In forced convection, the flow is driven by fans, pumps, and the like. A simple example is a fan that blows air past you in hot surroundings and cools you by replacing the air heated by your body with cooler air. A more complicated example is the cooling system of a typical car, in which a pump moves coolant through the radiator and engine to cool the engine and a fan blows air to cool the radiator.

In free or natural convection, the flow is driven by buoyant forces: hot fluid rises and cold fluid sinks because density decreases as temperature increases. The house in Figure.22 is kept warm by natural convection, as is the pot of water on the stove in Figure23. Clean currents and large-scale atmospheric circulation, which result from the buoyancy of warm air and water, transfer hot air from the tropics toward the poles and cold air from the poles toward the tropics. (Earth's rotation interacts with those flows, causing the observed eastward flow of air in the temperate zones.)

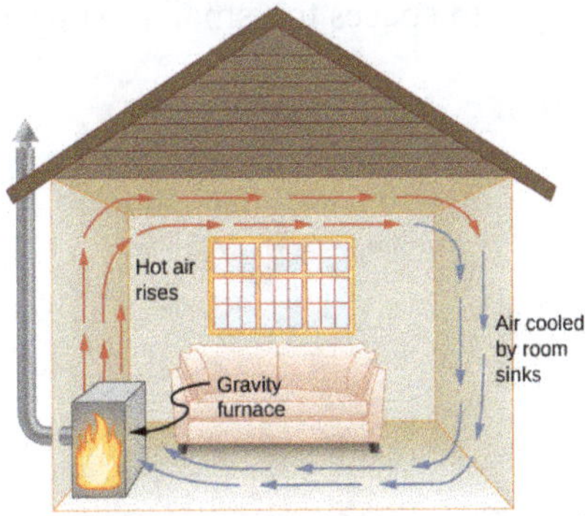

Fig. 22: Air heated by a so-called gravity furnace expands and rises, forming a convective loop that transfers energy to other parts of the room. As the air is cooled at the ceiling and outside walls, it contracts, eventually becoming denser than room air and sinking to the floor. A properly designed heating system using natural convection, like this one, can heat a home quite efficiently.

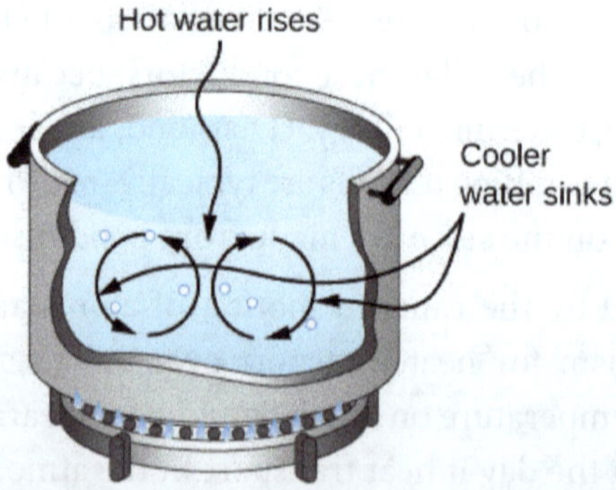

Fig. 23: Natural convection plays an important role in heat transfer inside this pot of water. Once conducted to the inside, heat transfer to other parts of the pot is mostly by convection. The hotter water expands, decreases in density, and rises to transfer heat to other regions of the water, while colder water sinks to the bottom. This process keeps repeating.

Convection is usually more complicated than conduction. Beyond noting that the convection rate is often approximately proportional to the temperature difference, we will not do any quantitative work comparable to the formula for conduction. However, we can describe convection qualitatively and relate convection rates to heat and time. However, air is a poor conductor. Therefore, convection dominates heat transfer by air, and the amount of available space for airflow determines whether air transfers heat rapidly or slowly. There is little heat transfer in a space filled with air with a small amount of other material that prevents flow. The space between the inside and outside walls of a typical American house, for example, is about 9 cm (3.5 in.)—large enough for convection to work effectively. The addition of wall insulation prevents airflow, so heat loss (or gain) is decreased. On the other hand, the gap between the two panes of a double-paned window is about 1 cm, which largely prevents convection and takes advantage of air's low conductivity reduce heat loss. Fur, cloth, and fiberglass also take advantage of the low conductivity of air by trapping it in spaces too small to support convection (Figure 8).

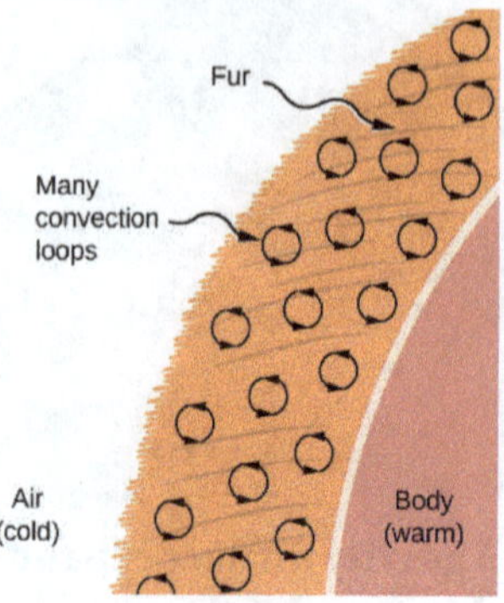

Fig. 24: Fur is filled with air, breaking it up into many small packets. Convection is very slow here, because the loops are so small. The low conductivity of air makes fur a very good lightweight insulator.

Some interesting phenomena happen when convection is accompanied by a phase change. The combination allows us to cool off by sweating even if the temperature of the surrounding air exceeds body temperature. Heat from the skin is required for sweat to evaporate from the skin, but without air flow, the air becomes saturated and evaporation stops. Air flow caused by convection replaces the saturated air by dry air and evaporation continues.

Another important example of the combination of phase change and convection occurs when water evaporates from the °Ceans. Heat is removed from the °Cean when water evaporates. If the water vapor condenses in liquid droplets as clouds form, possibly far from the °Cean, heat is released in the atmosphere. Thus, there is an overall transfer of heat from the °Cean to the atmosphere. This pr°Cess is the driving power behind thunderheads, those great cumulus clouds that rise as much as 20.0 km into the stratosphere (Figure 25). Water vapor carried in by convection condenses, releasing tremendous amounts of energy. This energy causes the air to expand and rise to colder altitudes. More condensation occurs in these regions, which in turn drives the cloud even higher. This mechanism is an example of positive feedback, since the pr°Cess reinforces and accelerates itself. It sometimes produces violent storms, with lightning and hail. The same mechanism drives hurricanes.

Fig. 25: Cumulus clouds are caused by water vapor that rises because of convection. The rise of clouds is driven by a positive feedback mechanism.

Radiation

You can feel the heat transfer from the Sun. The space between Earth and the Sun is largely empty, so the Sun warms us without any possibility of heat transfer by convection or conduction. Similarly, you can sometimes tell that the oven is hot without touching its door or looking inside—it may just warm you as you walk by. In these examples, heat is transferred by radiation (Figure 26). That is, the hot body emits electromagnetic waves that are absorbed by the skin. No medium is required for electromagnetic waves to propagate. Different names are used for electromagnetic waves of different wavelengths: radio waves, microwaves, infrared radiation, visible light, ultraviolet radiation, X-rays, and gamma rays.

Figure 26 : Most of the heat transfer from this fire to the observers °Occurs through infrared radiation. The visible light, although dramatic, transfers relatively little thermal energy. Convection transfers energy away from the observers as hot air rises, while conduction is negligibly slow here. Skin is very sensitive to infrared radiation, so you can sense the presence of a fire without looking at it directly.

The energy of electromagnetic radiation varies over a wide range, depending on the wavelength: A shorter wavelength (or higher frequency) corresponds to a higher energy. Because more heat is radiated at higher temperatures, higher temperatures produce more intensity at every wavelength but especially at shorter wavelengths. In visible light, wavelength determines color—red has the longest wavelength and violet the shortest—so a temperature change is accompanied by a color change. For example, an electric heating element on a stove glows from red to orange, while the higher-temperature steel in a blast furnace glows from yellow to white. Infrared radiation is the predominant form radiated by objects cooler than the electric element and the steel. The radiated energy as a function of wavelength depends on its intensity, which is represented in Figure 27 by the height of the distribution. (Electromagnetic Waves explains more about the electromagnetic spectrum, and Photons and Matter Waves discusses why the decrease in wavelength corresponds to an increase in energy.)

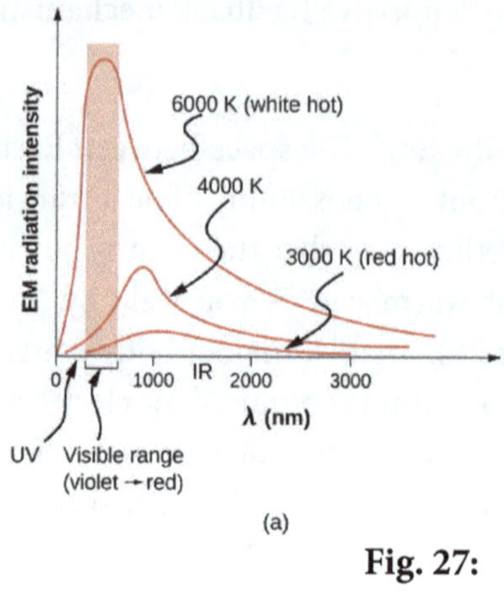

Fig. 27:

(a) A graph of the spectrum of electromagnetic waves emitted from an ideal radiator at three different temperatures. The intensity or rate of radiation emission increases dramatically with temperature, and the spectrum shifts down in wavelength toward the visible and ultraviolet parts of the spectrum. The shaded portion denotes the visible part of the spectrum. It is apparent that the shift toward the ultraviolet with temperature makes the visible appearance shift from red to white to blue as temperature increases. (b) Note the variations in color corresponding to variations in flame temperature.

The rate of heat transfer by radiation also depends on the object's color. Black is the most effective, and white is the least effective. On a clear summer day, black asphalt in a parking lot is hotter than adjacent gray sidewalk, because black absorbs better than gray (Figure 28). The reverse is also true—black radiates better than gray. Thus, on a clear summer night, the asphalt is colder than the gray sidewalk, because black radiates the energy more rapidly than gray. A perfectly black object would be an ideal radiator and an ideal absorber, as it would capture all the radiation that falls on it. In contrast, a perfectly white object or a perfect mirror would reflect all radiation, and a perfectly transparent object would transmit it all (Figure 29). Such objects would not emit any radiation. Mathematically, the color is represented by the emissivity e. A "blackbody" radiator would have an $e=1$, whereas a perfect reflector or transmitter would have $e=0$. For real examples, tungsten light bulb filaments have an e of about 0.5, and carbon black (a material used in printer toner) has an emissivity of about 0.95.

Fig. 28: The darker pavement is hotter than the lighter pavement (much more of the ice on the right has melted), although both have been in the sunlight for the same time. The thermal conductivities of the pavements are the same.

To see that, consider a silver object and a black object that can exchange heat by radiation and are in thermal equilibrium. We know from experience that they will stay in equilibrium (the result of a principle that will be discussed at length in Second Law of Thermodynamics). For the black object's temperature to stay constant, it must emit as much radiation as it absorbs, so it must be as good at radiating as absorbing. Similar considerations show that the silver object must radiate as little as it absorbs. Thus, one property, emissivity, controls both radiation and absorption.

Finally, the radiated heat is proportional to the object's surface area, since every part of the surface radiates. If you kn°Ck apart the coals of a fire, the radiation increases noticeably due to an increase in radiating surface area.

The rate of heat transfer by emitted radiation is described by the Stefan-Boltzmann law of radiation:

$P=\sigma AeT4$,(12) where $\sigma=5.67\times10-8J/s\cdot m2\cdot K4$ is the Stefan-Boltzmann constant, a combination of fundamental constants of nature; A is the surface area of the object; and T is its temperature in kelvins.

The proportionality to the fourth power of the absolute temperature is a remarkably strong temperature dependence. It allows the detection of even small temperature variations. Images called thermographs can be used medically to detect regions of abnormally high temperature in the body, perhaps indicative of disease. Similar techniques can be used to detect heat leaks in homes (Figure 14), optimize performance of blast furnaces, improve comfort levels in work environments, and even remotely map Earth's temperature profile.

Fig. 29 : A thermograph of part of a building shows temperature variations, indicating where heat transfer to the outside is most severe. Windows are a major region of heat transfer to the outside of homes.

The Stefan-Boltzmann equation needs only slight refinement to deal with a simple case of an object's absorption of radiation from its surroundings. Assuming that an object with a temperature T1 is surrounded by an environment with uniform temperature T2 , the net rate of heat transfer by radiation is

$$P==\sigma AeT^4 ,$$

where e is the emissivity of the object alone. In other words, it does not matter whether the surroundings are white, gray, or black: The balance of radiation into and out of the object depends on how well it emits and absorbs radiation. When $T_2>T_1$, the quantity Pnet is positive, that is, the net heat transfer is from hot to cold.

Before doing an example, we have a complication to discuss: different emissivities at different wavelengths. If the fraction of incident radiation an object reflects is the same at all visible wavelengths, the object is gray; if the fraction depends on the wavelength, the object has some other color. For instance, a red or reddish object reflects red light more strongly than other visible wavelengths. Because it absorbs less red, it radiates less red when hot. Differential reflection and absorption of wavelengths outside the visible range have no effect on what we see, but they may have physically important effects. Skin is a very good absorber and emitter of infrared radiation, having an emissivity of 0.97 in the infrared spectrum. Thus, in spite of the obvious variations in skin color, we are all nearly black in the infrared. This high infrared emissivity is why we can so easily feel radiation on our skin. It is also the basis for the effectiveness of night-vision scopes used by law enforcement and the military to detect human beings.

The average temperature of Earth is the subject of much current discussion. Earth is in radiative contact with both the Sun and dark space, so we cannot use the equation for an environment at a uniform temperature. Earth receives almost all its energy from radiation of the Sun and reflects some of it back into outer space. Conversely, dark space is very cold, about 3 K, so that Earth radiates energy into the dark sky. The rate of heat transfer from soil and grasses can be so rapid that frost may Occur on clear summer evenings, even in warm latitudes.

The average temperature of Earth is determined by its energy balance. To a first approximation, it is the temperature at which Earth radiates heat to space as fast as it receives energy from the Sun.

An important parameter in calculating the temperature of Earth is its emissivity (e). On average, it is about 0.65, but calculation of this value is complicated by the great day-to-day variation in the highly reflective cloud coverage. Because clouds have lower emissivity than either oceans or land masses, they reflect some of the radiation back to the surface, greatly reducing heat transfer into dark space, just as they greatly reduce heat transfer into the atmosphere during the day. There is negative feedback (in which a change produces an effect that opposes that change) between clouds and heat transfer; higher temperatures evaporate more water to form more clouds, which reflect more radiation back into space, reducing the temperature.

The often-mentioned greenhouse effect is directly related to the variation of Earth's emissivity with wavelength. The greenhouse effect is a natural phenomenon responsible for providing temperatures suitable for life on Earth and for making Venus unsuitable for human life. Most of the infrared radiation emitted from Earth is absorbed by carbon dioxide (CO_2) and water (H_2O) in the atmosphere and then re-radiated into outer space or back to Earth. Re-radiation back to Earth maintains its surface temperature about 40°C higher than it would be if there were no atmosphere. (The glass walls and roof of a greenhouse increase the temperature inside by blocking convective heat losses, not radiative losses.)

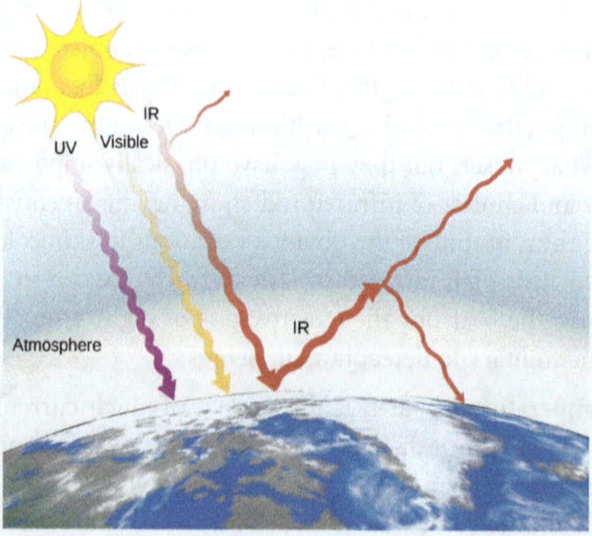

Fig. 30: The greenhouse effect is the name given to the increase of Earth's temperature due to absorption of radiation in the atmosphere. The atmosphere is transparent to incoming visible radiation and most of the Sun's infrared. The Earth absorbs that energy and re-emits it. Since Earth's temperature is much lower than the Sun's, it re-emits the energy at much longer wavelengths, in the infrared. The atmosphere absorbs much of that infrared radiation and radiates about half of the energy back down, keeping Earth warmer than it would otherwise be. The amount of trapping depends on concentrations of trace gases such as carbon dioxide, and an increase in the concentration of these gases increases Earth's surface temperature.

The greenhouse effect is central to the discussion of global warming due to emission of carbon dioxide and methane (and other greenhouse gases) into Earth's atmosphere from industry, transportation, and farming. Changes in global climate could lead to more intense storms, precipitation changes (affecting agriculture), reduction in rain forest biodiversity, and rising sea levels.

RELATED TERMINOLOGY

Action potential of nerve: impulses in the form voltage pulse formed by the cell potential changes from negative to positive and back again during depolarization-repolarization.

Activation energy: the amount of initial energy necessary for reactions to occur.

Active site: a specific region on the enzyme where the substrate binds .

Allosteric inhibition: the mechanism for inhibiting enzyme action in which a regulatory molecule binds to a second site (not the active site) and initiates a conformation change in the active site, preventing binding with the substrate.

Ametrope: People who have less than perfect vision.

Anabolic: describes the pathway that requires a net energy input to synthesize complex molecules from simpler ones.

Aqueous humor: fluid-like water that fills the front of the eye between the lens and the cornea.

Audiogram: Graph generated hearing test.

Auditory: perception of sound by the human ear-brain system.

Axons (nerve fibers): A kind of long tail that propagate signals from the cell body. Axons carry signals biolistrik or nerve impulses from the cell body to the muscles, glands, or other neurons.

Barometer: to measure air pressure tool

Bioenergetics: the concept of energy flow through living systems.

Brown, it was explained by Einstein as the result of the pollen grain being buffeted by the random motion

Brownian motion can be derived from the random walk equations by making the step size infinitely small along with a few other mathematical assumptions.

Brownian motion: Brownian motion is the seemingly random motion that a small particle.

Catabolic: describes the pathway in which complex molecules are broken down into simpler ones, yielding energy as an additional product of the reaction.

Catalyze: Some chemical reactions proceed much more quickly in the presence of a particular molecule than they do when that molecule is absent. The molecule, called a "catalyst," is said to catalyze the reaction.

Competitive inhibition: a general mechanism of enzyme activity regulation in which a molecule other than the enzyme's substrate is able to bind the active site and prevent the substrate itself from binding, thus inhibiting the overall rate of reaction for the enzyme.

Cone: photoreceptors in the retina used for daytime vision. Conjunctiva. Transparent: coating that protects the sclera.

Conformation distribution: The internal potential energy that a molecule has depends on its physical constantly evolving, and emergent behavior often appears. Biological and ecological systems are

Dendrites: input nerve ends are attached to the cell body. Dendrites carry signals from the sensor into the body's cells.

Depolarization: temporary reversal of the membrane potential of neurons.

Diastolic: minimum blood pressure.

ECG (electrocardiogram): potential recordings of heart on the skin.

EEG (electroencephalogram): Recording electrical signals from the brain.

EMG (electromyogram): potential recordings muscles during movement.

Emmetrope: People who have good vision.

Endergonic: describes a chemical reaction that results in products that store more chemical potential energy than the reactants.

Energy landscape: The energy of a physical system can be represented by a mathematical function that depends on several variables. The energy landscape that the system occupies is this function plotted asa hypersurface in space that is one dimension higher than the relevant number of variables. If the energy depends on one variable, then the energy landscape is a line drawn in a two-dimensional plane. If theenergy depends on two variables, the energy landscape is a two-dimensional surface embedded in threedimensionalspace that can look like mountains and valleys in a real landscape that one might encounteron the Earth's surface. The ground state of a system is the lowest point on the energy landscape.entropy: Entropy is a quantitative measure of the amount of order in a system. In statistical mechanics,

Enzymes: Enzymes are proteins that catalyze chemical reactions in biological systems.

EOG (Electrooculogram): Recording potential changes caused by eye movements.

ERG (Electroretinogram): Recording potential changes produced by the retina of the eye when exposed to a beam of light.

Exergonic: describes a chemical reaction that results in products with less chemical potential energy than the reactants, plus the release of free energy.

Extensor: muscles that causes the bones to move away from each other.

Eyepiece: The eyes are doing fine focusing, which can change its shape so as to have the ability to focus objects at various distances.

Feedback inhibition: a mechanism of enzyme activity regulation in which the product of a reaction or the final product of a series of sequential reactions inhibits an enzyme for an earlier step in the reaction series.

Fitness landscape: The fitness landscape is a visual representation of how well adapted different genotypes are to a set of environmental conditions. Each possible genotype occupies a point on the landscape. The distance between each pair of genotypes is related to how similar they are, and the height of each point indicates how well adapted that genotype is.

Flexor: muscles that causes the bones to move closer to each other.

Fluid: Substances that have the ability to flow, liquids and gases ie.

frustrated: A physical system is frustrated if it has no well-defined ground state because there are competing interactions among the pieces of the system that cannot simultaneously be at an energy minimum. A simple example is a system of three spins. If the interaction energy between two spins is lowest when they point in opposite directions, the ground state of a pair of spins is clearly for the two spins to point in opposite directions. If a third spin is added, it is pulled in opposite directions attempting to minimize its interaction with the other two.

Genome: An organism's genome is the complete set of genetic information required to reproduce and maintain that organism in a living state.

Ground state: The ground state of a physical system is the lowest energy state it can occupy. For example, a hydrogen atom is in its ground state when its electron occupies the lowest available energy level.

Handedness: Handedness, also called "chirality," is a directional property that physical systems may exhibit. A system is "right handed" if it twists in the direction in which the fingers of your right hand curl if your thumb is directed along the natural axis defined by the system. Most naturally occurring sugar molecules are right handed. Fundamental particles with spin also exhibit chirality. In this case, the twist is defined by the particle's spin, and the natural axis by the direction in which the particle is moving. Electrons produced in beta-decay are nearly always left handed.

Hearing cortex: part of the brain which is where the hearing process is formed.

Heat energy: the energy transferred from one system to another that is not work.

Infrasonic: sound waves having a frequency below 20 Hz.

Intensity: energy per unit time carried by a wave passing through a unit area perpendicular to the direction of energy flow. The amount related to the level of intensity of sound waves is expressed in decibels (dB).

Iris: Diaphragm muscular reflexes which regulate the light entering the eye.

Kinetic energy: the type of energy associated with objects in motion.

Laminar Flow: flow is smooth, so that the fluid layers adjacent to one another glide gently. In this flow of each fluid particles follow a smooth trajectory and particle trajectories do not intersect.

Land Retineks: theory that suggests that the cones do not operate completely independently, but rather organized into three independent systems, one for each cone. Sclera: tough membrane, white, and meetings that envelops the entire eye except the cornea.

Metabolism: all the chemical reactions that take place inside cells, including those that use energy and those that release energy.

Metastable states of atoms are put to use in atomic clocks because they are long lived, and therefore orzrespond to a clock frequency that can be known very precisely. In biological physics, valleys in the energy landscape correspond to metastable states, as do low-lying peaks in the fitness landscape.

Metastable: A metastable state has a higher energy than the ground state that a physical system can become trapped in for some length of time. A simple example is a ball sitting on a hilltop. The ball's energy would be lower if it rolled down the hill; but unless something disturbs it, it will remain where it is.

Missing: presbycusis hearing caused by age.

Monomer: A monomer is a small molecule that can bind to other like molecules to form a polymer. The amino acids that make up proteins are examples of monomers.

Myelin: Sheath jointed containing fatty substances and embungkus axons.

Neurons: Nerve cells that form a complex network in the body that receive, process and forward information from one part of the body to other body parts.

Noncompetitive Inhibition: a general mechanism of enzyme activity regulation in which a regulatory molecule binds to a site other than the active site and prevents the active site from binding the substrate; thus, the inhibitor molecule does not compete with the substrate for the active site; allosteric inhibition is a form of noncompetitive inhibition.

Paradoxical sleep or rapid eye movement (REM): high frequency pattern in the EEG during the sleep, the eyes move in this period.

Perception: of human vision to light through the eye-brain system. Perception consciousness through the senses.

Pitch (pitch): Perception of sound frequencies are determined by Titi tone frequency; higher the frequency, the higher the pitch titi.

Polar: A polar molecule has a nonzero electric dipole moment, so it has a side that is positively charged and a side that is negatively charged.

Polarizability: Some atoms and molecules that have no electric dipole moment in an electrically neutral environment will develop one in an electric field. The polarizability of an atom or molecule is a quantity that describes how susceptible it is to this effect.

Polarization: The polarization of a wave is the direction in which it is oscillating. The simplest type of polarization is linear, transverse polarization. Linear means that the wave oscillation is confined along a single axis, and transverse means that the wave is oscillating in a direction perpendicular to its direction of travel. Laser light is most commonly a wave with linear, transverse polarization. If the laser beam travels along the x-axis, its electric field will oscillate either in the y-direction or in the z-direction.

Pressure Gauge: manometer fluid.

Pressure Measurement (gauge pressure): Pressure melibihi atmospheric pressure.

Pupil: small hole in the center of the iris where light enters the eye.

Random walk: The random walk is the trajectory that arises when an object moves in steps that are all the same length, but in random directions. The path of a molecule in a gas follows a random walk, with the step size determined by how far (on average) the molecule can travel before it collides with something and changes direction. The behavior of many diverse systems can be modeled as a random walk, including the path of an animal searching for food, fluctuating stock prices, and the diffusion of a drop of food coloring placed in a bowl of water.

Reach Sounds (Audible Range): frequency range of sound waves that can be addressed human ear, between 20 Hz and 20,000 Hz.

Repolarization: Back to the rest of the state of polarity reversal potential.

Resting Potential: when the membrane potential at the axon does not conduct electrical pulses.

Resultant Vector: resulting from the addition or subtraction of two or more vectors.

Rod (rod): photoreceptors in the retina that is used for night vision and peripheral vision.

Second Law of Thermodynamics: The second law of thermodynamics states that the entropy of an isolated system will either increase or remain the same over time. This is why heat flows from a hot bject to a cold object, but not the other way; and why it's easy to dissolve salt in water, but not so easy to get the salt back out again.

Snellen: chart Chart used to test the acuity of the eye.

Sodium-Potassium Pump: pumping process that transports sodium ions out of the cell and bring potassium ions enter the same amount.

Sometimes there is not a single, unique minimum energy conformation. In this case, the conformation

Sphygmomanometer: Tools for menukur blood pressure.

Structure or conformation: Molecules tend toward structures that minimize their potential energy.

Substrate: a molecule on which the enzyme acts.

Synapses: Connections between neurons.

Systolic: maximum pressure in blood.

Tendons: stringy fibers in the muscle system that forwards styles performed by the muscles of the body to another.

The cornea: is transparent and crystal clear arch at the front of the eye that focuses light rays entering the eye.

The Disease: myasthenia gravis muscle weakness when it perform repetitive tasks.

The Light-Sensitive: retina half of the eyeball that covers the rear. Most of vision is limited to a small area called the macula lutea or yellow spots. Vision of the small parts of objects occurs in a very small area in the yellow spots called the fovea centralis (central fovea).

The Nodes of Ranvier: The gaps separating the axon segments of the myelin sheath.

The Propagation of Jumps (Saltatory): propagation of action potentials in axons that seems to jump from gap to gap sat next as the pace of the action potential in the axon wrapped in myelin so large compared to the cracks.

Thermody Namics: the science of the relationships between heat, energy, and work.

Turbulent Floe: flow characterized by small circles resembles whirlpools irregular and called eddy currents.

Turing Machine: In 1937, Alan Turing outlined the details of the Turing machine in a paper investigating the possibilities and limits of machine computation. The machine is an idealized computing device that consists, in its simplest form, of a tape divided up into cells that are processed by an active element called a "head." The cells can be in one of two states. The head moves along the tape, changing the cells from one state to the other and moving either forward or backward according to a set of predetermined instructions. Turing machines can be described with a set of simple mathematical equations that allowed scientists to understand many of the basic properties of digital computing long before the first modern computer was built.

Ultrasonic: sound waves having a frequency above 20,000 Hz.

Viscosity: friction in a fluid that prevents fluid moving freely and a friction force between the fluid layers adjacent: layers during the moving of one another.

Vision Cortex: part of the brain which is where the vision is formed.

Visual Acuity: Sharpness small parts of objects that can be detected by eye.

Vitreous humor: gelatinous fluid that fills the eyeball between the lens and the retina.

REFERENCES

A.L. Hodgkin and A.F. Huxley. Currents carried by sodium and potassium ions through the membrane of a giant axon of loligo. *Journal of Physiology,* 116:449{472, 1952.

A.L. Hodgkin and A.F. Huxley. The components of membrane conductance in the giant axon of loligo. *Journal of Physiology,* 116:473{496, 1952.

A.L. Hodgkin and B. Katz. The e_ect of sodium ions on the electrical activity of the giant axon of the squid. *Journal of Physiology*, 108:37{77, 1949.

A.L. Hodgkin, A.F. Huxley, and B. Katz. Measurements of current-voltage relations in the membrane of the giant axon of loligo. *Journal of Physiology*, 116:424{448, 1952.

Aho et al 1987: Retinal noise, the performance of retinal ganglion cells, and visual sensitivity in the dark00adapted frog. A–C Aho, K Donner, C Hyd´en, T Reuter & OY Orlov, J Opt Soc Am A 4, 2321–2329 (1987).

Aho et al 1988: Low retinal noise in animals with low body temperature allows high visual sensitivity. A–C Aho, K Donner, C Hyd´en, LO Larsen & T Reuter, Nature 334, 348–350 (1988).

Aho et al 1993: Visual performance of the toad (Bufo bufo) at low light levels: retinal ganglion cell responses and prey- catching accuracy. A–C Aho, K Donner, S Helenius, LO Larsen & T Reuter, J Comp Physiol A 172, 671–682 (1993).

Aidley 1998: The Physiology of Excitable Cells, 4th Edition DJ Aidley (Cambridge University Press, Cambridge, 1998).

Angeli V, Randolph GJ. Inflammation, lymphatic function, and dendritic cell migration. Lymphat Res Biol. 2006;4:217–228.

Annis, D. A., Nickbarg, E., Yang, X., Ziebell, M. R. & Whitehurst, C. E. Affinity selection-mass spectrometry screening techniques for small molecule drug discovery. Curr. Opin. Chem. Biol. 11, 518–526 (2007).

Antonova N, Todorov R, Exerowa D. Rheological behavior and parameters of the in vitro model of lung surfactant systems: The role of the main phospholipid component. Biorheology. 2003; 40(5):531–543

Aristotelous, T. et al. Discovery of β2 adrenergic receptor ligands using biosensor fragment screening of tagged wild-type receptor. ACS Med. Chem. Lett. 4, 1005–1010 (2013).

Aristotelous, T., Hopkins, A. L. & Navratilova, I. Surface plasmon resonance analysis of seventransmembrane receptors. Methods Enzymol. 556, 499–525 (2015).

Arkin, M. R., Tang, Y. & Wells, J. A. Small-molecule inhibitors of protein–protein interactions: progressing toward the reality. Chem. Biol. 21, 1102–1114 (2014).

Armer JM. The problem of post-breast cancer lymphedema: Impact and measurement issues. Cancer Invest. 2005;23:76–83.

B. Hille. Ionic Channels of excitable membranes. Sinaouer Associates, 1984.

B. Katz. Nerve, muscle and synapse. McGraw-Hill, 1966.

Badescu V (1997) Verification of some very simple clear and cloudy sky model to evaluate global solar irradiance. Sol Energy 61:251–264

Baell, J. & Walters, M. A. Chemistry: chemical con artists foil drug discovery. Nature 513, 481–483 (2014).

Bai, X. C. et al. An atomic structure of human gammasecretase. Nature 525, 212–217 (2015).

Bai, X. C., McMullan, G. & Scheres, S. H. How cryo-EM is revolutionizing structural biology. Trends Biochem. Sci. 40, 49–57 (2015).

Baksh, M. M., Kussrow, A. K., Mileni, M., Finn, M. G. & Bornhop, D. J. Label-free quantification of membrane-ligand interactions using backscattering interferometry. Nat. Biotechnol. 29, 357–360 (2011).

Barlow 1952: The size of ommatidia in apposition eyes. HB Barlow, J Exp Biol 29, 667–674 (1952).

Barlow 1956: Retinal noise and absolute threshold. HB Barlow, J Opt Soc Am 46, 634–639 (1956).

Barlow 1981: Critical limiting factors in the design of the eye and visual cortex. HB Barlow, Proc R Soc Lond Ser B 212, 1–34 (1981).

Bartesaghi, A. et al. 2.2 Å resolution cryo-EM structure of beta-galactosidase in complex with a cellpermeant inhibitor. Science 348, 1147–1151 (2015).

Basser PJ, McMahon TA, Griffith P. The mechanism of mucus clearance in cough. J Biomech Eng. 1989;111:288–297.

Baylor et al 1979a: The membrane current of single rod outer segments. DA Baylor, TD Lamb & K–W Yau, J Physiol (Lond) 288, 589–611 (1979).

Baylor et al 1979b: Rod responses to single photons. DA Baylor, TD Lamb & K–W Yau, J Physiol (Lond) 288, 613–634 (1979).

Baylor, G Matthews & K–W Yau, J Physiol (Lond) 309, 591–621 (1980).

Baylor et al 1984: The photocurrent, noise and spectral sensitivity of rods of the monkey Macaca fascicularis. DA Baylor, BJ Nunn & JF Schnapf, J Physiol (Lond) 357, 575–607 (1984).

Becker P, Erhardt DW, Smith AP (1989) Analysis of forest light environments Part I. Computerized estimation of solar radiation from hemispherical canopy photographs. Agric Forest Meteorol 44:217–232

Bellet S. Clinical disorders of the heart beat. Lea & Febiger; Philadelphia: 1963.

Bialek 1987: Physical limits to sensation and perception. W Bialek, Ann Rev Biophys Biophys Chem 16, 455–478 (1987).

Bilek AM, Dee KC, Gaver DP. Mechanisms of surface-tension-induced epithelial cell damage in a model of pulmonary airway reopening. *Journal of Applied Physiology.* 2003;94(2):770–783.

Björke, H. & Andersson, K. Measuring the affinity of a radioligand with its receptor using a rotating cell dish with in situ reference area. Appl. Radiat. Isot. 64, 32–37 (2006).

Blundell, T. L. & Patel, S. High-throughput X-ray crystallography for drug discovery. Curr. Opin. Pharmacol. 4, 490–496 (2004).

Bocquet, N. et al. Real-time monitoring of binding events on a thermostabilized human A2A receptor embedded in a lipid bilayer by surface Plasmon resonance. Biochim. Biophys. Acta 1848, 1224–1233 (2015).

Bohr 1933: Light and life. N Bohr, Nature 131, 421–423 (1933). Cairns et al 1966: Phage and the Origins of Molecular Biology, J Cairns, GS Stent & JD Watson, eds (Cold Spring Harbor Press, Cold Spring Harbor NY, 1966).

Bornhop, D. J. et al. Free-solution, label-free molecular interactions studied by back-scattering interferometry. Science 317, 1732–1736 (2007).

Bouman 1961: History and present status of quantum theory in vision. MA Bouman, in Sensory Communication, W Rosenblith, ed, pp 377–401 (MIT Press, Cambridge, 1961).

Bradkte F (1951) Katathermometrische Feststellung der mittleren Strahlungstemperatur der Umgebung. GI 72:3–7

Bradshaw, J. M. et al. Prolonged and tunable residence time using reversible covalent kinase inhibitors. Nat. Chem. Biol. 11, 525–531 (2015).

Braissant, O. et al. Isothermal microcalorimetry accurately detects bacteria, tumorous microtissues, and parasitic worms in a label-free well-plate assay. Biotechnol. J. 10, 460–468 (2015).

Brough, P. A. et al. 4,5-diarylisoxazole Hsp90 chaperone inhibitors: potential therapeutic agents for the treatment of cancer. J. Med. Chem. 51, 196–218.

Brough, P. A. et al. Combining hit identification strategies: fragment-based and in silico approaches to orally active 2-aminothieno[2,3-d]pyrimidine inhibitors of the Hsp90 molecular chaperone. J. Med. Chem. 52, 4794–4809 (2009).

Brühl Ch, Zdunkowski W (1983) An approximate calculation method for parallel and diffuse solar irradiances on inclined surfaces in

Bruse M, Fleer H (1998) Simulating surface-plant-air interactions inside urban environments with a three dimensional numerical model. Environ Model Softw 13:373–384

Bruylants, G., Wouters, J. & Michaux, C. Differential scanning calorimetry in life science: thermodynamics, stability, molecular recognition and application in drug design. Curr. Med. Chem. 12, 2011–2020 (2005).

Bunnage, M. E. Getting pharmaceutical R&D back on target. Nat. Chem. Biol. 7, 335–339 (2011).

C.M. Armstrong and L. Binstock. Anomalous recti_cation in the squid giant axon injected with tetraethylammonium chloride. *Journal of General Physiology*, 48:859{872, 1965}.

Cala, O., Guilliere, F. & Krimm, I. NMR-based analysis of protein-ligand interactions. Anal. Bioanal. Chem. 406, 943–956 (2014).

Cameron, J. R., J.G. Skofronick. (1978). Medical Physics. New York: John Wiley & Sons, Inc.

Campana D, Salta FA. Surfactant effects on the Rayleigh instability in capillary tubes - non-ideal systems. International *Journal of Multiphase Flow.* 2007;33(11):1153–1171.

Cassidy KJ, et al. Surfactant effects in model airway closure experiments. *Journal of Applied Physiology.* 1999;87(1):415–427.

Ceballos JC, De A Moura GB (1997) Solar irradiation assessment using meteosat 4-Vis imagery. Sol Energy 60:209–219

Chaires, J. B. Calorimetry and thermodynamics in drug design. Annu. Rev. Biophys. 37, 135–151 (2008).

Chalmers, M. J., Busby, S. A., Pascal, B. D., West, G. M. & Griffin, P. R. Differential hydrogen/ deuterium exchange mass spectrometry analysis of protein–ligand interactions. Expert Rev. Proteom. 8, 43–59 (2011).

Chapman L, Thornes JE, Bradley AV (2001) Rapid determination of canyon geometry parameters for use in surface radiation budgets. Theor Appl Climatol 69:81–89

Chen JM, Black TA (1991) Measuring leaf area index of plant canopies with branch architecture. Agric Forest Meteorol 57:1–12 Clark RP, Edholm OG (1985) Man and his thermal environment. E. Arnold, London

Chen, D. et al. Fragment screening of GPCRs using biophysical methods: identification of ligands of the adenosine A2A receptor with novel biological activity. ACS Chem. Biol. 7, 2064–2073 (2012).

Cheng, Y., Grigorieff, N., Penczek, P. A. & Walz, T. A primer to single-particle cryo-electron microscopy. Cell 161, 438–449 (2015).

Chodera, J. D. & Mobley, D. L. Entropy-enthalpy compensation: role and ramifications in biomolecular ligand recognition and design. Annu. Rev. Biophys. 42, 121–142 (2013).

Chr. Koch. Biophysics of computation. Oxford University Press, 1999.

Christopeit, T. et al. A surface plasmon resonancebased biosensor with full-length BACE1 in a reconstituted membrane. Anal. Biochem. 414, 14–22 (2011).

Christopher, J. A. et al. Biophysical fragment screening of the β1-adrenergic receptor: identification of high affinity arylpiperazine leads using structure-based drug design. J. Med. Chem. 56, 3446–3455 (2013).

Christopher, J. A. et al. Discovery of HTL6641, a dual orexin receptor antagonist with differentiated pharmacodynamic properties. MedChemComm 6, 947–955 (2015).

Chu EK, et al. Bronchial epithelial compression regulates epidermal growth factor receptor family ligand expression in an autocrine manner. American *Journal of Respiratory Cell and Molecular Biology.* 2005;32(5):373–380.

Chu EK, et al. Induction of the plasminogen activator system by mechanical stimulation of human bronchial epithelial cells. American *Journal of Respiratory Cell and Molecular Biology.* 2006 ;35(6):628–638.

Chu, R., Reczek, D. & Brondyk, W. Capture-stabilize approach for membrane protein SPR assays. Sci. Rep. 4, 7360 (2014).

Claridge, S. A., Schwartz, J. J. & Weiss, P. S. Electrons, photons, and force: quantitative single-molecule measurements from physics to biology. ACS Nano. 5, 693–729 (2011).

Codd SL, et al. Tensile stiffness of ovine tracheal wall. *Journal of Applied Physiology.* 1994; 76 (6):2627–2635.

Concepcion, J. et al. Label-free detection of biomolecular interactions using BioLayer interferometry for kinetic characterization. Comb. Chem. High Throughput Screen. 12, 791–800 (2009).

Congreve, M. et al. Fragment screening of stabilized G-protein-coupled receptors using biophysical methods. Methods Enzymol. 493, 115–136 (2011).

Cooper, M. A. & Singleton, V. T. A survey of the 2001 to 2005 quartz crystal microbalance biosensor literature: applications of acoustic physics to the analysis of biomolecular interactions. J. Mol. Recognit 20, 154–184 (2007).

Cooper, M. A. Optical biosensors in drug discovery. Nat. Rev. Drug Discov. 1, 515–528 (2002).

Copeland, R. A. The drug-target residence time model: a 10-year retrospective. Nat. Rev. Drug Discov. 15, 87–95 (2016).

Copeland, R. A., Pompliano, D. L. & Meek, T. D. Drug-target residence time and its implications for lead optimization. Nat. Rev. Drug Discov. 5, 730–739 (2006).

Craggs C, Conway EM, Pearsall NM (2000) Statistical investigation of the optimal averaging time for solar irradiance on horizontal and vertical surfaces in the UK. Sol Energy 68:79–187

Cusack, K. P. et al. Design strategies to address kinetics of drug binding and residence time. Bioorg. Med. Chem. Lett. 25, 2019–2027 (2015).

Czeplak G, Kasten F (1987) Parametrisierung der atmosphärischen Wärmestrahlung bei bewölktem Himmel. Meteorol Rndsch 40:184–187.

D. Purves, G.J. Augustine, D. Fitzpatrick, L.C. Katz, A-S. LaMantia, and J.O. McNamara. Neuroscience. Sinauer associates, 1997.

Dahl, G. & Åkerud, T. Pharmacokinetics and the drugtarget residence time concept. Drug Discov. Today 18, 697–707 (2013).

Danielson, U. H. Fragment library screening and lead characterization using SPR biosensors. Curr. Top. Med. Chem. 9, 1725–1735 (2009).

Darby, J. F. et al. Discovery of selective small-molecule activators of a bacterial glycoside hydrolase. Angew. Chem. Int. 53, 13419–13423 (2014).

Davis, B. J. & Erlanson, D. A. Learning from our mistakes: the 'unknown knowns' in fragment screening. Bioorg. Med. Chem. Lett. 23, 2844–2852 (2013).

Dawson, R. J. et al. Structure of the acid-sensing ion channel 1 in complex with the gating modifier Psalmotoxin 1. Nat. Commun. 3, 936 (2012).

Daylight, sunlight and solar gain in the urban environment. Sol Energy 70:177–185

de Ruyter van Steveninck & Laughlin 1996a: The rate of information transfer at graded-potential synapses. RR de Ruyter van Steveninck & SB Laughlin, Nature 379, 642–645 (1996).

de Ruyter van Steveninck & Laughlin 1996b: Light adaptation and reliability in blowfly photoreceptors. R de Ruyter van Steveninck & SB Laughlin, Int. J. Neural Syst. 7, 437- 444 (1996)

de Vries M, Vonkeman WG, van Ginkel RJ, Hoekstra HJ. Morbidity after inguinal sentinel lymph node biopsy and completion lymph node dissection in patients with cutaneous melanoma. Eur J Surg Oncol. 2006;32:785–789.

Delbr¨uck 1949: A physicist looks at biology. M Delbr¨uck, Trans Conn Acad Arts Sci 38, 173–190 (1949). Reprinted in Cairns et al (1966), pp 9–22.

Delbr¨uck 1970: A physicist's renewed look at biology: twenty years later. M Delbr¨uck, Science 168, 1312–1315 (1970).

Dennis Kunkel. (2004). http://rabi.phvs.virainia.edu/HTW/book.html. www.Cellsbio.com

Diag GR, Bland WL, Mecikalski JR, Anderson MC (2000) Satellitebased estimates of longwave radiation for agricultural applications. Agric Forest Meteorol 103:349–355

Dixon JB, Gashev AA, Zawieja DC, Moore JE, Jr, Cote GL. Image correlation algorithm for measuring lymphocyte velocity and diameter changes in contracting microlymphatics. Ann Biomed Eng. 2007;35:387–396.

Dixon JB, Greiner ST, Gashev AA, Cote GL, Moore JE, Zawieja DC. Lymph flow, shear stress, and lymphocyte velocity in rat mesenteric prenodal lymphatics. Microcirculation. 2006;13:597–610.

Dixon JB, Zawieja DC, Gashev AA, Cote GL. Measuring microlymphatic flow using fast video microscopy. J Biomed Opt. 2005;10:064016.

DM Green, & JA Swets (Wiley, New York, 1966).

Dominguez, J. L. et al. Effect of the protonation state of the titratable residues on the inhibitor affinity to BACE-1. Biochemistry 49, 7255–7263 (2010).

Egloff, P. et al. Structure of signaling-competent neurotensin receptor 1 obtained by directed evolution in Escherichia coli. Proc. Natl Acad. Sci. USA 111, E655–E662 (2014).

Elinder, M. et al. Inhibition of HIV-1 by non-nucleoside reverse transcriptase inhibitors via an induced fit mechanism-Importance of slow dissociation and relaxation rates for antiviral efficacy. Biochem. Pharmacol. 80, 1133–1140 (2010).

Erlanson, D. A. & Zartler, E. Fragments in the clinic: 2015 edition. Practical Fragments, http:// practicalfragments.blogspot.fr/2015/01/fragments-inclinic- 2015-edition.html (2015).

Errey, J. C., Dore, A. S., Zhukov, A., Marshall, F. H. & Cooke, R. M. Purification of stabilized GPCRs for structural and biophysical analyses. Methods Mol. Biol. 1335, 1–15 (2015).

F Rieke & DA Baylor, Rev Mod Phys 70, 1027– 1036 (1998).

Fanger PO (1972) Thermal comfort. McGraw-Hill, New York

Ferenczy, G. G. & Keseru, G. M. Thermodynamics guided lead discovery and optimization. Drug Discov. Today 15, 919–932 (2010).

Feynman et al 1963: The Feynman Lectures on Physics. RP Feynman, RB Leighton & M Sands (Addison–Wesley, Reading, 1963).

Folmer, R. H. Integrating biophysics with HTS-driven drug discovery projects. Drug Discov. Today 21, 491–498 (2016).

Frank SF, Gerding RB, O'Rourke PA, Terhung WH (1981) An urban radiation obstruction model. Boundary-Layer Meteorology 20:259–264

Frazer GW, Fournier RA, Trofymow JA, Hall RJ (2001) A comparison of digital and film fisheye photography for analysis of forest canopy structure and gap light transmission. Agric

Freeman. (2004). The Science of Biology, 4th Edition, by Sinauer Associates (www.sinauer.com) and (www.whfreeman.com).

Friberg, A. et al. Discovery of potent myeloid cell leukemia 1 (Mcl-1) inhibitors using fragment-based methods and structure-based design. J. Med. Chem. 56, 15–30 (2013).

Fruh, V. et al. Application of fragment-based drug discovery to membrane proteins: identification of ligands of the integral membrane enzyme DsbB. Chem. Biol. 17, 881–891 (2010).

Fujioka H, Grotberg JB. Steady propagation of a liquid plug in a two-dimensional channel. J Biomech Eng. 2004;126(5):567–577.

Fujioka H, Grotberg JB. The steady propagation of a surfactant-laden liquid plug in a two dimensional channel. Physics of Fluids. 2005;17:Art. 082102.

Fujioka H, Takayama S, Grotberg JB. Unsteady propagation of a liquid plug in a liquid-lined straight tube. Physics of Fluids. 2008;20(6)

Fuortes & Yeandle 1964: Probability of occurrence of discrete potential waves in the eye of Limulus. MGF Fuortes & S Yeandle, J Gen Physiol 47, 443–463 (1964).

Gagge AP, Fobelets AP, Berglund LG (1986) A standard predictive index of human response to the thermal environment. ASHRAE Trans 92:709–731

Gashev AA, Davis MJ, Delp MD, Zawieja DC. Regional variations of contractile activity in isolated rat lymphatics. Microcirculation. 2004;11:477–492.

Gashev AA, Davis MJ, Zawieja DC. Inhibition of the active lymph pump by flow in rat mesenteric lymphatics and thoracic duct. J Physiol. 2002b;540:1023–1037.

Gashev AA, Zawieja DC. Physiology of human lymphatic contractility: A historical perspective. Lymphology. 2001;34:124–134.

Gashev AA. Physiologic aspects of lymphatic contractile function: Current perspectives. Ann N Y Acad Sci. 2002a; 979:178–187. discussion 188–196.

Gaver DP, et al. The steady motion of a semi-infinite bubble through a flexible-walled channel. *Journal of Fluid Mechanics.* 1996;319:25–65.

Geitmann, M. et al. Interaction kinetic and structural dynamic analysis of ligand binding to acetylcholinebinding protein. Biochemistry 49, 8143–8154 (2010).

Genick, C. C. et al. Applications of biophysics in highthroughput screening hit validation. J. Biomol. Screen 19, 707–714 (2014).

Geschwindner, S., Carlsson, J. F. & Knecht, W. Application of optical biosensors in small-molecule screening activities. Sensors (Basel) 12, 4311–4323

Geschwindner, S., Ulander, J. & Johansson, P. Ligand binding thermodynamics in drug discovery: still a hot tip? J. Med. Chem. 58, 6321–6335 (2015).

Ghadiali SN, Gaver DP. The influence of non-equilibrium surfactant dynamics on the flow of a semi-infinite bubble in a rigid cylindrical capillary tube. *Journal of Fluid Mechanics.* 2003;478:165–196.

Giancoli, D. C. (1998). Physics, Fifth Edition. London: Prentice-Hall International (UK) Limited.

Giannetti, A. M. From experimental design to validated hits a comprehensive walk-through of fragment lead identification using surface Plasmon resonance. Methods Enzymol. 493, 169–218 (2011).

Glaser, R. (2001). Biophysics. Berlin: Springer-Verlag.

Gnepp DR. Lymphatics. In: Staub Taylor., editor. Edema. New York: Raven; 1984.

Gopinathan KK (1992) Estimation of hourly global radiation and diffuse solar radiation from hourly sunshine duration. Solar Energy 48:3–5

Gossas, T. et al. The advantage of biosensor analysis over enzyme inhibition studies for slow dissociating inhibitors – characterization of hydroxamate-based matrix metalloproteinase-12 inhibitors. MedChemComm 4, 432–442 (2013).

Granger HJ, Laine GA, Barnes GE, Lewis RE. Dynamics and control of transmicrovascular fluid exchange. In: Taylor, editor. Staub. Edema; New York, Raven: 1984.

Green & Swets 1966: Signal Detection Theory and Psychophysics.

Gregg DE. Coronary circulation in health and disease. Lea & Febiger; Philadelphia: 1950.

Grimaldi A, Moriondo A, Sciacca L, Guidali ML, Tettamanti G, Negrini D. Functional arrangement of rat diaphragmatic initial lymphatic network. Am J Physiol Heart Circ Physiol. 2006;291:H876–885.

Grohmann, D., Werner, F. & Tinnefeld, P. Making connections — strategies for single molecule fluorescence biophysics. Curr. Opin. Chem. Biol. 17, 691–698 (2013).

Gueymard C (2000) Prediction and performance assessment of mean hourly global radiation. Sol Energy 68:285–303

Gul MS, Muneer T, Kambezidis HD (1998) Models for obtaining solar radiation from other meteorological data. Sol Energy 64:99–108

Gunnarsson, A. et al. Drug discovery at the single molecule level: inhibition-in-solution assay of membrane-reconstituted beta-secretase using singlemolecule

Guyton AC. Medical Physiology. Saunders; Philadelphia: 1971.

Hajduk, P. J. & Greer, J. A decade of fragment-based drug design: strategic advances and lessons learned. Nat. Rev. Drug Discov. 6, 211–219 (2007).

Halpern D, et al. Liquid and surfactant delivery into pulmonary airways. Respiratory Physiology & Neurobiology. 2008;163(1–3):222–231.

Halpern D, et al. Unsteady bubble propagation in a flexible channel: predictions of a viscous stick-slip instability. *Journal of Fluid Mechanics.* 2005;528:53–86.

Halpern D, Grotberg JB. Fluid-elastic instabilities of liquid-lined flexible tubes. *Journal of Fluid Mechanics.* 1992;244:615–632.

Halpern D, Grotberg JB. Nonlinear saturation of the Rayleigh instability in a liquid-lined tube due to oscillatory flow. J Fluid Mech. 2003;492:251 – 270.

Halpern D, Grotberg JB. SURFACTANT EFFECTS ON FLUID-ELASTIC INSTABILITIES OF LIQUID-LINED FLEXIBLE TUBES - A MODEL OF AIRWAY-CLOSURE. *Journal of Biomechanical Engineering-Transactions of the ASME.* 1993;115(3):271–277.

Harner, M. J., Frank, A. O. & Fesik, S. W. Fragmentbased drug discovery using NMR spectroscopy. J. Biomol. NMR 56, 65–75 (2013).

Hazel AL, Heil M. The influence of gravity on the steady propagation of a semi-infinite bubble into a flexible channel. Physics of Fluids. 2008;20(9)

Hazel AL, Heil M. Three-dimensional airway reopening: the steady propagation of a semi-infinite bubble into a buckled elastic tube. *Journal of Fluid Mechanics.* 2003;478:47–70.

Hecht et al 1942: Energy, quanta and vision. S Hecht, S Shlaer & MH Pirenne, J Gen Physiol 25, 819–840 (1942).

Heil M, Hazel AL, Smith JA. The mechanics of airway closure. Respiratory Physiology & Neurobiology. 2008;163(1–3):214–221.

Heil M. Finite Reynolds number effects in the propagation of an air finger into a liquid-filled flexible-walled channel. *Journal of Fluid Mechanics.* 2000;424:21–44.

Henderson, R. The potential and limitations of neutrons, electrons and X-rays for atomic resolution microscopy of unstained biological molecules. Q. Rev. Biophys. 28, 171–193 (1995).

Hennig, M., Ruf, A. & Huber, W. Combining biophysical screening and X-ray crystallography for fragment-based drug discovery. Top. Curr. Chem. 317, 115–143 (2012).

Heym, R. G., Hornberger, W. B., Lakics, V. & Terstappen, G. C. Label-free detection of smallmolecule binding to a GPCR in the membrane environment. Biochim. Biophys. Acta 1854, 979–986 (2015).

Higueruelo, A. P., Jubb, H. & Blundell, T. L. Protein–protein interactions as druggable targets: recent technological advances. Curr. Opin. Pharmacol.

Hirst, E. R., Yuan, Y. J., Xu, W. L. & Bronlund, J. E. Bond-rupture immunosensors — a review. Biosens. Bioelectron. 23, 1759–1768 (2008).

Hofstadler, S. A. & Sannes-Lowery, K. A. Applications of ESI-MS in drug discovery: interrogation of noncovalent complexes. Nat. Rev. Drug Discov. 5, 585–595 (2006).

Holdgate, G. A. & Gill, A. L. Kinetic efficiency: the missing metric for enhancing compound quality? Drug Discov. Today 16, 910–913 (2011).

Holdgate, G. A. Thermodynamics of binding interactions in the rational drug design process. Expert Opin. Drug Discov. 2, 1103–1114 (2007).

Holdgate, G. A., Anderson, M., Edfeldt, F. & Geschwindner, S. Affinity-based, biophysical methods to detect and analyze ligand binding to recombinant proteins: matching high information content with high throughput. J. Struct. Biol. 172, 142–157 (2010).

Holmer B (1992) A simple operative method for determination of sky view factors in complex urban canyon from fisheye photographs. Meteorol Zeitschrift, NF 1:236–239

Holmer B, Postgärd U, Eriksson M (2001) Sky view factors in forest canopies calculated with IDRISI. Theor Appl Climatol 68:33–40

Höppe P (1984) Die Energiebilanz des Menschen. Wiss Mitt Meteorol Inst Univ München No. 49

Höppe P (1992) Ein neues Verfahren zur Bestimmung der mittleren Strahlungstemperatur im Freien. Wetter und Leben 44:147–151

Höppe P (1993) Heat balance modelling. Experientia 49:741–746

Höppe P (1999) The physiological equivalent temperature – a universal index for the biometeorological assessment of the thermal environment. Int J Biometeorol 43:71–75

Hossler FE, Monson FC. Structure and blood supply of intrinsic lymph nodes in the wall of the rabbit urinary bladder--studies with light microscopy, electron microscopy, and vascular corrosion casting. Anat Rec. 1998;252:477–484.

Hubbard, R. E. & Murray, J. B. Experiences in fragment-based lead discovery. Methods Enzymol. 493, 509–531 (2011).

Hubbard, R. E., Davis, B., Chen, I. & Drysdale, M. J. The SeeDs approach: integrating fragments into drug discovery. Curr. Top. Med. Chem. 7, 1568–1581 (2007).

Huber, W. & Mueller, F. Biomolecular interaction analysis in drug discovery using surface Plasmon resonance technology. Curr. Pharm. Des. 12,

Hughes JMB, Rosenzweig DY, Kivitz PB. Site of airway closure in excised dog lungs: histologic demonstration. *Journal of Applied Physiology.* 1970;29:340–344.

Hughes, W. (1979). Aspects of Biophysics. New York: John Wiley & Sons, Inc.

Huh D, et al. Acoustically detectable cellular-level lung injury induced by fluid mechanical stresses in microfluidic airway systems. Proceedings of the National Academy of Sciences. 2007;104(48):18886–18891. doi: 10.1073/pnas.0610868104.

Iversen, P. W. et al. in Assay Guidance Manual (eds Sittampalam, G. S. et al.) (Bethesda (MD), 2004).

J. Hertz, A. Krogh, and R. Palmer. Introduction to the theory of neural computation, volume 1 of Santa Fe Institute. Addison-Wesley, Redwood City, 1991.

J.W. Moore, M.P. Blaustein, N.C. Anderson, and T. Narahashi. Basis of tetrodotoxin's selectivity in blockage of squid axons. *Journal of General Physiology,* 50:1401{1411, 1967.

Jacob AM, Gaver DP. An investigation of the influence of cell topography on epithelial mechanical stresses during pulmonary airway reopening. Physics of Fluids. 2005;17(3).

Jahnke, W. et al. Allosteric non-bisphosphonate FPPS inhibitors identified by fragment-based discovery. Nat. Chem. Biol. 6, 660–666 (2010).

JD Kalma, TR Oke and A Auliciems (eds) Biometeorology and Urban Climatology at the Turn of the Millenium : Selected Papers from the Conference ICB-ICUC'99, Sydney, WCASP-50,

Jendritzky G, Menz H, Schirmer H, Schmidt-Kessen W (1990)

Jendritzky G, Nübler W (1981) A model analysing the urban thermal environment in physiologically significant terms. Arch Met Geoph Biokl, Ser B29:313–326

Jensen OE, et al. The steady propagation of a bubble in a flexible-walled channel: Asymptotic and computational models. Physics of Fluids. 2002;14(2):443–457.

Jessel W (1983) Die diffuse Himmelstrahlung. Eine vergleichende Darstellung der Bestrahlungsstärke bezogen auf eine kugelförmige und eine ebene horizontale Empfangsfläche. Arch Met Geoph Biokl, Ser B32:23–52

Jhoti, H., Williams, G., Rees, D. C. & Murray, C. W. The 'rule of three' for fragment-based drug discovery: where are we now? Nat. Rev. Drug Discov. 12, 644–645 (2013).

Johnson ID, Watson ID (1984) The determination of view-factors in urban canyons. J Clim Appl Meteorol 23:329–335

Judson 1979: The Eighth Day of Creation HF Judson (Simon and Schuster, New York, 1979).

Kaempfert W (1949) Zur Frage der Besonnung enger Strassen. Meteorol Rdsch 2:222–227

Kaempfert W (1951) Ein Phasendiagramm der Besonnung. Meteorol Rdsch 4:141–144

Kamm RD, Schroter RC. Is airway closure caused by a thin liquid instability? Respiration Physiology. 1989;75:141–156.

Kanda M, Kawai T, Nagakawa K (2004) A simple theoretical radiation scheme for regular building arrays. Boundary Layer Meteorology 114:71–90

Kapanidis, A. N. & Strick, T. Biology, one molecule at a time. Trends Biochem. Sci. 34, 234–243 (2009).

Karpanen T, Alitalo K. Molecular biology and pathology of lymphangiogenesis. Annu Rev Pathol. 2007.

Kasten F (1980) A simple parametrization of the pyrheliometric formula for determining the Linke turbidity factor. Meteorol Rdsch 33:124–127.

Kay SS, et al. Pressure gradient, not exposure duration, determines the extent of epithelial cell damage in a model of pulmonary airway reopening. *Journal of Applied Physiology.* 2004;97(1):269–276.

Kemmoku Y, Orita S, Nakagawa S, Sakakibara T (1999) Daily insolation forecasting using a multi-stage neural network. Solar Energy 66:193–199.

Kern, J., Yachandra, V. K. & Yano, J. Metalloprotein structures at ambient conditions and in real-time: biological crystallography and spectroscopy using X-ray free electron lasers. Curr. Opin. Struct. Biol. 34, 87–98 (2015).

Kerslake Mc K (1972) The stress of hot environments. Cambridge University Press, Cambridge Littlefair P (2001).

Kesters, D. et al. Structural basis of ligand recognition in 5-HT3 receptors. EMBO Rep. 14, 49–56 (2013).

Kimbal, J. W. (1983). Biologi, Jilid 2, Edisi ke-5, Terjemahan. Jakarta: Erlangga.

King M, Macklem PT. Rheological properties of microliter quantities of normal mucus. *Journal of Applied Physiology.* 1977;42:797–802.

Kitova, E. N., El-Hawiet, A., Schnier, P. D. & Klassen, J. S. Reliable determinations of proteinligand interactions by direct ESI-MS measurements. Are we there yet? J. Am. Soc. Mass Spectrom. 23, 431–441 (2012).

Klebe, G. Applying thermodynamic profiling in lead finding and optimization. Nat. Rev. Drug Discov. 14, 95–110 (2015).

Klebe, G. The use of thermodynamic and kinetic data in drug discovery: decisive insight or increasing the puzzlement? ChemMedChem. 10, 229–231 (2015).

Koller A, Mizuno R, Kaley G. Flow reduces the amplitude and increases the frequency of lymphatic vasomotion: Role of endothelial prostanoids. Am J Physiol. 1999;277:R1683–1689.

Konermann, L., Pan, J. & Liu, Y. H. Hydrogen exchange mass spectrometry for studying protein structure and dynamics. Chem. Soc. Rev. 40, 1224–1234 (2011).

Kozma, P., Hamori, A., Kurunczi, S., Cottier, K. & Horvath, R. Grating coupled optical waveguide interferometer for label-free biosensing. Sens. Actuators B: Chem. 155, 446–450 (2011).

Kozma, P., Kehl, F., Ehrentreich-Forster, E., Stamm, C. & Bier, F. F. Integrated planar optical waveguide interferometer biosensors: a comparative review. Biosens. Bioelectron. 58, 287–307 (2014).

Krueger MA, Gaver DP. A theoretical model of pulmonary surfactant multilayer collapse under oscillating area conditions. *Journal of Colloid and Interface Science.* 2000;229:353–364.

Kuo, L. C. Fragment-based drug design — tools, practical approaches, and examples. Preface. Methods Enzymol. 493, xxi–xxii (2011).

Ladbury, J. E., Klebe, G. & Freire, E. Adding calorimetric data to decision making in lead discovery: a hot tip. Nat. Rev. Drug Discov. 9, 23–27 (2010).

Land & Nilsson 2002: Animal Eyes MF Land & D–E Nilsson (Oxford University Press, Oxford, 2002).

Langer, A. et al. Protein analysis by time-resolved measurements with an electro-switchable DNA chip. Nat. Commun. 4, 2099 (2013).

Larson 2003: The Complete Far Side. G Larson (Andrews Mc- Neel Publishing, Kansas City, 2003).

Lawson & Uhlenbeck 1950: Threshold Signals. MIT Radiation Laboratory Series vol 24 JL Lawson & GE Uhlenbeck (McGraw–Hill, New York, 1950).

Leake, M. C. The physics of life: one molecule at a time. Phil. Trans. R. Soc. B 368, 20120248 (2013).

Leeson, P. D. & Springthorpe, B. The influence of drug-like concepts on decision-making in medicinal chemistry. Nat. Rev. Drug Discov. 6, 881–890 (2007).

Liao, M., Cao, E., Julius, D. & Cheng, Y. Structure of the TRPV1 ion channel determined by electron cryomicroscopy. Nature 504, 107–112 (2013).

Lighthill 1958: Introduction to Fourier Analysis and Generalized Functions. MJ Lighthill (Cambridge University Press, Cambridge, 1958)

Lighthill 1958: Introduction to Fourier Analysis and Generalized Functions. MJ Lighthill (Cambridge University Press, Cambridge, 1958)

Lipp MM, et al. Coexistence of buckled and flat monolayers. Physical Review Letters. 1998; 81(8):1650–1653.

Lo, M. C. et al. Evaluation of fluorescence-based thermal shift assays for hit identification in drug discovery. Anal. Biochem. 332, 153–159 (2004).

Lubert, Styer. (2000). Biokomia. Vol I. Edisi 4. Penerbit Buku Kedokteran EGC. Jakarta.

Luchinat, E. & Banci, L. A. Unique tool for cellular structural biology: in-cell NMR. J. Biol. Chem. 291, 3776–3784 (2016).

Ludbrook J. Regulation of arterial blood flow, pressure and resistance in the systemic circulation. In: Schwartz CJ, Werthessen NT, Wolf S, editors. Structure and Function of the circulation. Vol. 1. Plenum, NY: 1980.

Mallock 1894: Insect sight and the defining power of composite eyes. A Mallock, Proc R Soc Lond 55, 85–90 (1894).

Marcus ML. The coronary circulation in health and disease. McGraw-Hill; New York: 1983.

Markgren, P. O., Hämäläinen, M. & Danielson, U. H. Kinetic analysis of the interaction between HIV-1 protease and inhibitors using optical biosensor technology. Anal. Biochem. 279, 71–78 (2000).

Marki A, Antonić O (1999) Annual models of monthly mean hourly direct, diffuse, and global radiation at ground. Meteorol Zeitschrift, NF 8:91–95

Martinez Molina, D. & Nordlund, P. The cellular thermal shift assay: a novel biophysical assay for in situ drug target engagement and mechanistic biomarker studies. Annu. Rev. Pharmacol. Toxicol. 56, 141–161 (2016).

Martinez Molina, D. et al. Monitoring drug target engagement in cells and tissues using the cellular thermal shift assay. Science 341, 84–87 (2013).

Marx, V. Structural biology: 'seeing' crystals the XFEL way. Nat. Methods 11, 903–908 (2014).

Marzinzik, A. L. et al. Discovery of novel allosteric nonbisphosphonate inhibitors of farnesyl pyrophosphate synthase by integrated lead finding. ChemMedChem. 10, 1884–1891 (2015).

Matzarakis A, Mayer H (1996) Another Kind of Environmental Stress: Thermal Stress. NEWSLETTERS No. 18, 7–10.

Maurer, T. et al. Small-molecule ligands bind to a distinct pocket in Ras and inhibit SOS-mediated nucleotide exchange activity. Proc. Natl Acad. Sci. USA 109, 5299–5304 (2012).

Mayer H (1993) Urban bioclimatology. Experientia 49:957–963

McMahon & Bonner 1983: On Size and Life. TA McMahon & JT Bonner (WH Freeman, New York, 1983).

Meek DW (1997) Estimation of maximum possible daily global radiation. Agric Forest Meteorol 87:223–241

Milhorn HT. The application of control theory to physiological systems. In: Mohrman DE, Heller LJ, editors. 1997 Cardiovascular Physiology. Saunders; philadelphia: McGraw-Hill; 1966.

Mohrman DE, Heller LJ. Cardiovascular Physiology. McGraw-Hill; 1997.

Mohsen MA (1979) Solar radiation and courtyard house forms – I. A mathematical model. Build Environ 14:89–106

Mora-Lopez LL, Sidrach-de-Cardona M (1998) Multicaptive arma models to generate hourly series of global irradiation. Sol Energy 63:283–291

Moree, B. et al. Protein conformational changes are detected and resolved site specifically by secondharmonic generation. Biophys. J. 109, 806–815 (2015).

Moree, B. et al. Small molecules detected by second-harmonic generation modulate the conformation of monomeric α-synuclein and reduce its aggregation in cells. J. Biol. Chem. 290, 27582–27593 (2015).

Motulsky, H. J. & Mahan, L. C. The kinetics of competitive radioligand binding predicted by the law of mass action. Mol. Pharmacol. 25, 1–9 (1984).

Murray, C. W., Verdonk, M. L. & Rees, D. C. Experiences in fragment-based drug discovery. Trends Pharmacol. Sci. 33, 224–232 (2012).

Murray, J. B., Roughley, S. D., Matassova, N. & Brough, P. A. Off-rate screening (ORS) by surface plasmon resonance. An efficient method to kinetically sample hit to lead chemical space from unpurified reaction products. J. Med. Chem. 57, 2845–2850 (2014).

Navratilova, I., Dioszegi, M. & Myszka, D. G. Analyzing ligand and small molecule binding activity of solubilized GPCRs using biosensor technology. Anal. Biochem. 355, 132–139 (2006).

Navratilova, I., Sodroski, J. & Myszka, D. G. Solubilization, stabilization, and purification of chemokine receptors using biosensor technology. Anal. Biochem. 339, 271–281 (2005).

Neumann, T., Junker, H. D., Schmidt, K. & Sekul, R. SPR-based fragment screening: advantages and applications. Curr. Top. Med. Chem. 7, 1630–1642.

Ng CP, Helm CL, Swartz MA. Interstitial flow differentially stimulates blood and lymphatic endothelial cell morphogenesis in vitro. Microvasc Res. 2004;68:258–264.

Nienaber, V. L. et al. Discovering novel ligands for macromolecules using X-ray crystallographic screening. Nat. Biotechnol. 18, 1105–1108 (2000).

Niesen, F. H., Berglund, H. & Vedadi, M. The use of differential scanning fluorimetry to detect ligand interactions that promote protein stability. Nat. Protoc. 2, 2212–2221 (2007).

Niewienda A, Heidt FD (1996) Sombrero: A pc-tool to calculate shadows on arbitrarily oriented surfaces. Sol Energy 58:253–363.

Nisimaru Y. Summary of our studies concerning the structure and function of lymphatic system. Hiroshima J Med Sci. 1982;31:145–160.

Nordström, H. et al. Identification of MMP-12 inhibitors by using biosensor-based screening of a fragment library. J. Med. Chem. 51, 3449–3459 (2008).

Nunez M, Eliasson I, Lindgren J (2000) Spatial variation of incoming longwave radiation in Göteborg, Sweden. Theor Appl Climatol 67:181–192.

Oberg B. Overall cardiovascular regulation. Annual Reviews of Physiology. 1976;38:537–570.

Olaru, A., Bala, C., Jaffrezic-Renault, N. & Aboul-Enein, H. Y. Surface plasmon resonance (SPR) biosensors in pharmaceutical analysis. Crit. Rev. Anal. Chem. 45, 97–105 (2015).

Olseth JA, Skartveit A (1993) Characteristics of hourly global irradiance modelled from cloud data. Sol Energy 51:197–204

Oltersdorf, T. et al. An inhibitor of Bcl-2 family proteins induces regression of solid tumours. Nature 435, 677–681 (2005).

Osborn MJ. Sudden cardiac death: A. Mechanism, incidence, and prevention of sudden cardiac death. In: Giuliani ER, Gersh BJ, McGoon MD, Hayes DL, Schaff HV, editors. Mayo Clinic Practice of Cardiology. Mosby; St. Luis: 1996.

Ostrem, J. M., Peters, U., Sos, M. L., Wells, J. A. & Shokat, K. M. K-Ras(G12C) inhibitors allosterically control GTP affinity and effector interactions. Nature 503, 548–551 (2013).

Otis DR, et al. ROLE OF PULMONARY SURFACTANT IN AIRWAY-CLOSURE - A COMPUTATIONAL STUDY. *Journal of Applied Physiology.* 1993;75(3):1323–1333.

P. Peretto. An introduction to the modeling of neural networks. Cambridge University Press, 1992.

Pan, A. C., Borhani, D. W., Dror, R. O. & Shaw, D. E. Molecular determinants of drug-receptor binding kinetics. Drug Discov. Today 18, 667–673 (2013).

Pantoliano, M. W. et al. High-density miniaturized thermal shift assays as a general strategy for drug discovery. J. Biomol. Screen 6, 429–440 (2001).

Patching, S. G. Surface plasmon resonance spectroscopy for characterisation of membrane proteinligand interactions and its potential for drug discovery. Biochim. Biophys. Acta 1838, 43–55 (2014).

Pellecchia, M. et al. Perspectives on NMR in drug discovery: a technique comes of age. Nat. Rev. Drug Discov. 7, 738–745 (2008).

Pereira FOR, Silva CAN, Turkienikz B (2001) A methodology for sunlight urban planning: A computer-based solar and sky vault obstruction analysis. Sol Energy 70:217–226.

Perpeet, M. et al. SAW sensor system for marker-free molecular interaction analysis. Anal. Lett. 39, 1747–1757 (2006).

Perspicace, S. et al. Isothermal titration calorimetry with micelles: thermodynamics of inhibitor binding to carnitine palmitoyltransferase 2 membrane protein. FEBS Open Bio. 3, 204–211 (2013).

Pickhardt, M. et al. Identification of small molecule inhibitors of tau aggregation by targeting monomeric tau as a potential therapeutic approach for tauopathies. Curr. Alzheimer Res. 12, 814–828 (2015).

Power H (2001) Estimating atmospheric turbidity from climate data. Atmos Environ 35:125–134.

Pries AR, Secomb TW. Control of blood vessel structure: Insights from theoretical models. Am J Physiol Heart Circ Physiol. 2005;288:H1010–1015.

Purves D, Augustine GJ, Fitzpatrick D, Hall WC, LaMantia AS, McNamara JO, White LE. Neurosciences. Sinauer Associates; Sunderland MA: 2008.

Quraishi MS, Jones NS, Mason J. The rheology of nasal mucus: a review. Clinical Otolaryngology. 1998;23 (5):403–413.

Randall WC. Nervous control of cardiovascular function. Oxford University Press; NY: 1984.

Recht, M. I. et al. Identification and optimization of PDE10A inhibitors using fragment-based screening by nanocalorimetry and X-ray crystallography. J. Biomol.Screen 19, 497–507 (2014).

Reddy NP, Krouskop TA, Newell PH., Jr A computer model of the lymphatic system. Comput Biol Med. 1977;7:181–197.

Reinhard, F. B. et al. Thermal proteome profiling monitors ligand interactions with cellular membrane proteins. Nat. Methods 12, 1129–1131 (2015).

Renaud, J. P., Neumann, T. & Van Hijfte, L. Small Molecule Medicinal Chemistry: Strategies and Technologies. (eds Czechtizky, W. & Hamley, P.) 221–249 (John Wiley & Sons Inc., 2015).

Revfeim KJA (1997) On the relationship between radiation and mean daily sunshine. Agric Forest Meteorol 86:183–191.

Rich PM, Clark DB, Clark DA, Oberbauer SF (1993) Long-term study of solar radiation regimes in a tropical wet forest using quantum sensors and hemispherical photography. Agric Forest Meteorol 65:107–127

Rieke & Baylor 1998: Single photon detection by rod cells of the retina. F Rieke & DA Baylor, Rev Mod Phys 70, 1027– 1036 (1998).

Roberts N, Kloos B, Cassella M, Podgrabinska S, Persaud K, Wu Y, Pytowski B, Skobe M. Inhibition of vegfr-3 activation with the antagonistic antibody more potently suppresses lymph node and distant metastases than inactivation of vegfr-2. Cancer Res. 2006;66:2650–2657.

Röckle R, Richter CJ, Höfl HC, Steinicke W, Streifeneder M, Matzarakis A (2003) Klimaanalyse Stadt Freiburg. Auftraggeber Stadtplanungsamt der Stadt Freiburg. November 2003.

Roddy, T. P. et al. Mass spectrometric techniques for label-free high-throughput screening in drug discovery. Anal. Chem. 79, 8207–8213 (2007).

Roderick ML (1999) Estimating the diffuse component from daily and monthly measurements of global radiation. Agric Forest Meteorol 95:169–185.

Roy, R., Hohng, S. & Ha, T. A practical guide to singlemolecule FRET. Nat. Methods 5, 507–516 (2008).

Rushmere RF. Cardiovascular dynamics. Saunders; Philadelphia: 1970.

Saalau-Bethell, S. M. et al. Discovery of an allosteric mechanism for the regulation of HCV NS3 protein function. Nat. Chem. Biol. 8, 920–925 (2012).

Sackner MA, Kim CS. Phasic flow mechanisms of mucus clearance. European *Journal of Respiratory Diseases - Supplement.* 1987;153:159–164.

Salafsky, J. S. Detection of protein conformational change by optical second-harmonic generation. J. Chem. Phys. 125, 074701 (2006).

Salsibury JW, D'Aria DM (1992) Emissivity of terrestrial material in the 8–14 μm atmospheric window. Remote Sens Environ 42:83–106.

Savitski, M. M. et al. Tracking cancer drugs in living cells by thermal profiling of the proteome. Science 346, 1255784 (2014).

Schiebel, J. et al. One question, multiple answers: biochemical and biophysical screening methods retrieve deviating fragment hit lists. ChemMedChem. 10, 1511–1521 (2015).

Schiele, F., Ayaz, P. & Fernandez-Montalvan, A. A universal homogeneous assay for high-throughput determination of binding kinetics. Anal. Biochem. 468, 42–49 (2014).

Schlichting, I. Serial femtosecond crystallography: the first five years. IUCrJ 2, 246–255 (2015).

Schmid-Schonbein GW. Microlymphatics and lymph flow. Physiol Rev. 1990;70:987–1028.

Schr¨odinger 1944: What is Life? E Schr¨odinger (Cambridge University Press, Cambridge, 1944).

Seeger, C. et al. Histaminergic pharmacology of homooligomeric β3 γ-aminobutyric acid type A receptors characterized by surface plasmon resonance biosensor technology. Biochem. Pharmacol. 84, 341–351 (2012).

Seidel, S. A. et al. Label-free microscale thermophoresis discriminates sites and affinity of protein–ligand binding. Angew. Chem. Int. 51, 10656–10659 (2012).

Seidel, S. A. et al. Microscale thermophoresis quantifies biomolecular interactions under previously challenging conditions. Methods 59, 301–315 (2013).

Seidler, J., McGovern, S. L., Doman, T. N. & Shoichet, B. K. Identification and prediction of promiscuous aggregating inhibitors among known drugs. J. Med. Chem. 46, 4477–4486 (2003).

Sen Z (1998) Fuzzy algorithm for estimation of solar radiation from sunshine duration. Sol Energy 63:39–49.

Shah, N. B. & Duncan, T. M. Bio-layer interferometry for measuring kinetics of protein–protein interactions and allosteric ligand effects. J. Vis. Exp. 18, e51383 (2014).

Sharff, A. & Jhoti, H. High-throughput crystallography to enhance drug discovery. Curr. Opin. Chem. Biol. 7, 340–345 (2003).

Shuker, S. B., Hajduk, P. J., Meadows, R. P. & Fesik, S. W. Discovering high-affinity ligands for proteins: SAR by NMR. Science 274, 1531–1534 (1996).

Silver MD. Cardiovascular Pathology. University Park Press; Baltimore: 1983.

Śledź, P., Abell, C. & Ciulli, A. NMR of Biomolecules: Towards Mechanistic Systems Biology. 1st edn (eds Bertini, I., McGreevy, K. S. & Parigi, G.) 265–280 (Wiley-VCH Verlag GmbH & Co. KGaA, 2012).

Smiley, R. D. & Hammes, G. G. Single molecule studies of enzyme mechanisms. Chem. Rev. 106, 3080–3094 (2006).

Snyder 1977: Acuity of compound eyes: Physical limitations and design. AW Snyder, J Comp Physiol 116, 161–182 (1977).

Snyder et al 1977: Information capacity of compound eyes. AW Snyder, DS Stavenga & SB Laughlin, J Comp Physiol 116, 183–207 (1977).

Souers, A. J. et al. ABT-199, a potent and selective BCL-2 inhibitor, achieves antitumor activity while sparing platelets. Nat. Med. 19, 202–208 (2013).

Stavenga & Hardie 1989: Facets of Vision. DG Stavenga & RC Hardie, eds (Springer–Verlag, Berlin, 1989).

Steadman RG (1971) Indices of windchill of clothed persons. J Appl Meteorology 10:674–683

Strausfeld 1976: Atlas of an Insect Brain. N Strausfeld (Springer–Verlag, Berlin, 1976).

Sun, Q. et al. Discovery of small molecules that bind to K-Ras and inhibit Sos-mediated activation. Angew. Chem. Int. 51, 6140–6143 (2012).

Sussman, F., Villaverde, M. C., Dominguez, J. L. & Danielson, U. H. On the active site protonation state in aspartic proteases: implications for drug design. Curr. Pharm. Des. 19, 4257–4275 (2013).

Swinney, D. C. The role of binding kinetics in therapeutically useful drug action. Curr. Opin. Drug Discov. Devel 12, 31–39 (2009).

Szuba A, Shin WS, Strauss HW, Rockson S. The third circulation: Radionuclide lymphoscintigraphy in the evaluation of lymphedema. J Nucl Med. 2003;44:43–57. 9. Ferrell RE. Research perspectives in inherited lymphatic disease. Ann N Y Acad Sci. 2002;979:39–51. discussion 76–39. 10. Oliver G, Alitalo K. The lymphatic vasculature: Recent progress and paradigms. Annu Rev Cell Dev Biol. 2005;21:457–483.

Talibov, V. O., Linkiviene, V., Matulis, D. & Danielson, U. H. Kinetically selective inhibitors of human carbonic anhydrase isoenzymes I, II, VII, IX, XII and XIII. J. Med. Chem. 59, 2083–2093 (2016).

Taylor CA, Draney MT, KU JP, Parker D, Steele BN, Wang K, Zarins CK. Predictive medicine: Computational techniques in theraputic decision-making. Computer Aided Surgery. 1999;4:231–247.

Terjung WH, Louie S (1974) A climatic model of urban energy budgets. Geogr Anal 6:341–367

the presence of obstructing mountain or buildings. Arch Met Geoph Biocl, Ser B 32:111–129

Thom EC (1959) The Discomfort Index. Weatherwise 12:57–60

Trzewik J, Mallipattu SK, Artmann GM, Delano FA, Schmid-Schonbein GW. Evidence for a second valve system in lymphatics: Endothelial microvalves. Faseb J. 2001;15:1711–1717.

Tsagaris TJ, Sutton RB, Kuida H. Hemodynamic effects of varying pacemaker sites. American *Journal of Physiology*. 1970;218:246–250.

Tschumperlin DJ, et al. Mechanical stress triggers selective release of fibrotic mediators from bronchial epithelium. American *Journal of Respiratory Cell and Molecular Biology*. 2003;28(2):142–149.

Tschumperlin DJ, et al. Mechanotransduction in the lung - Bronchial epithelial compression regulates MAP kinase signaling and HB-EGF-like growth factor expression. AMERICAN *JOURNAL OF PHYSIOLOGY-LUNG CELLULAR AND MOLECULAR PHYSIOLOGY*. 2002;282(5):L904–L911.

Tuukkanen, A. T. & Svergun, D. I. Weak protein-ligand interactions studied by small-angle X-ray scattering. FEBS J. 281, 1974–1987 (2014).

Underwood CR, Ward EJ (1966) The solar radiation area on man. Ergonomics 9:155–168

Urone, P. P. (1986). Physics with Health Science Application. New York: John Wiley & Sons, Inc.

Valko P (1966) Die Himmelsstrahlung in ihrer Beziehung zu verschiedenen Parametern. Arch Met Geoph Biocl B14: 337–359

VDI (1998) VDI 3787, Part I: Environmental Meteorology, Methods for the human biometeorological evaluation of climate and air quality for the urban and regional planning at regional level. Part I: Climate. Beuth, Berlin, p 29

VDI (2001) VDI 3789, Part 3: Environmental Meteorology, Interactions between Atmosphere and Surfaces; Calculation of spectral irradiances in the solar wavelength range. Beuth, Berlin, p 77

Vestergaard, B. & Sayers, Z. Investigating increasingly complex macromolecular systems with small-angle X-ray scattering. IUCrJ 1, 523–529 (2014).

Vivat Hannah, V., Atmanene, C., Zeyer, D., Van Dorsselaer, A. & Sanglier-Cianferani, S. Native MS: an 'ESI' way to support structure- and fragmentbased drug discovery. Future Med. Chem. 2, 35–50 (2010).

von der Weid PY, Crowe MJ, Van Helden DF. Endothelium-dependent modulation of pacemaking in lymphatic vessels of the guinea-pig mesentery. J Physiol. 1996;493 (Pt 2):563–575. 22. Fortes ZB, Scivoletto R, Garcia-Leme J. Endothelin-1 induces potent constriction of lymphatic vessels in situ. Eur J Pharmacol. 1989;170:69–73.

Wachter H (1950) Strahlungsmessung für bioklimatische Zwecke. Meteorol Rdsch 3:65–68.

Walter, N. G., Huang, C. Y., Manzo, A. J. & Sobhy, M. A. Do-it-yourself guide: how to use the modern single-molecule toolkit. Nat. Methods 5, 475–489 (2008).

Wartchow, C. A. et al. Biosensor-based small molecule fragment screening with biolayer interferometry. J. Comput. Aided Mol. Des. 25, 669–676 (2011).

Watson ID, Johnson GT (1987) Graphical estimation of sky viewfactors in urban environments. J Climatology 7:193–197

Watson ID, Johnson GT (1988) Estimating person view factors from fish-eye lens photographs. Int J Biometeorol 32:123–128

Wei HH, Halpern D, Grotberg JB. Linear stability of a surfactant-laden annular film in a time-periodic pressure-driven flow through a capillary. *Journal of Colloid and Interface Science.* 2005;285(2):769–780.

WHO Colloborating Centre for Air Quality Management and Air Pollution Control

Widdicombe J. Airway and alveolar permeability and surface liquid thickness: Theory. *Journal of Applied Physiology.* 1997;82(1):3–12.

Wigle JT, Harvey N, Detmar M, Lagutina I, Grosveld G, Gunn MD, Jackson DG, Oliver G. An essential role for prox1 in the induction of the lymphatic endothelial cell phenotype. Embo J. 2002;21:1505–1513.

Winslow CEA, Herrington LP, Gagge AP (1936) A new method of particional calorimetry. Amer J Physiology 116:641–655

Winter, A. et al. Biophysical and computational fragment-based approaches to targeting proteinprotein interactions: applications in structure-guided drug discovery. Q. Rev. Biophys. 45, 383–426 (2012).

Wittstein IS, Thiemann DR, Lima JAC, Baughman KL, Schulman SP, Gerstenblith G, Wu KC, Rade JJ, Bivalacqua TJ, Champion HC. Neurohumoral features of myocardial stunning due to sudden emotional stress. N Engl J Med. 2005;352:539–548.

Wu, B. et al. High-throughput screening by nuclear magnetic resonance (HTS by NMR) for the identification of PPIs antagonists. Curr. Top. Med. Chem. 15, 2032–2042 (2015).

Wu, B. et al. HTS by NMR of combinatorial libraries: a fragment-based approach to ligand discovery. Chem. Biol. 20, 19–33 (2013).

Yager D, et al. Airway surface liquid thickness as a function of lung volume in small airways of the guinea pig. *Journal of Applied Physiology.* 1994;77(5):2333–40.

Yalcin HC, Perry SF, Ghadiali SN. Influence of airway diameter and cell confluence on epithelial cell injury in an in vitro model of airway reopening. *Journal of Applied Physiology.* 2007;103(5):1796–1807.

Yap DYK, Gaver DP. The influence of surfactant on two-phase flow in a flexible- walled channel under bulk equilibrium conditions. Physics of Fluids. 1998;10(8):1846–1863.

Yeates DB. Mucus rheology. In: Crystal RG, West JB, editors. The Lung: Scientific Foundations. Raven Press, Ltd; New York: 1990. pp. 197–203.

Yin YB, Zhu KQ. Oscillating flow of a viscoelastic fluid in a pipe with the fractional Maxwell model. Applied Mathematics and Computation. 2006;173(1):231–242.

Zamir M. Mechanics of blood supply to the heart:Wave reflection effects in a right coronary artery. Proceedings of the Royal Society of London B. 1998;265:439–444.

Zamir M. The Physics of Coronary Blood flow. Springer; New York: 2005.

Zasadzinski JA, et al. The physics and physiology of lung surfactants. Current Opinion in Colloid & Interface Science. 2001;6(5–6):506–513.

Zawieja DC, Davis KL, Schuster R, Hinds WM, Granger HJ. Distribution, propagation, and coordination of contractile activity in lymphatics. Am J Physiol. 1993;264:H1283–1291.

Zdunkowski W, Brühl Ch (1983) A fast approximate method for the calculation of the infrared radiation balance within city street cavities. Arch Met Geoph Biocl, Ser B 33:237–241.

Zhang, J. et al. Targeting Bcr–Abl by combining allosteric with ATP-binding-site inhibitors. Nature 463, 501–506 (2010).

Zhao, Y. et al. Lab-on-a-chip technologies for singlemolecule studies. Lab. Chip 13, 2183–2198 (2013).

Zimmer ME, Williams HAR, Gaver DP. The pulsatile motion of a semi-infinite bubble in a channel: flow fields, and transport of an inactive surface-associated contaminant. *Journal of Fluid Mechanics.* 2005;537:1–33.

INDEX

D

E

F

www.ingramcontent.com/pod-product-compliance
Lightning Source LLC
LaVergne TN
LVHW062218200926
848484LV00006B/1423